MAKERS OF AMERICAN MACHINIST'S TOOLS

A HISTORICAL DIRECTORY OF MAKERS AND THEIR TOOLS

Kenneth L. Cope

ASTRAGAL PRESS
Mendham, New Jersey

Library of Congress Catalog Card Number 94-78585
International Standard Book Number 1-879335-55-7

Published by
THE ASTRAGAL PRESS
5 Cold Hill Road, Suite 12
P.O. Box 239
Mendham, New Jersey 07945-0239

Manufactured in the United States of America

DEDICATION

This book is fondly dedicated to our four grandchildren, Ben and Rebecca Hess, and Tim and Sam Hays. I wonder what they will think of the drawers full of odd looking tools so treasured by their grandfather.

ACKNOWLEDGEMENTS

This book could never have been written by one person. Only the unselfish help of a number of tool collectors, many of whom exposed valuable catalogs and tools to the hazards of the U.S. Post Office and other shipping agencies, made it possible to include so much information. Special thanks are due to:

Martin J. Donnelly, Bath, NY
Emil Pollak, Mendham, NJ
William Robertson, Kansas City, MO

Ken Cope
Milwaukee, WI
May, 1994

TABLE OF CONTENTS

INTRODUCTION

The writing of this introduction marks the end of a task begun in earnest some five years ago. I had, at that time, become a serious collector of American machinist's tools and was painfully aware of the nearly complete lack of information regarding them. For some years prior to starting a collection, I had been interested in machinist's tools, largely as a result of the 30 years spent in the machine tool industry. As chief engineer of a large machine tool builder, I was aware of the use of machinist's tools but knew nothing of their history.

This began to change when I found a stack of bound *American Machinist* magazines in the archives of another machine tool builder. The owner of the company was kind enough to allow me to take the 1885 through 1903 volumes on loan. The advertisements and articles were, of course, fascinating and I began to learn of the history of these important tools.

With that beginning, I soon found that the Milwaukee public library had the *American Machinist* magazine from 1881 to the current issue, *Machinery* magazine from 1901 into the 1960's and *Iron Age* magazine from 1886 to the 1960's. The information found in these magazines forms the basis for this book.

The second great source of information was contained in the patent records. At first, I had intended to include the patent data as an appendix to this book. When I had finished accumulating the data, however, it was obvious that it was a book in itself. In fact, it became a book, AMERICAN MACHINIST'S TOOLS: AN ILLUSTRATED DIRECTORY OF PATENTS, published by Astragal Press in September, 1993. The reader will find many references to patents in the text of this book, but only a few patent drawings. It would have been much too cumbersome to include the more than one thousand drawings found in the patent book. It is my hope, therefore, that the patent book will serve you as a companion volume to this one.

The third great source was found in contemporary maker and dealer catalogs and, of course, in the tools themselves. Other collectors, met at shows and through correspondence, generously allowed me to copy their catalogs and study their collections, even at some risk of damage, especially to the fragile catalogs.

The time period covered by this book is generally from about 1830 to 1920. A few makers who began after 1920 are included if the tools are of special interest. Some information on post-1920 tools from the larger makers, such as the L.S. Starrett Co., is also included.

Generally, however, development of general purpose tools for the individual machinist ended soon after World War I. Machine shops had become production facilities and the demand for tools was concentrated in special measuring gauges, dedicated to a single part, which could be used by nearly unskilled labor. The American machinist is still with us, but in smaller and smaller numbers. High production machine tools, since the first decade of this century, have been designed to make parts without the need for operator skill.

Machine tools for producing low volume continued to require highly skilled machinists for their operation until the 1960's. The development of electronic Numerical Control systems, which began to be applied to machine tools at that time, has now nearly eliminated the need for operator skill and the need for his use of precision tools. Few skills die out completely, but I fear that the use of machinist's tools is rapidly becoming an art known only to a small and now aging group. Perhaps this book will help preserve a small part of our manufacturing history.

BROWN & SHARPE

BROWN, JOSEPH R. Providence, RI

Joseph Rogers Brown (1810-1876) was born in Warren, RI, January 26, 1810. He learned the machinist's trade in 1827, working in Valley Falls, RI. In 1828, Brown joined his father, who was a maker of tower clocks and, in 1833. father and son began doing business as D. BROWN & SON in Providence, RI.

On April 22, 1833, the following advertisement appeared in the *Providence Daily Journal:*

> D. BROWN & SON-Watch, Clock Makers, etc. have taken the stand No. 43 South Main street, sign of the Turret Clock, where they offer their services in the above business.
>
> They will give occular demonstration that they are prepared, and by assiduity will show themselves qualified to supply many of the wants of others to their satisfaction.
>
> They will pay particular attention to repairing Watches and all other Time Pieces that may be offered, and make such as are called for. They have for sale, 1 Turret Clock, and 1 large cutting engine, on an improved plan.
>
> They have a dividing engine for making the most accurate graduations for Mathematical and Nautical Instruments and will attend to making and repairing such as are ordered.
>
> Dividing plates for all sized Engines graduated in the most perfect manner.
>
> Spur, Spiral and Bevel Gear, and Screws for Worm Gear, Taps, etc. for repairs of Spinning Machinery, turned and cut as directed. After having been to great research and expense, and arrived to do the above work with despatch by manual labor only and individually, they think they need only to be known to have steady business.
>
> DAVID BROWN
> JOSEPH R. BROWN

In 1841, David Brown left the business and Joseph Brown began operating under his own name. For the next nine years most of Brown's business consisted of repairing clocks and watches.

In 1850, a new business was begun when Brown invented and built a linear dividing engine which is thought to be the first automatic machine for graduating rules in the United States. This machine, and two others made in 1854 and 1859 were reported as still in use as late as 1916.

Using his new dividing engine, Brown began making steel machinist's rules in 1850. In 1851 he introduced the vernier caliper, reading to thousandths of an inch and, the following year, applied the vernier to protractors. Brown's introduction of the vernier caliper made it possible, for the first time, for accurate measurements to be made in the every day work of the machine shop. This development was one of the most significant advancements in the history of precision manufacturing.

Sales of the vernier caliper were slow, the records showing that only four had been made by January, 1852. At that time Brown wrote his New York agent and asked him to return the vernier caliper the agent had on exhibition because he "needed it for some fine work and did not have another in the shop." The firm had now expanded to 1,800 square feet of floor space and, in 1853 Lucien Sharpe (1830-1899), who had been Brown's apprentice for five years, became Brown's partner. The name was then changed to:

BROWN & SHARPE, J.R., Providence, RI

The new partnership continued to produce steel rules, vernier calipers and protractors. Apparently, it did not expect to sell great numbers of the vernier calipers as Lucien Sharpe wrote their agent at Newark, NJ, in 1854, that: "it could not be expected there would be a market for many more tools in that neighborhood, as $500 worth had already been sold there".

The new vernier calipers while, for the first time putting a means of measuring to thousandths of an inch into the hands of the individual machinist, were not cheap. Selling price in 1868 was $18.00 for the smallest (6 inch) size. This would be close to a month's wages for the average machinist of the time.

In 1855, Lucien Sharpe began work on a new wire gauge standard. In January, 1856, his correspondence shows that the idea had been conceived of making a wire gauge with a regular progression of sizes, a table of sizes being made in February, 1856. In February, 1857, the Waterbury Brass Association passed resolutions adopting the new standard. The new gauge was introduced to the public in March, 1857. By 1868, the B.& S. standard gauge was widely adopted both by wire makers and sheet brass makers. The B.& S. wire gauge, also known as the American Wire Gauge is still in common use. On May 24, 1866, Brown and Sharpe formed a partnership with their largest competitor in steel rules, Samuel Darling of Bangor, ME. Darling, along with his stock, six workmen, grinder, and dividing machine, moved to Providence in the spring of 1868.

Henry Sharpe, grandson of Lucien Sharpe, recalls that the Darling dividing machine was made of old saw blades from the sawmills in Maine. Darling had not even bothered to remove the teeth from the blades except where necessary. According to him the Darling machine was superior to the J.R. Brown machines and was used for a longer time.

The new partnership was named DARLING, BROWN & SHARPE and operated as a separate company, although sharing factory space with J.R. Brown & Sharpe. Darling's end of the business was to make tools which required graduating, while Brown & Sharpe continued to make all other small tools and carried on its main business of making machine tools and sewing machines. This arrangement continued until January 1, 1892, when Samuel Darling's interest was bought out. DARLING, BROWN & SHARPE was absorbed into Brown & Sharpe Mfg. Co. in late 1895.

On January 1, 1868, Fred W. Howe and Thomas McFarlane were admitted as partners and the company was reorganized as:

BROWN & SHARPE MFG. CO. Providence, RI

Products are often marked B.& S. MFG. CO., PROV. R.I.

The new firm quickly became involved in the development of a new screw thread standard which had been proposed in 1864 by William Sellers, a prominent machine tool builder in Philadelphia, PA, and promoted by the Franklin Institute. As first adopted, the standard called for a sharp V thread which proved very difficult to make in production. The thread form was modified to allow for a "flat" on the top and bottom of the thread and is the system still in use. The Brown & Sharpe Mfg. Co. was the first to make screw thread gauges for grinding the proper angle and flat on thread-cutting tools. An illustration and description of the gauge was published in *American Artisan* magazine of November 25, 1868.

Joseph R. Brown and Lucien Sharpe, on a visit to the Paris Exposition in 1867, saw a Palmer micrometer, patented in France September 7, 1848. They were impressed with the potential of this tool and brought one back with them. In 1868, they introduced an improved version of the Palmer design as their "pocket sheet metal gauge". An illustration of the tool and a lengthy description appeared in the *American Artisan* issue of December 30, 1868. The development of this gauge is detailed in Appendix VI.

The "pocket sheet metal gauge" was designed for the purpose of accurately measuring sheet metal and had a range of only ¼". By 1877 the tool had evolved into the micrometer very much as we know it today and had a range of one inch. Brown & Sharpe used the term "micrometer caliper" for the first time in its 1877 catalog.

The new micrometer was quick to gain acceptance. Luther Burlingame, then chief draftsman for Brown & Sharpe, states in a 1915 article that: "In 1880, the Brown & Sharpe Mfg. Co.was regularly making one-inch micrometers in lots of five hundred and had been making them in lots of this size for several years prior to that date. A year or two later one-inch micrometers were made in lots of one thousand."

Sales were boosted in 1882, when the Railway Master Mechanics committee on Standard Wire Gauge recommended adoption of the Brown & Sharpe micrometer as the standard by which the new Wire Gauge was to be measured.

In 1883, Brown & Sharpe bought the micrometer business of the VICTOR SEWING MACHINE CO. of Middletown, CT. The purchase included all stock and patents, including the important Pratt patent of April 23, 1878 (No. 202866), which covered an inserted anvil. This was a crucial patent as it had proven impossible to properly harden anvils which were integral with the frame.

As part of the deal they also got the services of Frank Spalding, who had been in charge of the micrometer department at Victor. Spalding headed the micrometer department at Brown & Sharpe for many years and developed nearly all of the later features of Brown & Sharpe micrometers, including the first technically successful direct reading micrometer.

In 1884 Brown & Sharpe introduced an improved version of the "Victor" micrometer incorporating the January 22, 1884 (No. 292409), Cyrus Carleton patent for an improved method for taking up wear. It also acquired Merrick M. Barnes' patents of January 30, 1883 (No. 271401), and July 3, 1883 (No. 280339), for a means of covering the screw threads and the Jerrey Moffitt patent of September 6, 1887 (No. 369357), to avoid openings exposed to the dirt. Moffitt is also credited with the idea for stamping decimal equivalents on micrometer frames, a practice still very common. By 1885 the Brown & Sharpe micrometer had reached its final form. Many small improvements were added in manufacture and design but the micrometer of 1885 would be very familiar to the machinist of 1994.

On April 11, 1893, a U.S. patent (No. 495379) was issued to J. Ciceri Smith, of Edinburgh, Scotland, for a direct reading micrometer caliper. This patent was acquired by Brown & Sharpe but the tool was never made by them.

In 1888, the entire line of micrometers was expanded and given numbers. Tables of decimal equivalents were stamped on the frame, verniers reading to .0001" were offered, and the number 4 model was made with the outer end of the frame the same size as the spindle.

The size of the plant continued to expand. Beginning with 1,800 square feet in 1853, it had several thousand square feet in 1872, expanding to 482,473 square feet in 1903 and, by 1933, it had 32 acres under roof.

Brown & Sharpe introduced, in 1894, a micrometer in which the graduations, rather than being in inches, were in oz. per sq.ft. of steel sheet. Each graduation of the barrel equaled 20 ounces per square foot of steel of corresponding thickness while each graduation of the thimble equaled one ounce per square foot. The company also began supplying a new design of ratchet stop based on the H.V. Bernhardt patent of November 6, 1894 (No. 528759), replacing the earlier design patented December 6, 1892 (No. 487410), by George Sylvia.

The company introduced, in 1895, a new screw pitch gauge with leaves narrow enough to enter a nut. The tool was patented by Samuel Darling in his final tool patent, granted December 31, 1895 (No. 552238).

The presently used trademark (B S inside the "L" of a square) was adopted about 1895. Vernier calipers "in 3", 6", 12" and 24" sizes, larger sizes made to order" were advertised under the BROWN & SHARPE MFG. CO. name in December, 1895. This ad marks the end of tool production under the DARLING, BROWN & SHARPE name, although tools in stock must have been shipped over many years.

It appears that all of the DARLING, BROWN & SHARPE tools were continued by BROWN & SHARPE MFG. CO., although many were gradually replaced with newer designs. Brown & Sharpe introduced, in 1901, lines of spring calipers nearly identical to the Starrett "Fay" and "Yankee" types but with its own design of speed nut covered by the Frank Spalding patent of October 5, 1897 (No. 591363). It also offered the new #57 micrometer for measuring "all sizes from 6" to 12"." The micrometer was supplied with 6 anvils in one inch increments.

The first surface gauges offered by Brown & Sharpe were introduced in 1902. They were offered in the #860, #861 and #862 pattern with a rectangular base and pivoting stem and a #864 and #865 pattern with a round base and fixed stem. The surface gauges were covered by the Charles J. W. Hayes patent of July 11, 1905 (No. 794500). An improved 29 degree screw thread tool gauge was also introduced.

The 1904 catalog shows rules with end graduations marked with the August 25, 1885 (No. 325096), Coffin & Leighton patent. Brown & Sharpe evidently bought either the patent or a license at some time when the patent was still in effect.

The company introduced, in 1904, a line of toolmaker's calipers which are still in production, and the #614, #615, #616 and #617 "hammerless center punches" (we would now call them automatic center punches) patented by James Hartley and Joseph Stryhal on February 7, 1905 (No. 781947), and February 28, 1905 (No. 783749). The #612 spacing attachment for the punches was introduced in 1907.

1905 offerings included new large size try squares, 24" to 36", patented June 7, 1904 (No. 761929), by Charles L. Weld; the #467 lathe test indicator: #1160 combination caliper and divider; #253 tool maker's rule and holder set. The Weld design for large squares replaced an earlier design patented January 17, 1882 (No. 252436), by Samuel Darling.

In 1906, Brown & Sharpe began offering a set of knife edge straight edges which had been developed by Frederick N. Gardner in 1883. The straight edges were first made by the HARTFORD TOOL CO. and later by the PRATT & WHITNEY CO., both of Hartford, CT.

1907 offerings included the new #806 and #807 inside micrometer calipers, patented January 29, 1907 (No. 842263), by Frank Spalding; and #870 surface gauges which were a redesign of the #864 and #865 round base pattern which had been introduced in 1902. The new pattern surface gauges were made under the same C.J.W. Hayes patent issued July 11, 1905 (No. 794500).

In 1909, it added lines of heavy micrometers and tubular inside micrometers; speed indicators patented March 8, 1910 (No. 951491), by Frank Spalding; beveled edge hardened squares; an indicator based on Byron F. Stowell's patent of July 24, 1906 (No. 826971); and a new style surface gauge with a fine adjustment built into a block which supported the stem. This surface gauge was probably quite expensive to make and was soon replaced by a more conventional design with a pivoting arm for the fine adjustment.

Brown & Sharpe introduced a direct reading micrometer in 1913, based on a patent issued to Frank Spalding on May 5, 1914 (No. 1095745). This was never a commercial success, probably due to the price, which at $12.00 was more than twice the $5.50 price of a standard micrometer in 1915.

Catalog #26, issued in 1915, offered micrometers, steel rules, machinist's squares, combination squares, bevel protractors, straight edges, vernier calipers, screw pitch gauges, #620, #621 and #622 surface gauges with the pivoting arm adjustment, thickness gauges, standard reference disks, limit gauges, wire gauges, indicators, and spring calipers.

Micrometers offered in the 1915 catalog included the sheet metal micrometer #220 and micrometer caliper #221 which had remained nearly unchanged since 1868; #250 inside micrometers and #249 gear tooth micrometers, based on Frank Spalding's patent of January 29, 1907 (No. 842263); #222 micrometer caliper, an odd sheet metal gauge which had been patented

June 18, 1901 (No. 676637), by Frank Spalding and which continued to be offered for many years; #227 and #228 micrometer calipers for electricians which were stamped with conversions from thousandths of an inch to wire sizes and ohms per hundred feet. These electrician's micrometers had been made since about 1900, and continued to be made well after 1915. Examples, however, are rather scarce.

New tools in 1915 included the #243 bench micrometers and the low cost "REX" micrometers which sold for $3.50 compared with $5.50 for a micrometer from its standard line. An improved direct reading micrometer, based on the J.W. Parker patent of November 28, 1922 (No. 1437352), was introduced in 1924. This new design also sold poorly, probably for the same reason as the first design. The micrometer was priced at $20.00, or more than twice the $8.50 charged for a standard model in 1924.

Other new tools offered in the number 29 catalog, issued in1924, included: #226 ball anvil attachment; #350 stainless steel rules; #377 key seat clamps; #500 combination bevel; #552 die maker's square; #554 adjustable square; #607 depth micrometer; #625 planer and shaper gauge; #672 taper parallel gauges; #748 speed indicator, patented November 28, 1922 (No. 1437353), by John W. Parker; #749 V blocks. It is interesting to note that the key seat clamps and the combination bevel were nearly identical to tools made by the L.S. STARRETT CO., who in turn had copied the designs of both tools from those made by the SAWYER TOOL MFG. CO.

New style hook rules, based on John W. Parker's patent of March 10, 1925 (No. 1529210), were introduced in catalog number 30, issued in 1926. Other new tools included: #263 and #265 inside micrometers; #900 cutter clearance gauge, patented December 23, 1924 (No. 1520450), by Walter F. Smith.

Catalog number 31, issued in 1929 offered a number of redesigned tools and some new tools including: new combination squares incorporating the John W. Parker patent of March 10, 1925 (No. 1529209); #465 height gauge attachments; #616 rule depth gauge and protractor; #746 vest pocket speed indicator; #756 toolmaker's clamps, patented May 14, 1929 (No. 1713239), by Malcolm K. Parkhurst.

A supplement included with the number 31 catalog offered more new tools, including: #24 micrometer calipers, reading by ten-thousandths directly; #468 depth gauge attachment; #590 telescoping gauges; #614 rod depth gauge; #617 drill point gauge and depth gauge; #627 radius gauges, patented April 15, 1930 (No. 1755152), by John W. Parker. The #24 micrometers look to be a useful development and should have been popular with machinists. However, they do not seem to have sold well, probably for the same reason as the direct reading micrometers. Price of the new design was $18.00 versus $11.75 for the old design where ten-thousandths were read off a vernier.

For a detailed look at the tools offered by the BROWN & SHARPE MFG. CO., a collection of catalogs, *Brown & Sharpe Mfg. Co. and Darling, Brown & Sharpe Catalogs for 1868, 1884 and 1891* is available from: The Astragal Press, P.O. Box 239, Mendham, NJ, 07945. Original catalogs from the 1930's and later are easily found and contain many of the earlier tools.

AMES' PATENT UNIVERSAL SQUARE.

"As a **CENTER-SQUARE** alone, it is invaluable to every mechanic. * * * In short, it combines, in a most convenient form, so many useful instruments, no mechanic's list of tools can well be complete without a Universal Square."—*Sc. Amer. Sep.* 22, 1855.

6 in. blade, $3.00 –8 in. $4.00.–10 in. $5.00.–12 in. $6.00.

This Square combines, in the most convenient form *five different instruments*: viz: The TRY SQUARE, the MITER, the T-Square, the GRADUATED RULE, and (what is entirely new) the CENTER SQUARE for finding the center of a circle.

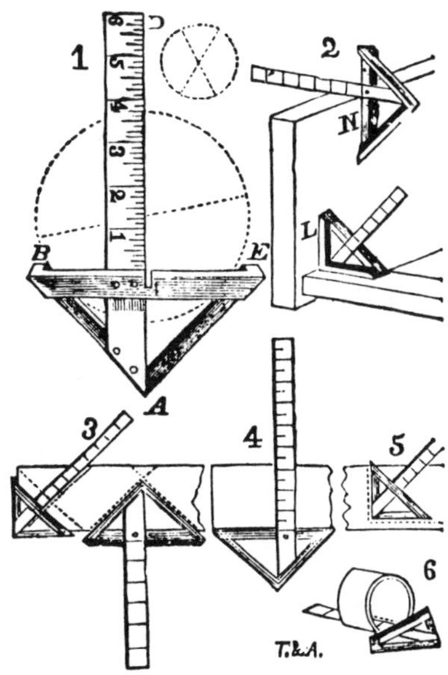

The tongue, D A, (Fig. 1,) being fastened, as it is, into the triangular frame, B A E, cannot be moved or knocked from its place,—in this respect, constituting a great improvement over the carpenters' Try-Square, T-Square and Miter, in common use. The instruments are made of the best material, neatly finished, and perfectly true.

Fig. 1, explains its application as a CENTER-SQUARE. Put the instrument over the circle, as the end of a bolt or shaft, with the arms, B A, E A, resting against the circumference, in which position one edge of the rule, A D, will cross the center. Mark a straight line in this position; apply the instrument again to another part of the circumference, and mark another line crossing the first. The point where the two lines cross each other, will be the center of the circle. The whole is the work of a moment. *Fig.* 2, explains the application of the instrument, as a carpenter's TRY-SQUARE, N, and an OUTSIDE-SQUARE, L;—*Fig.* 3, as a MITER;—*Fig.* 4, as a T-SQUARE and a GRADUATED RULE;—*Figs.* 5, and 6, as an OUTSIDE-SQUARE for drawing, and a T-SQUARE for machinists.

STEEL TRIANGLES.

OPEN STEEL TRIANGLES FOR DRAUGHTSMEN.

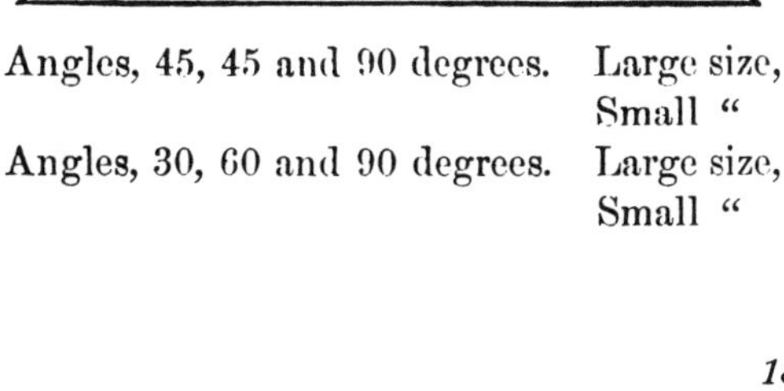

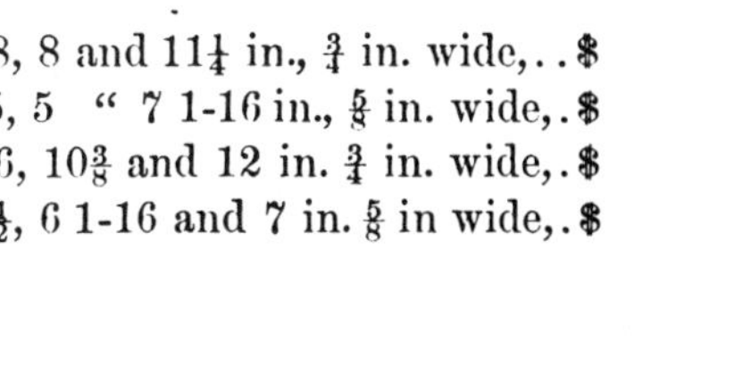

Angles, 45, 45 and 90 degrees.	Large size, sides 8, 8 and 11¼ in., ¾ in. wide,.. $
	Small " " 5, 5 " 7 1-16 in., ⅝ in. wide,. $
Angles, 30, 60 and 90 degrees.	Large size, sides 6, 10⅜ and 12 in. ¾ in. wide,. $
	Small " " 3½, 6 1-16 and 7 in. ⅝ in wide,. $

1868

AMERICAN ARTISAN

STANDARD SCREW-THREADS.

We take pleasure in laying before our readers any information we may be able to obtain concerning the progress our engineers are making in establishing a uniform system of screw-threads for bolts and nuts. The publication of the report of the committee of the Franklin Institute has caused numerous inquiries for the necessary gauge to form the flat end upon the cutting tools, for the want of such gauges made itself felt at the very outset in the attempt to introduce the proposed system. We now inform our readers that this gauge can be obtained from the Brown & Sharpe Manufacturing Company, of Providence, R. I., who have been for some time past providing the necessary machinery for its construction. We here give an illustration of the gauge.

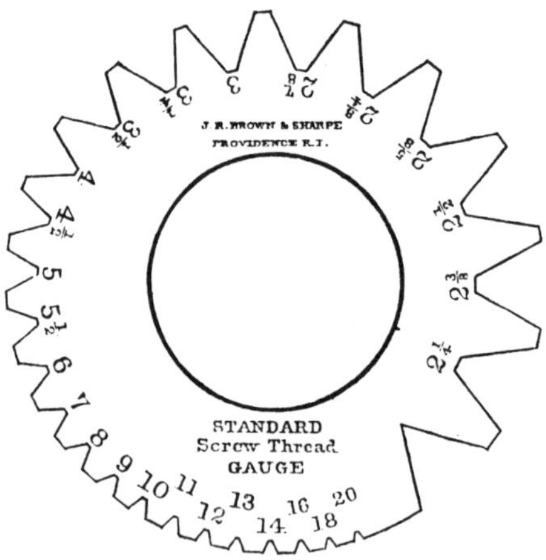

A change of system in the screw-cutting machinery of any country must necessarily be slow ; but the advantages to be obtained from the adoption of any system were so palpable that many of our large machine-works and one or two of our navy-yards made preparations to adopt the proposed standard at once, and have so far progressed that a few of the larger shops are now prepared and do execute all new work to the standard thread, and it is also extensively used on some of our principal railways.

Nov. 25, 1868

VERNIER CALIPERS.

IMPROVED PATTERN.

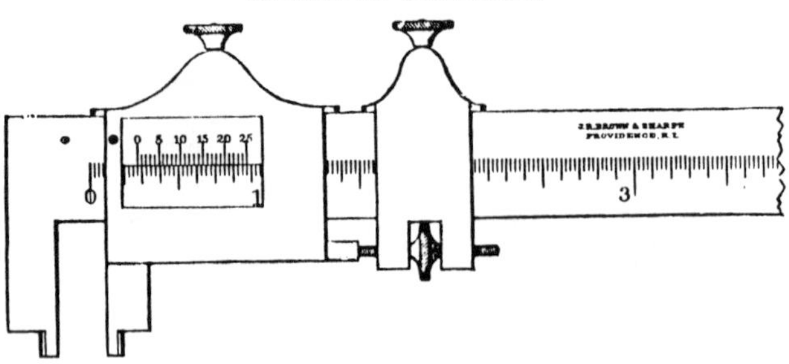

Front View.

1868

The above cut is a fac-simile of one side of the Vernier Caliper, which reads to thousandths of inches. On the other side are 64ths or 50ths of inches, to read without a vernier. This instrument is furnished with both inside and outside calipers and points to transfer the distance with dividers. An explanation of the Vernier accompanies each instrument. These instruments are made of steel, and have the points tempered and jaws ground.

AMERICAN ARTISAN

A POCKET SHEET-METAL GAUGE.

IN the operations of many different trades it is essential that some reliable and convenient means should be at hand for accurately measuring the thickness of sheet-metal and other sheet material. Such an appliance is represented in the adjoining cut, and its uses and construction will be readily understood from the subjoined description :—

The piece in the form of the letter   has a projecting hub, *a*, on one end. Through the two ends are tapped holes, in one of which is the adjusting screw, B, and in the other the gauge screw, C. Attached to the screw, C, is a thimble, D, which fits over the exterior of the hub, *a*. The end of this thimble is beveled, and the beveled edge graduated into twenty-five parts and figured, 0, 5, 10, 15, 20. A line of gradations 40 to the inch is also made upon the outside of the hub, *a*, the line of these divisions running parallel with the center of the screw, C, while the gradations on the thimble are circular. The pitch of the screw, C, being 40 to the inch, one revolution of the thimble opens the gauge $\frac{1}{40}$ or $\frac{25}{1000}$ of an inch. The divisions on the thimble are then read off for any additional part of the revolution of the thimble, and the number of such

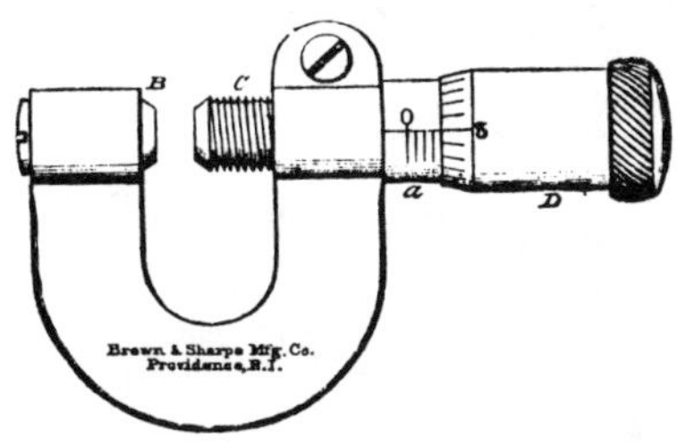

divisions are added to the turn or turns already made by the thimble, allowing $\frac{25}{1000}$ for each gradation on the hub, *a*. For example suppose the thimble to have made four revolutions and one-fifth. It will then be noticed that the beveled edge has passed four of the gradations on the hub, *a*, and opposite the line of gradation will be found on the thimble the line marked 5. Add this number to the amount of the four gradations, which is $\frac{100}{1000}$, and it equals $\frac{105}{1000}$, which is the measurement shown by the gauge.

The advantages claimed for this over gauges of the ordinary construction may be stated thus :—The edges of metal being often imperfect, ordinary gauges are prevented from going on readily, and it also usually happens that the extreme edges are thinner than the rest of the sheet and cannot therefore be relied upon to give the thickness correctly ; both of these objections are obviated in the gauge under consideration. Furthermore, in selecting sheets for many purposes, it is desirable to have a gauge to indicate the exact thickness in parts of an inch, and this end is effectually secured by the means above described, the gauge being capable of showing the thickness of a piece of metal up to three-tenths of an inch in thousandths of an inch, and at some distance from the edge of the sheet. The small size, lightness, and suitability for being carried in the pocket are characteristics which will be readily appreciated by those having occasion to use a device of this kind.

Tables of decimal and fractional parts of an inch, and the thickness in decimal parts of an inch of the various numbers of the American and Birmingham wire gauges, are furnished with this gauge By the aid of these tables the measurement obtained by this gauge may be made to show the numbers according to the American or Birmingham gauge.

These improved sheet-metal gauges are manufactured and for sale by the Brown & Sharpe Manufacturing Company, of Providence, R. I.

December 30, 1868

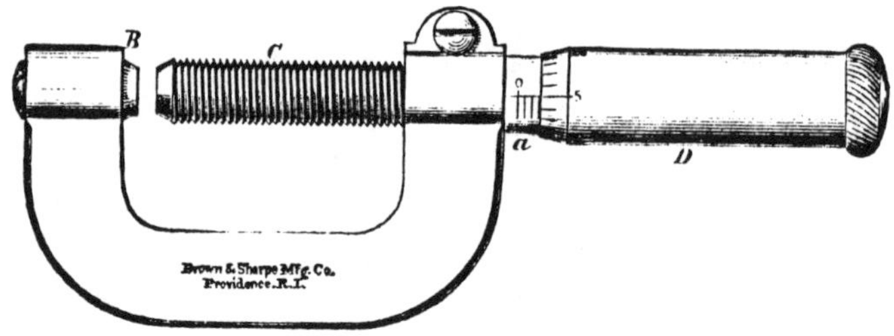

STANDARD MEASURING INSTRUMENT ADOPTED BY THE MASTER MECHANICS' ASSOCIATION.

1882

VICTOR MICROMETER CALIPER,

MADE BY

BROWN & SHARPE M'F'G CO.,

PROVIDENCE, R. I.

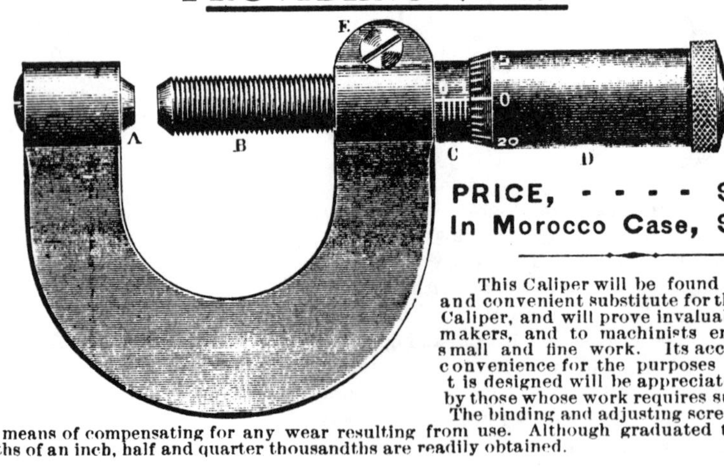

PRICE, - - - - $6.00.
In Morocco Case, $7.00.

This Caliper will be found a reliable and convenient substitute for the Vernier Caliper, and will prove invaluable to tool makers, and to machinists engaged on small and fine work. Its accuracy and convenience for the purposes for which t is designed will be appreciated at once by those whose work requires such a tool. The binding and adjusting screws furnish the ready means of compensating for any wear resulting from use. Although graduated to read to thousandths of an inch, half and quarter thousandths are readily obtained.

1883

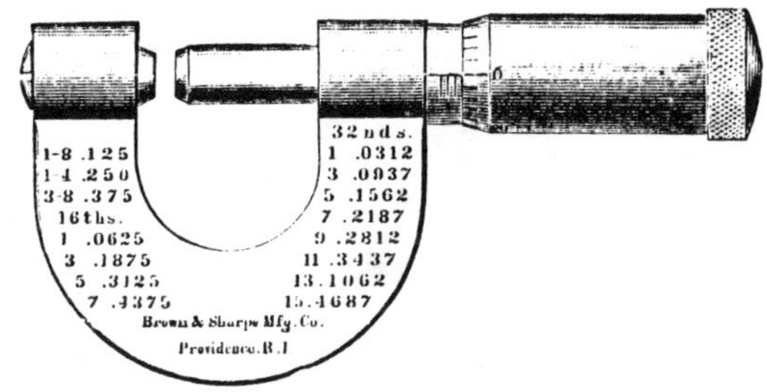

No. 2.

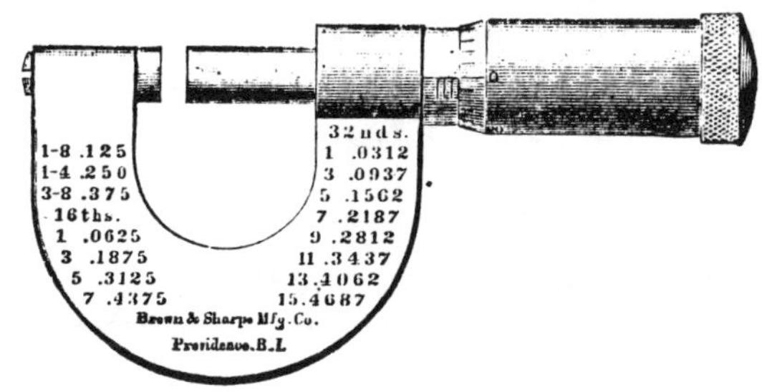

No. 4.

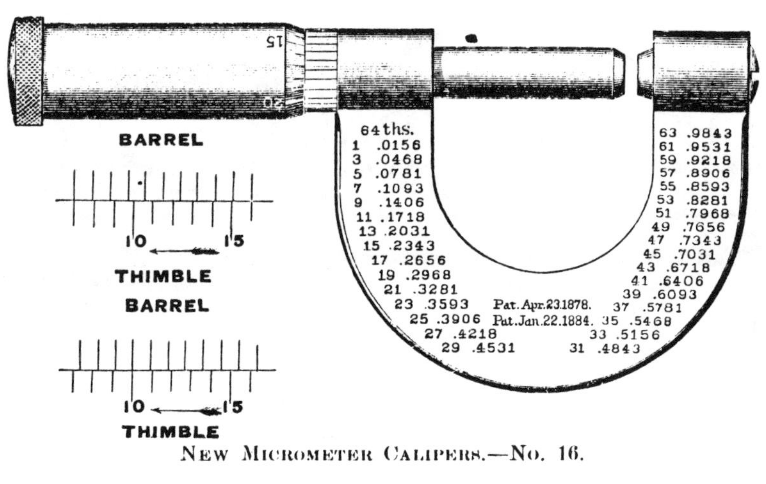

New Micrometer Calipers.—No. 16.

1888

1888

1889

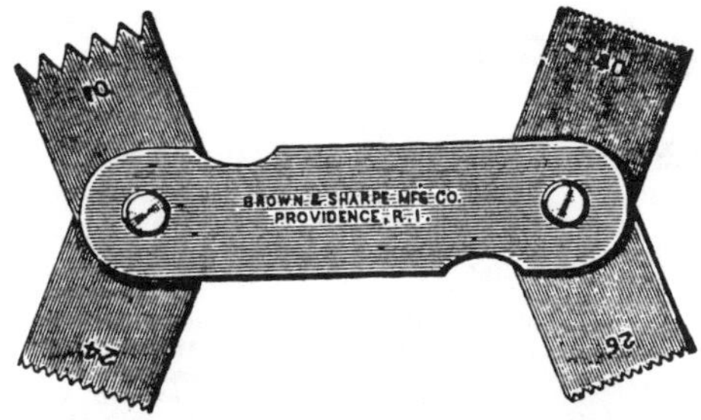

Screw Pitch Gauge.

For 16 Pitches.

Price, - $ .75

(Cut Full Size.)

1891

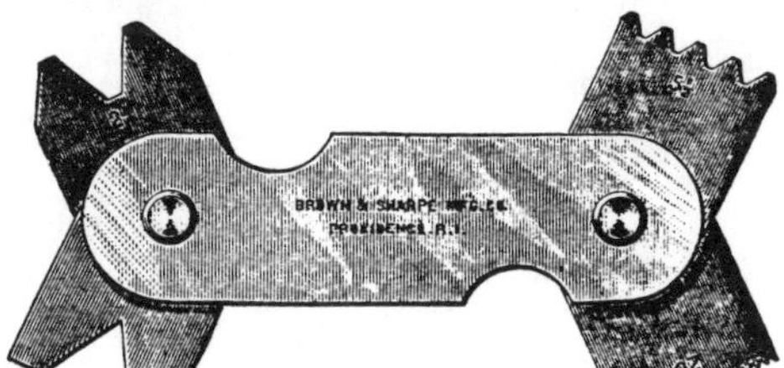

Two-Thirds Size.

Screw Pitch Gauge.

For 25 Pitches.

FRANKLIN INSTITUTE THREAD.

Price, $1.75

MICROMETER CALIPERS WITH LARGE MEASURING SURFACES.

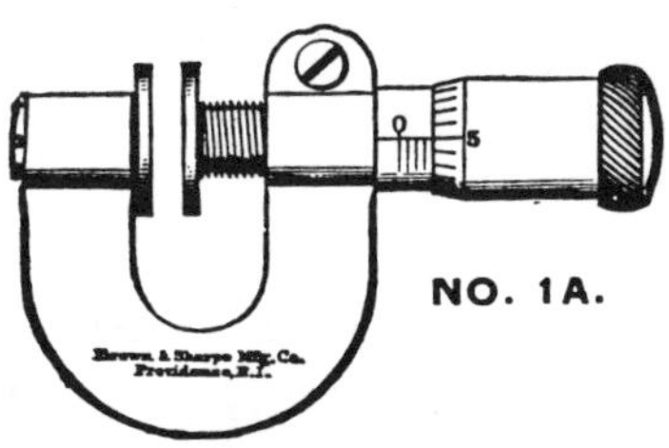

NO. 1A.

For measuring the thickness of cardboard, paper, or other yielding substances, it is often found of advantage to substitute discs or washers (usually ½″ diameter), in place of the beveled ends of Measuring Spindles and Adjusting Screws. The comparatively large surfaces have less tendency to compress the objects measured, and in many hands enable accurate measurements to be more quickly obtained. No. 1 Micrometer Caliper fitted with these discs.

Price, $5.50. In Morocco Case, $6.00.

1891

BROWN & SHARPE MFG. CO.,

PROVIDENCE, R. I.

CAST IRON STRAIGHT EDGES.

18 in. x 1 1-2 in.	7 lbs.	$7.00.
24 in. x 1 5-8 in.	10 "	9.50.
30 in. x 1 3-4 in.	16 "	12.00.
36 in. x 1 7-8 in.	19 "	15.00.
48 in. x 2 in.	36 "	20.50.
60 in. x 2 1-8 in.	51 "	26.50.
72 in. x 2 1-4 in.	78 "	33.00.
96 in. x 2 1-2 in.	153 "	39.00.
120 in. x 2 5-16 in.	175 "	45.00.

1891

S. A. SMITH,

Western Representative,

23 So. Canal Street, CHICAGO, ILLS.

BROWN & SHARPE MFG. CO.,

PROVIDENCE, R. I.

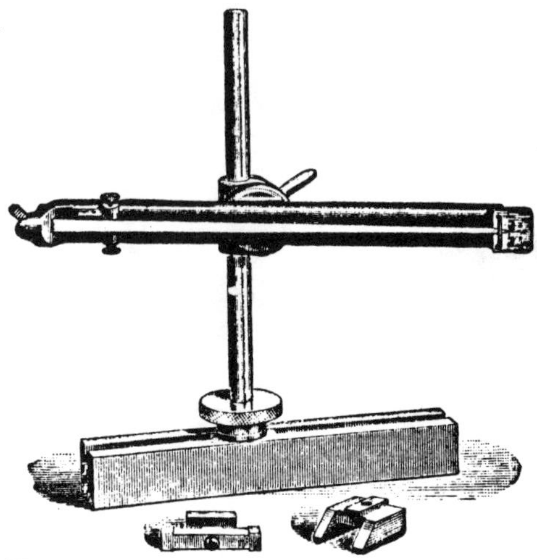

TEST INDICATOR.

This Indicator is especially useful to those erecting or inspecting machines. It is possible by its use to readily determine the degree of inaccuracy of a plane surface on the top, bottom or side of a piece of work, or to easily ascertain the amount of end movement, for example, of a spindle, or the extent to which a spindle runs out of true.

The upright post, or stand, may be clamped at any point upon the base by the knurled nut. The sleeve which carries the arm may be fastened at any height on the post or turned around the post to bring the arm on either side. The arm turns in the sleeve and may be set at any angle relative to the base, or may be inverted so that the point brought in contact with the work will be downward rather than in position shown in cut.

S. A. SMITH,

Western Representative,

23 So. Canal St., CHICAGO, ILL.

1891

BROWN & SHARPE MFG. CO.,

PROVIDENCE, R. I.

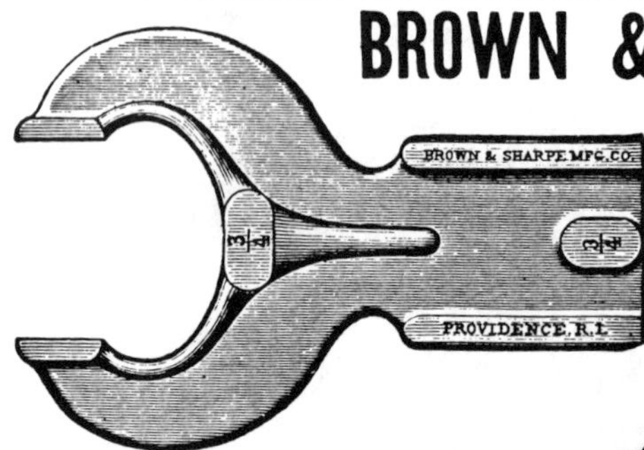

STANDARD CALIPER GAUGES.

Convenient for general shop use.

Gauges 3 inch and larger are made in two parts.

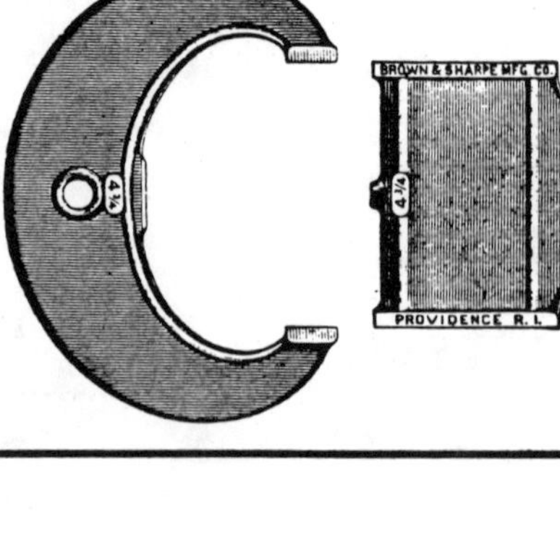

ENGLAND—BUCK & HICKMAN, 280 Whitechapel Road, London, E.

GERMANY—SCHUCHART & SCHUTTE, 59 Spanoauerstrasse, Berlin, C. (Small Tools).

GERMANY—G. DIECHMANN, Ansbacherstr, 5 Berlin, W. 62.

FRANCE—FENWICK FRERES & CO., 21 Rue Martel, Paris.

FRANCE—F. G. KREUTZBERGER, 140 Rue de Neuilly Puteaux (Seine).

CHICAGO, ILL.—FRED. A. RICH, 23 South Canal St.

1894

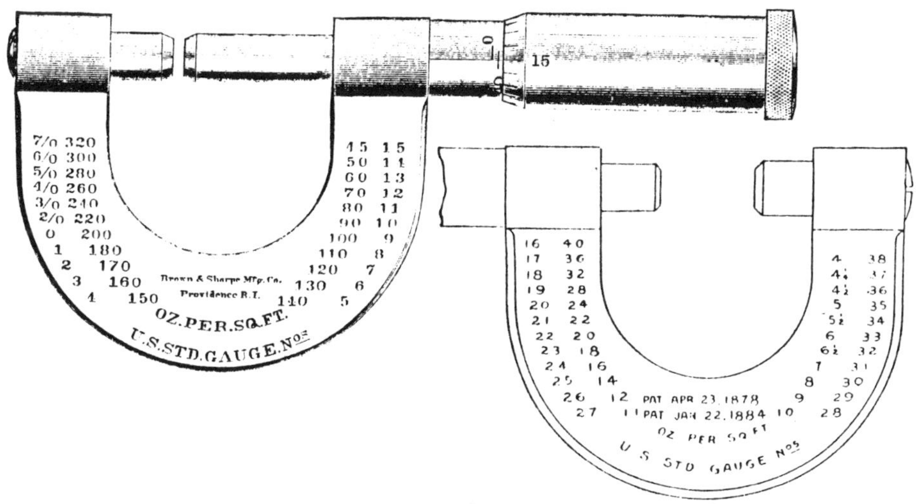

MICROMETER CALIPER FOR SHEET AND PLATE IRON AND STEEL.

1895

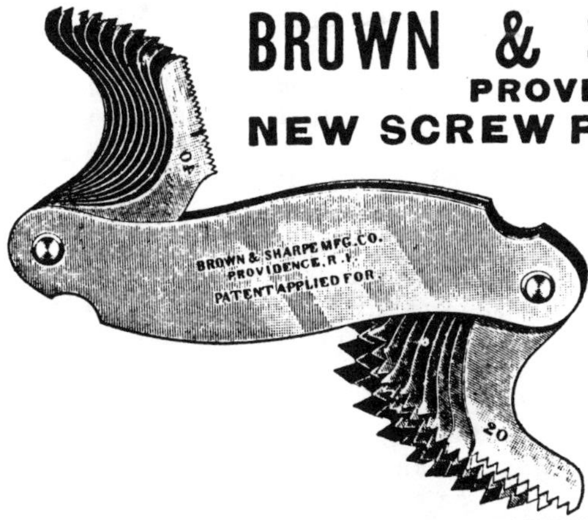

BROWN & SHARPE MFG. CO.

PROVIDENCE, R. I.

NEW SCREW PITCH GAUGE NO. 20,

22 PITCHES.

It will measure the threads of nuts as well as of screws, and contains the pitches 9, 10, 11, **11 1-2,** 12, 13, 14, 15, 16, 18, 20, on one end, and 22, 24, 26, **27,** 28, 30, 32, 34, 36, 38 and 40 on the other end.

It contains the 11½ and 27 pitches for pipe threads.

It contains practically two complete gauges as heretofore made. Price, $1.

We make no additional charge for the extra pitches.

REPRESENTATIVES AT HOME:

CHICAGO, ILL.—FRED. A. RICH, 23 So. Canal St.
NEW YORK CITY—F. G. KRETSCHMER, 136 Liberty St., Room 507.
BUFFALO, N. Y.—R. HOFFELD & CO., 61 Carroll St.
CINCINNATI, OHIO—E. A. KINSEY & CO., 227 W. Fourth St.
PITTSBURGH, PA.—U. BAIRD MACHINERY CO., 74 Water St.

REPRESENTATIVES ABROAD:

ENGLAND—BUCK & HICKMAN, 280 Whitechapel Road, London, E.
ENGLAND—CHAS. CHURCHILL & CO., Ltd., 21 Cross St., Finsbury, London, E. C.
GERMANY—SCHUCHARDT & SCHUTTE, 59 Spandauerstrasse, Berlin C. (Small Tools).
GERMANY—GUSTAV DIECHMANN & SOHN, Zimmer-Strasse 87, Berlin, S. W. 12, Germany.
FRANCE—FENWICK FRERES & CO., 21 Rue Martel, Paris.
FRANCE—F. G. KREUTZBERGER, 140 Rue de Neuilly Puteaux (Seine).

1895

Universal Surface Gauges.

PATENT APPLIED FOR.

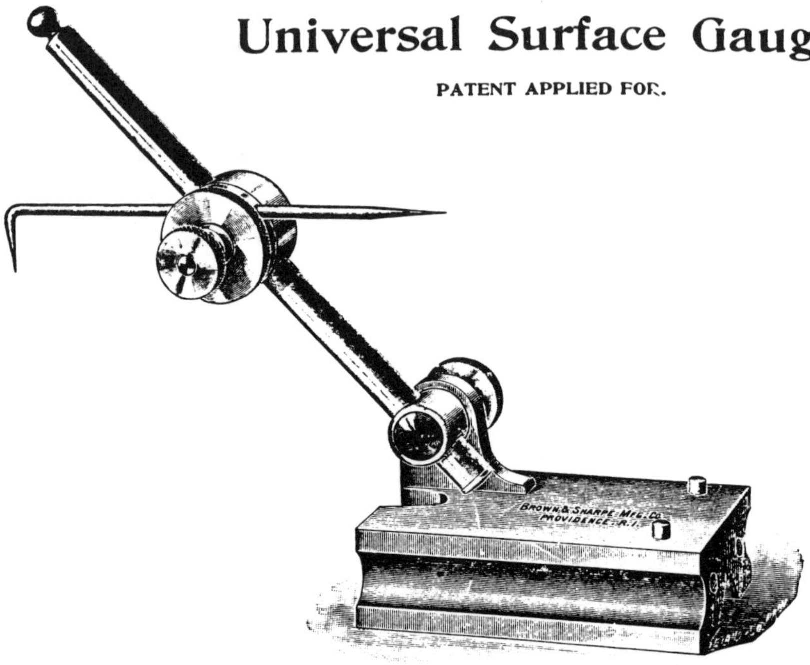

No. 860, Tool Makers, Price, $2 50.

No. 861, 9 inch, Price, $2 50. No. 862, 12 inch, Price, $3 00.

This Gauge, shown full size, is especially designed to meet the requirements of tool makers and those desiring a reliable tool for accurate and complicated measurements.

The base is solid, carefully case-hardened, handsomely mottled and of a form most convenient to handle.

The bases of the 9″ and 12″ gauges have in addition to the V shaped groove in the end, a corresponding groove in the bottom, adapting the gauges for use on cylindrical work. These sizes are also provided with two gauge pins, in the rear end of the base, that can be pushed down and used against the edge of a plate or side of a T slot.

The post swivels can be set and rigidly clamped in any position from the vertical to the horizontal and the scriber used below the base as a depth gauge.

The scriber has a fine adjustment that can be used after the sliding block is set at the approximate height. The adjustment is made by means of the large knurled nut, shown in cut, which, when turned, revolves the scriber clasp slowly and continuously and admits of the scriber being set at any position within its range.

1907

Surface Gauges.

PATENT APPLIED FOR.

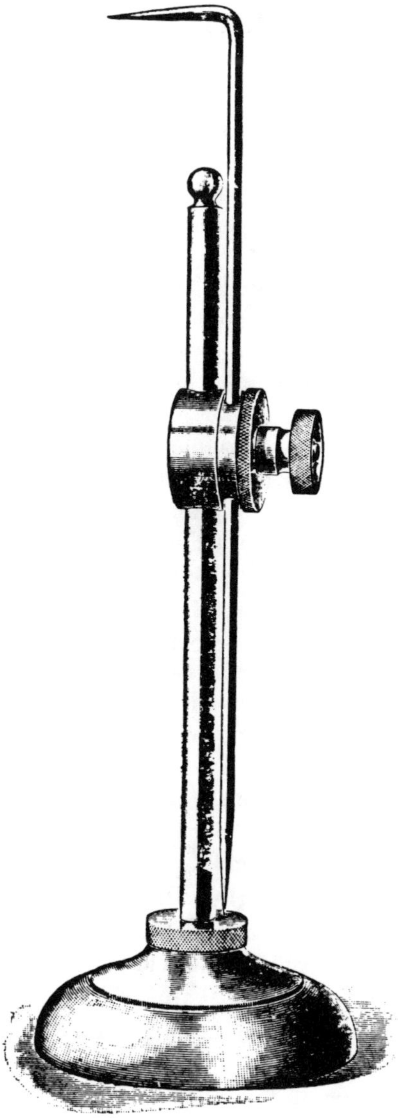

No. 864.	**9″ high.**	**Price, $2 50.**
No. 865.	**12″ high.**	**Price, $3 50.**

This Gauge presents new features that greatly enhance its value as a convenient and desirable tool.

The post is of steel and the base is carefully case-hardened and handsomely mottled.

The sleeve and scriber clasp are entirely new in design and extremely simple in construction. When loosened for adjustment they are both held in position by a spring friction and can be rigidly clamped by a single knurled nut.

The scriber has a fine adjustment that can be used after the sliding block is set at the approximate height. This device is simple and cannot get out of order. The adjustment is made by means of the large knurled disk, shown in cut, which, when turned, revolves the scriber clasp slowly and continuously and admits of the scriber being set at any position within its range.

1903

Universal Depth Gauge.

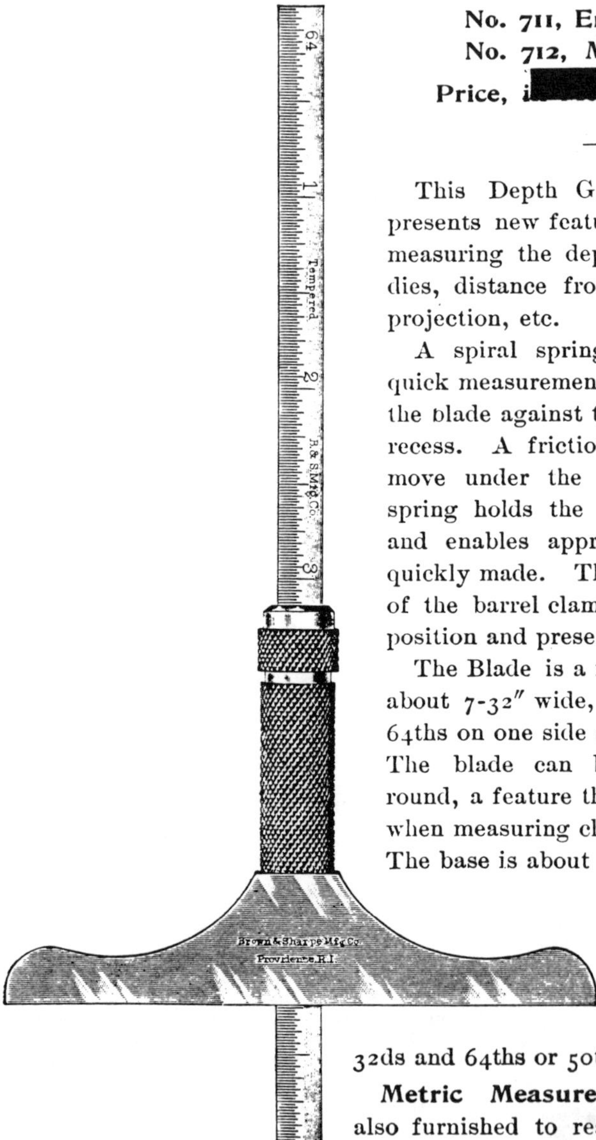

No. 711, English Measure.
No. 712, Metric Measure.
Price, [illegible] **$3 00.**

This Depth Gauge, shown full size, presents new features and is designed for measuring the depth of holes, recesses in dies, distance from a plain surface to a projection, etc.

A spiral spring in the barrel makes quick measurements possible, as it forces the blade against the bottom of the hole or recess. A friction clutch that is free to move under the pressure of the spiral spring holds the blade without clamping and enables approximate settings to be quickly made. The clamp nut at the top of the barrel clamps the blade securely in position and preserves the setting.

The Blade is a narrow 6″ tempered rule about 7-32″ wide, accurately graduated to 64ths on one side and 100ths on the other. The blade can be swiveled completely round, a feature that is of great advantage when measuring close to a corner or angle. The base is about 3″ long, 7-16″ wide and carefully hardened and ground.

Other Blades. We furnish, when desired, blades divided into 32ds and 64ths or 50ths and 100ths of inches.

Metric Measure. This Depth Gauge is also furnished to read to 1-50 m/m on one side and to 1-2 m/m on the other.

1903

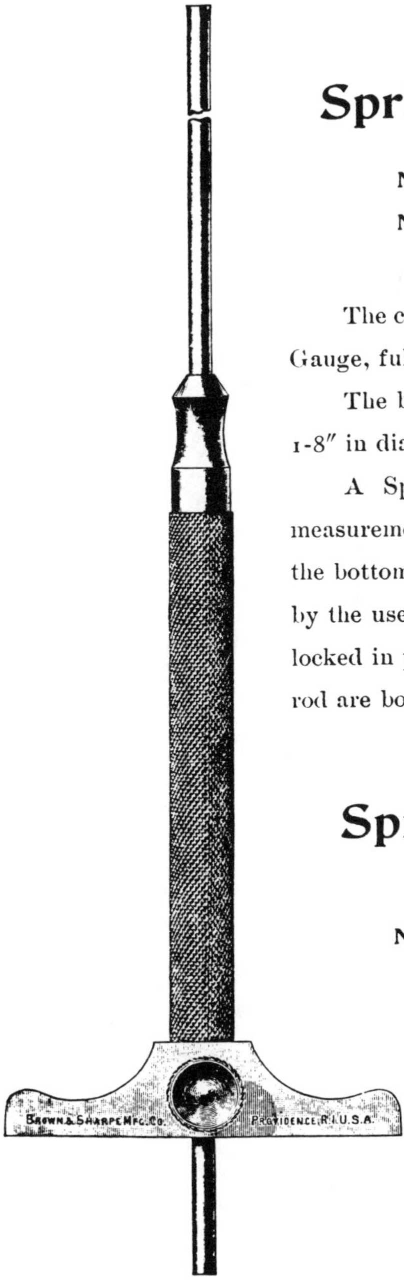

Spring Depth Gauges.

No. 713, 2″ Base, Price, $1 50.

No. 717, 4″ Base, Price, $2 00.

The cut shows the head and barrel of the Depth Gauge, full size, together with a portion of the rod.

The base is about 7-16″ wide and the rod about 1-8″ in diameter and 6″ long.

A Spiral Spring in the barrel makes quick measurements possible, as it forces the rod against the bottom of the hole or recess to be measured and by the use of the clamp screw, the rod is securely locked in position. The base and lower end of the rod are both hardened.

Spring Depth Gauge

WITH FRICTION.

No. 725, 3″ Base, Price, $2 25.

This gauge differs from No. 713 only in that the rod is held by a friction clutch that is free to move under pressure of the spiral spring and enables approximate settings to be quickly made.

1903

Combination Caliper and Divider.

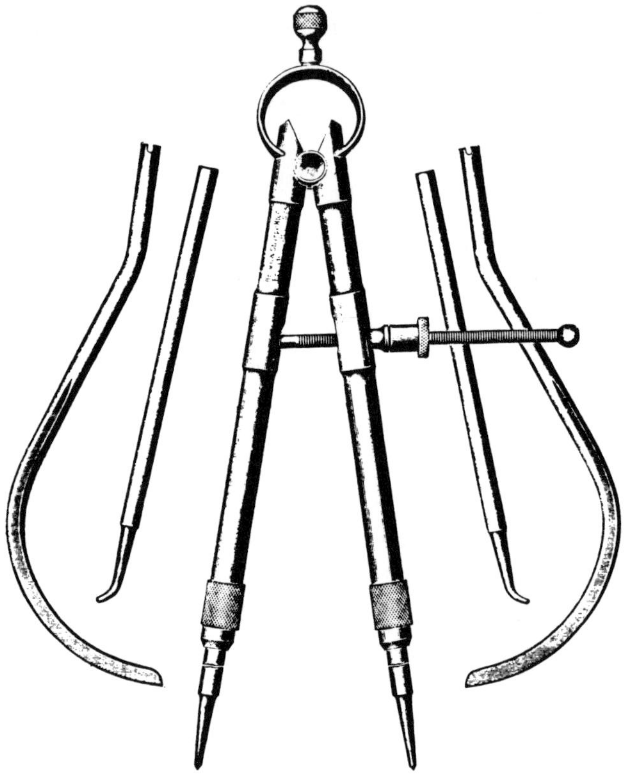

This Combination Caliper and Divider is entirely new in design. The arms or holders are provided with split chucks to receive the auxiliary legs, which are held firmly by the simple turn of the knurled nut that closes the chuck concentrically. A pencil can be substituted for one of the legs if desired.

The tool is of steel carefully finished and sharp corners are practically eliminated.

Used as a divider it will describe a circle 21 1-2″ in diameter.

PRICES.

No. 1160, Set Complete, $3 00. **No. 1161, with Divider Legs only, $2 00.**

1905

Universal Dividers.

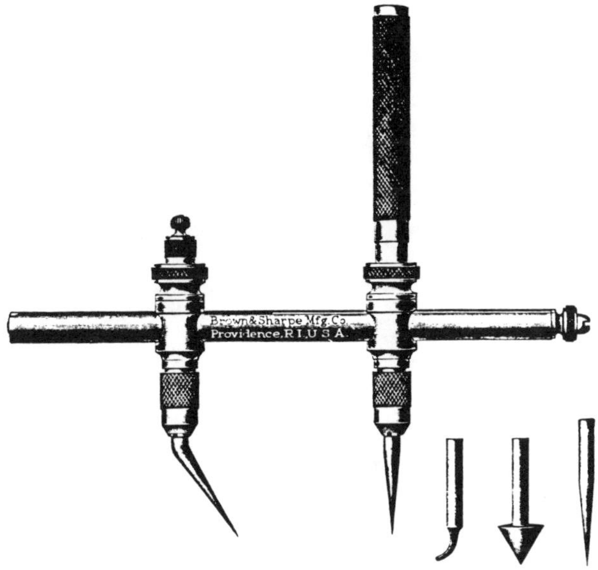

No. 856. Price, $3 00.

1905

Lathe Test Indicator.

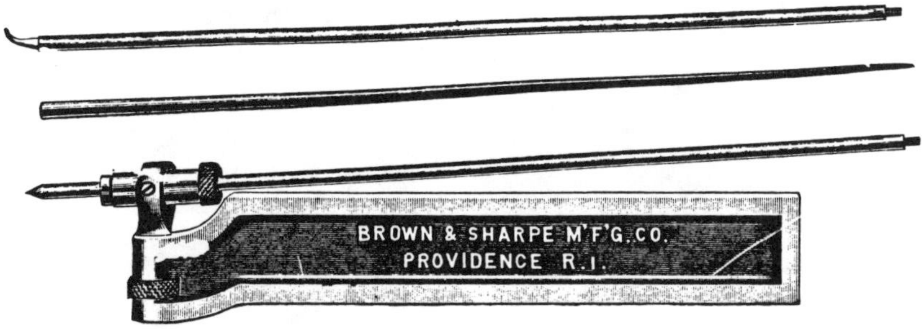

No. 467. Price, $3 00.

1905

Tempered Hook Rules.

Brown & Sharpe Mfg. Company, Providence, R. I., is offering tempered hook rules, as shown in the accompanying cuts. The narrow rule illustrated in Fig. 2 is of the same construction as that in Fig. 1, differing only in that it is narrow enough to allow measurements to be taken through a hole ⅜ inch in diameter. The rules are designed for measuring diameters of flanges or circular

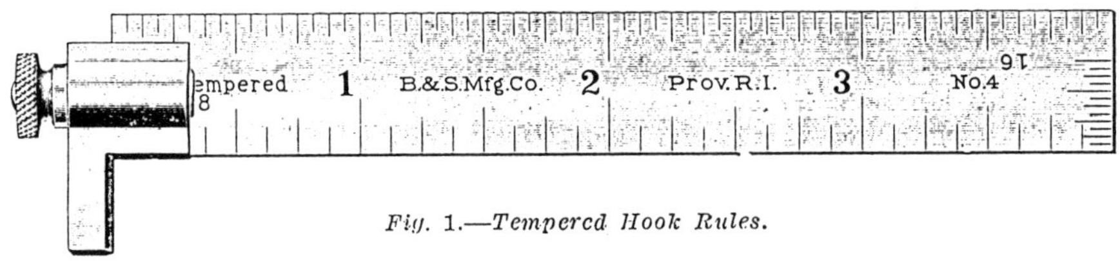

Fig. 1.—Tempered Hook Rules.

pieces, through the hubs of pulleys, setting calipers and dividers and work of a similar character. The hook is held rigidly in position by a clamp that is closed by the simple tightening of the knurled nut shown at the left. This construction requires no hole for attaching the hook and allows it to be clamped in any position on the rule, or

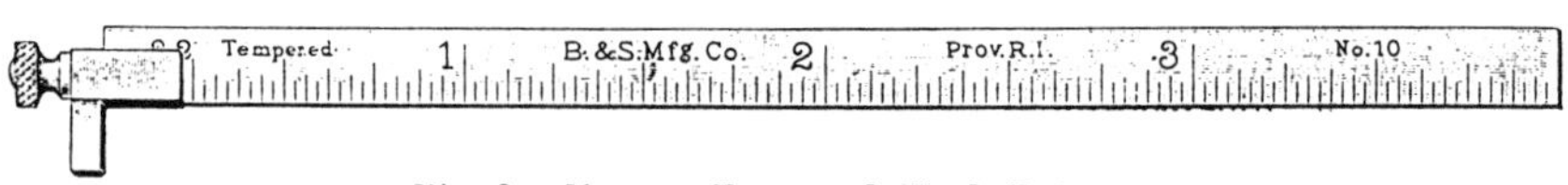

Fig. 2.—Narrow Tempered Hook Rules.

it can be used as a slide in measuring against a shoulder or the depth of a hole. The hooks of the rules are carefully hardened. The wide rule is made 4, 6, 9, 12, 18, 24 and 36 inch lengths, while the narrow rule is made only in the first four lengths.

1905

Steel Rules with Holder

The rules and holder shown herewith are convenient where the ordinary rule

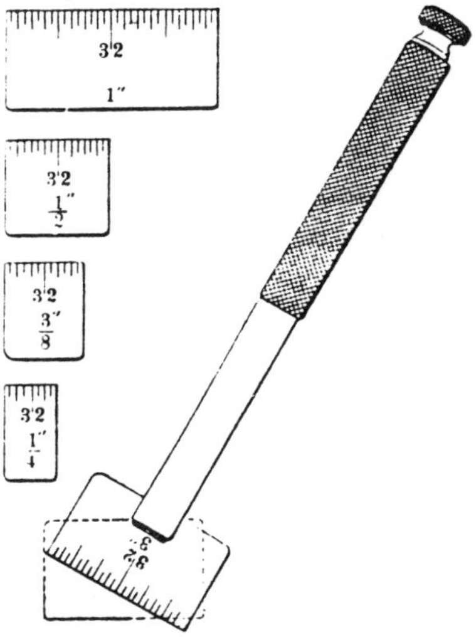

STEEL RULES WITH HOLDER

cannot be used, as in measuring a recess or keyway, as well as in tool and die work. The holder takes any of the five sizes of rules.

1907

Universal Dividers.

The scriber point of the Universal Dividers, shown herewith, has both a fine and a quick adjustment, the fine adjustment being obtained by a screw inclosed in the beam, which engages the nut on the scriber point holder. By pulling up the small knurled knob shown at the top of the post the screw is released and the post can be quickly adjusted, the knob springing into place as soon as released. The scriber point is adjustable either side of the center, as shown by the dotted lines, and can be

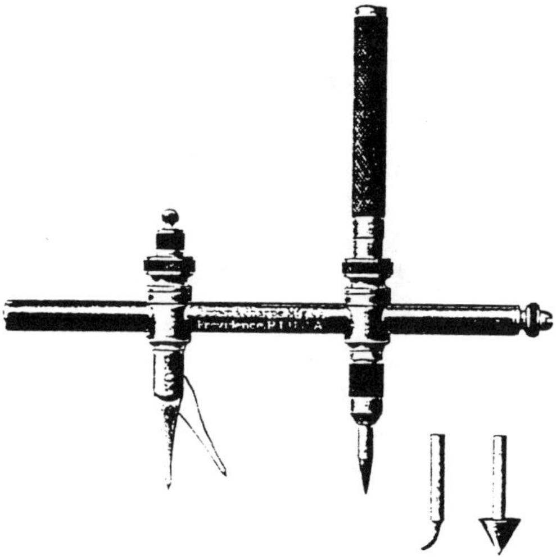

Universal Dividers.

set for scribing small circles or for working close to a shoulder. The adjustable center point is held by a spring chuck and can be easily removed and replaced by pencil or other special points. The posts are clamped by knurled nuts and held in place by spring friction when the nuts are unclamped for setting the points. A V-point is furnished for use in describing a circle about a hole already drilled. A caliper point is also included. The beam is 4 inches long and the points can be set to describe a circle 8 inches in diameter. The dividers are put on the market by the Brown & Sharpe Mfg. Company, Providence, R. I.

1905

Interchangeable Steel Rules.

The Interchangeable steel rules with holder shown in the accompanying cut are convenient when the ordinary rule cannot be used, as in measuring a recess or keyway, as well as in the general class of tool and die work. The holder takes either of the five sizes of rules, the lengths being ¼, ⅜, ½, ¾ and 1 in. The barrel is knurled for finger grip. The rules are held in a split chuck ad-

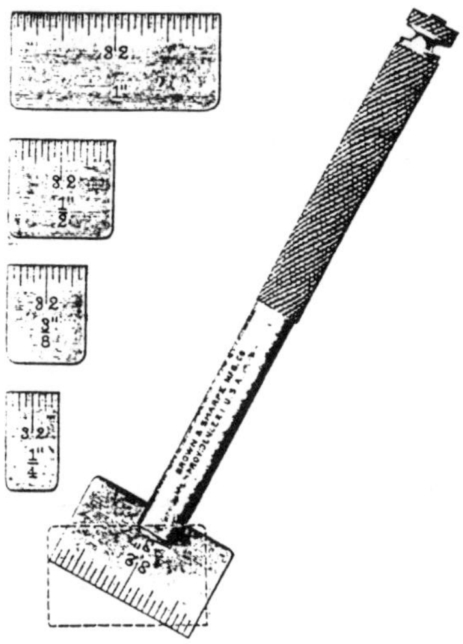

Interchangeable Steel Rules with Holder.

justed by a knurled nut at the top of the barrel, and can be set at various angles according to the work. The rules are of tempered steel, graduated on both sides, as follows: 32ds on one side and 64ths on the other, and 50ths on one side and 100ths on the other. The equipment is manufactured by the Brown & Sharpe Mfg. Company, Providence, R. I.

Spacing Attachment for Automatic Center Punch

The illustration shows an attachment the Brown & Sharpe Manufacturing Company, of Providence, R. I., has recently designed for use in connection with their automatic center punch, the arrangement being useful for spacing and laying out work to be machined or drilled. When adjusted, the attachment is screwed on to the center punch in place of the remov-

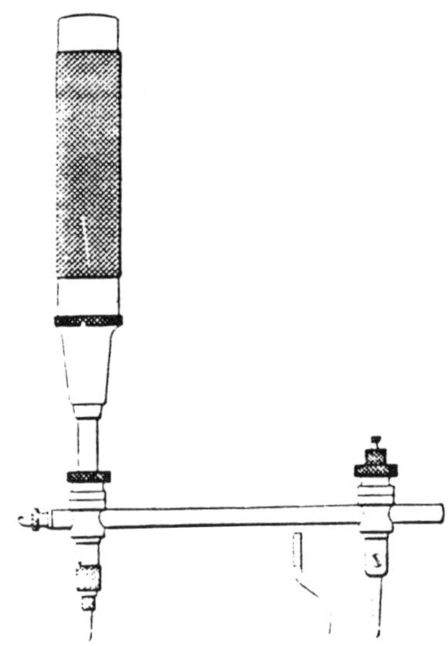

SPACING ATTACHMENT FOR AUTOMATIC CENTER PUNCH

able point. Fine adjustment of the locating point is obtained by means of the screw at the end of the beam and quick adjustment by pulling out the knob at the top of the post. The point is held by a knurled check nut and can be adjusted to the desired length.

1907

Universal Surface Gauges.

HEAVY BASE.

Patented July 11, 1905.

No. 870, with 12″ Spindle. Price, $3 00.

No. 871, with 12″ and 18″ Spindles.
Price, $3 50.

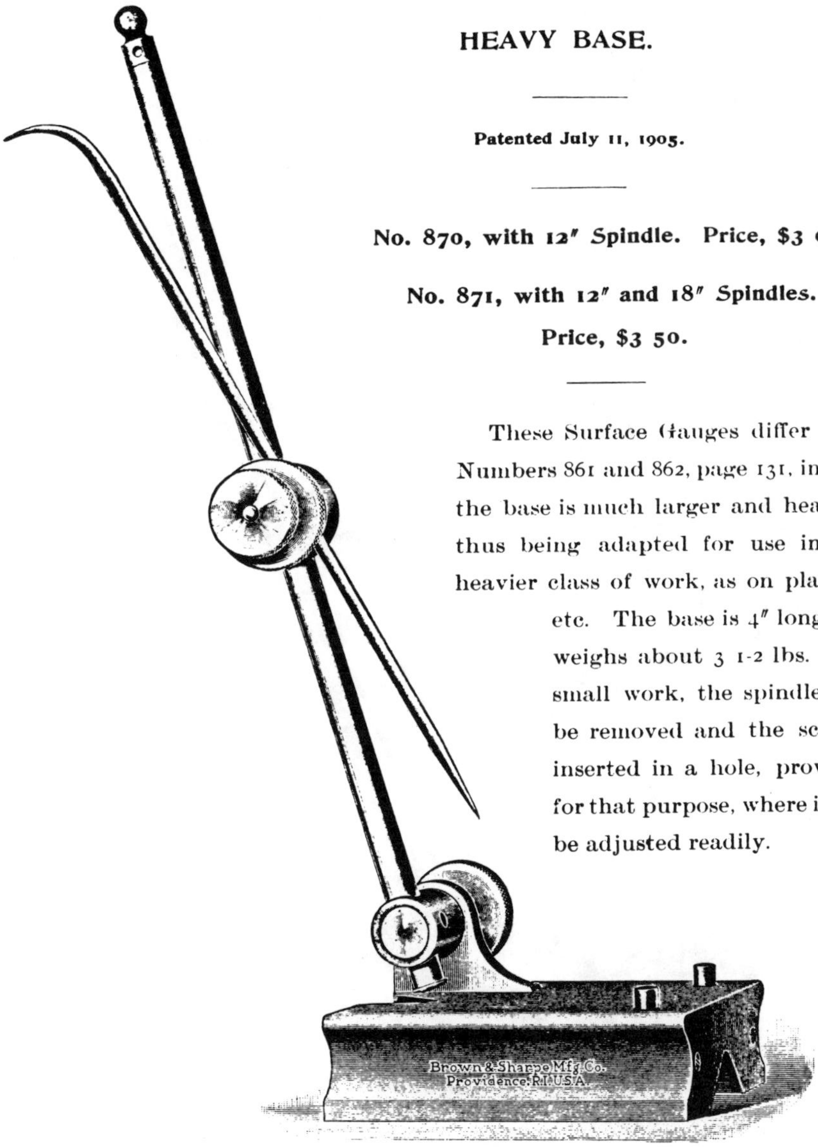

These Surface Gauges differ from Numbers 861 and 862, page 131, in that the base is much larger and heavier, thus being adapted for use in the heavier class of work, as on planers, etc. The base is 4″ long and weighs about 3 1-2 lbs. For small work, the spindle can be removed and the scriber inserted in a hole, provided for that purpose, where it can be adjusted readily.

1907

Universal Surface Gauges.

This line of Surface Gauges has been designed so that a wide range of adjustments can be quickly obtained by means of the knurled adjusting screw. The maximum adjustment can be quickly obtained, and at the same time the adjustment is so arranged that all moving parts can be firmly clamped in position when the desired setting has been obtained.

The spindle passes through a bushing and bolt which are clamped by a knurled nut firmly to the boss on the base thereby removing any possibility of play in the working parts when the spindle has been clamped in position.

The base is of a form most convenient to handle. A V-shaped groove in the bottom especially adapts it for cylindrical work. It has two gauge pins in the rear end that can be pushed down and used against the edge of a plate or the side of a T slot.

The spindle swivels, can be securely clamped in any position from the vertical to the horizontal, and the scriber may be used below the base as a depth gauge. For small work the spindle may be removed and the scriber inserted in a hole in the spindle swivelling bolt, where it is readily adjusted.

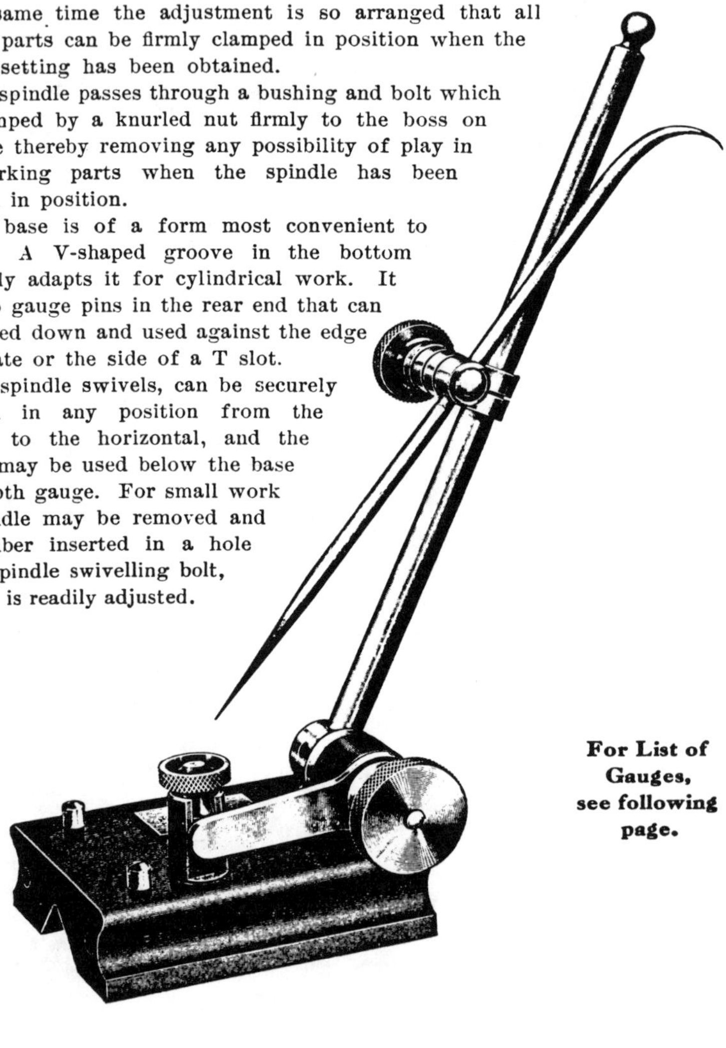

For List of Gauges, see following page.

1915

Direct Reading

Micrometer Caliper No. 25.

Patented Aug. 22, 1911; May 5, 1914.

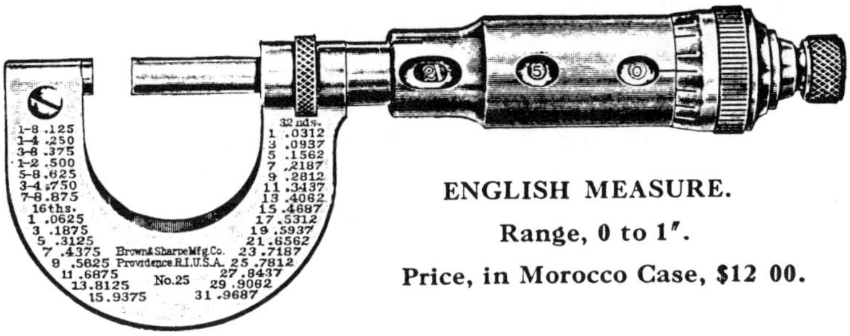

ENGLISH MEASURE.

Range, 0 to 1″.

Price, in Morocco Case, $12 00.

Direct Reading
Micrometer Caliper No. 26

Range, 0 to 1″ by thousandths of an inch

Price, $20.00 Leather Case, **$1.25**

Patented November 28, 1922

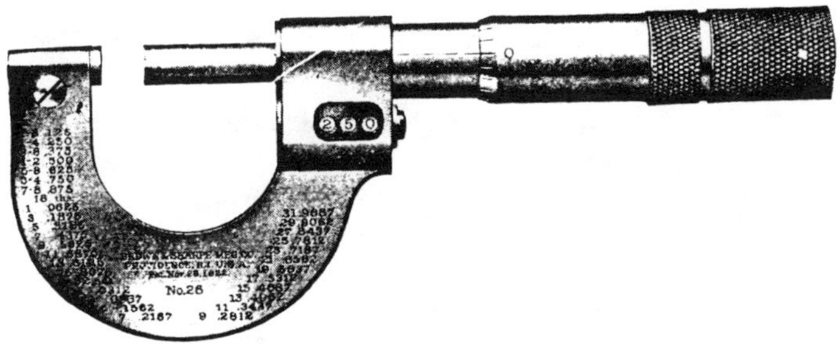

Micrometer Caliper No. 30.

ENGLISH OR METRIC MEASURE.

Range, 0 to 2″ or 0 to 50 m/m.

Price, $7 50. With Ratchet Stop, $8 00. Morocco Case, $0 75.

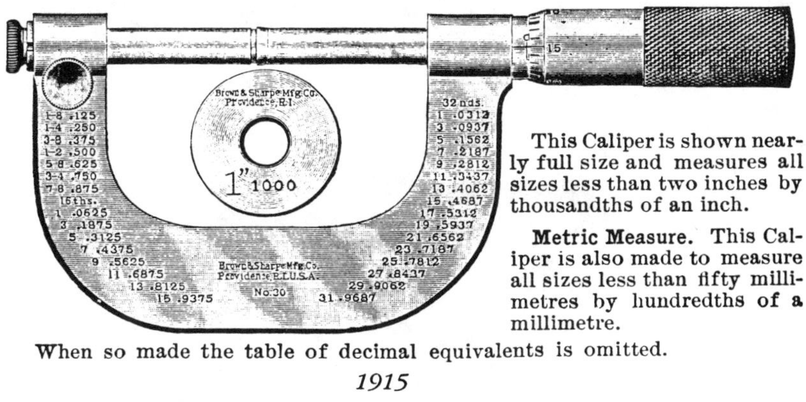

This Caliper is shown nearly full size and measures all sizes less than two inches by thousandths of an inch.

Metric Measure. This Caliper is also made to measure all sizes less than fifty millimetres by hundredths of a millimetre. When so made the table of decimal equivalents is omitted.

1915

Sheet Metal Micrometer No. 220.

Price, $10 00.

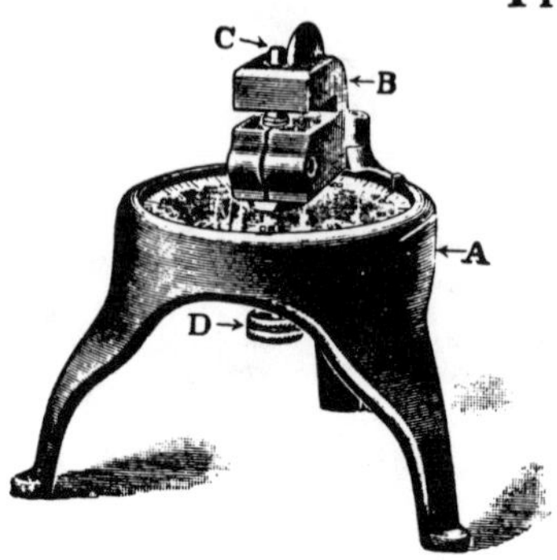

This Micrometer measures to 1-4″ by thousandths of an inch. It is a convenient tool for Jewelers, Silversmiths, Sheet Metal Rollers and Workers, Rubber and Paper Manufacturers, Type Founders, etc.

The frame A supports the measuring mechanism. The arm B holds the measuring screw D and the adjusting screw C. The knurled thumb screw D operates the measuring screw and the movable dial.

The movable dial is of German silver and the graduations are read by means of the pointer shown at the right of arm B.

Provision is made for compensation for wear.

Micrometer Caliper No. 221.

(POCKET SHEET METAL GAUGE.)

Range, 0 to 3-10″.

Price, $4 00. With Ratchet Stop, $4 50. Morocco Case, $0 50.

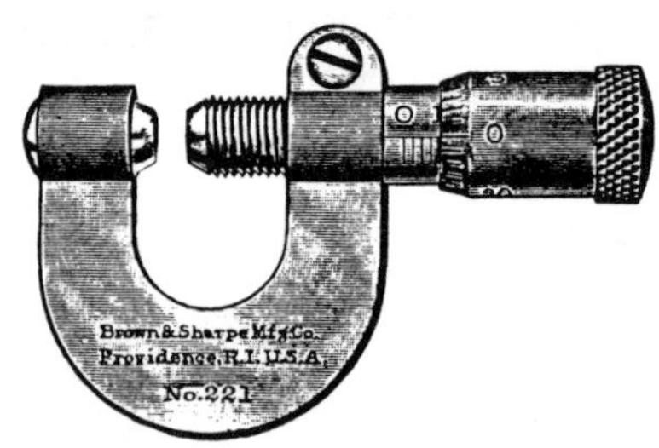

This Caliper is shown full size and measures all sizes less than three-tenths of an inch by thousandths of an inch.

Paper Gauge Micrometer Caliper, No. 231.

ENGLISH MEASURE.

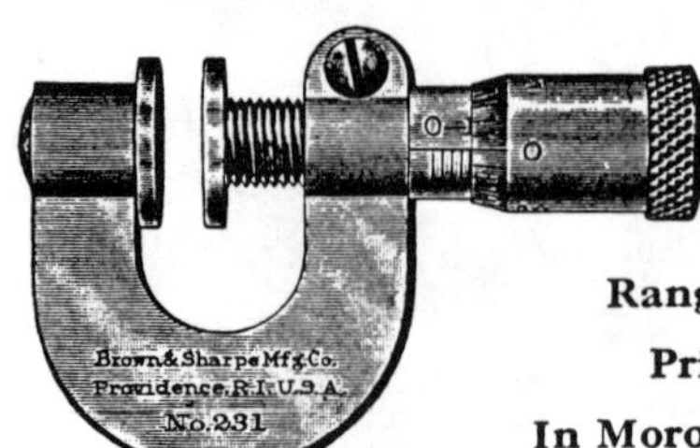

1915

Range, 0 to 1-4″.

Price, $5 00.

In Morocco Case, $5 50.

Micrometer Caliper No. 222.

ENGLISH OR METRIC MEASURE.

Sheet Metal Gauge.

Range, 0 to 1-2″ or 0 to 13 m/m.

Patented June 18, 1901.

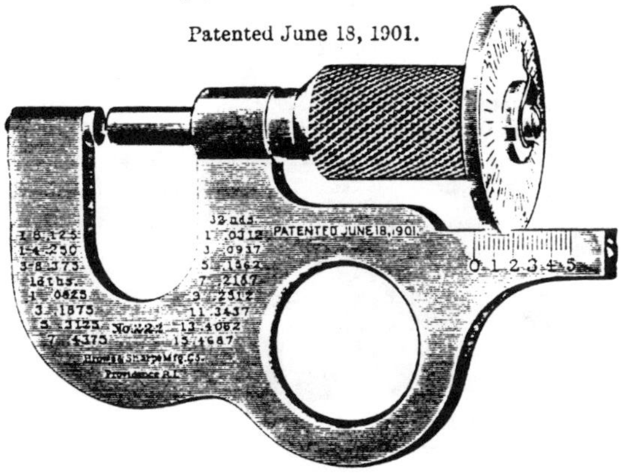

Price, $5 50. **Morocco Case, $0 50.**

Micrometer Caliper No. 227.

FOR ELECTRICIANS.

Range, 0 to 1-2″.

Price, $5 50. With Ratchet Stop, $6 00. Morocco Case, $0 50.

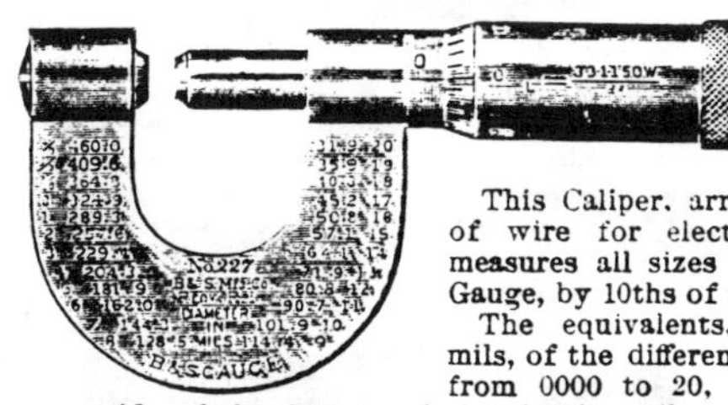

This Caliper, arranged for users of wire for electrical purposes, measures all sizes to 0000, B & S Gauge, by 10ths of mils.

The equivalents, expressed in mils, of the different sizes of wire from 0000 to 20, B & S Gauge, are stamped on one side of the frame and the circular mils of the same size on the other.

Three formulas are stamped on the thimble: one for the weight, length in feet and diameter being known; one for length in feet, weight and diameter being known; and one for resistance of commercial copper wire, in ohms per hundred feet at 75° F., length and diameter being known.

Micrometer Caliper No. 228.

FOR ELECTRICIANS.

Range, 0 to 1-2″.

Price, $5 50. With Ratchet Stop, $6 00. Morocco Case, $0 50.

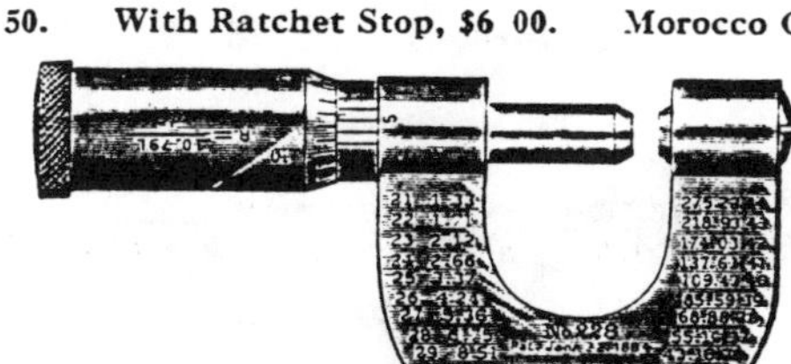

This Caliper differs from Micrometer Caliper No. 227, only in that the equivalents stamped on one side of the frame are for wire from 21 to 44, B & S Gauge, and the resistance of commercial copper wire, in ohms per hundred feet at 75° F., of the same sizes on the other. *1915*

Bench Micrometer Caliper, No. 243.

ENGLISH OR METRIC MEASURE.

Range, 0 to 1" or 0 to 25 m/m.

Price, $7 50.

Patented December 30, 1902.

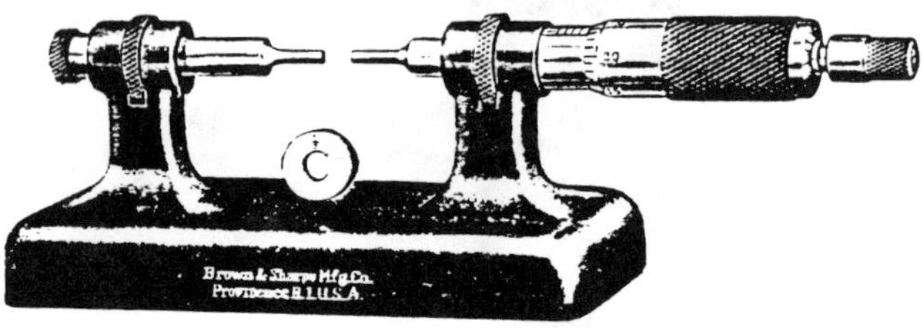

Lathe Test Indicator, No. 736.

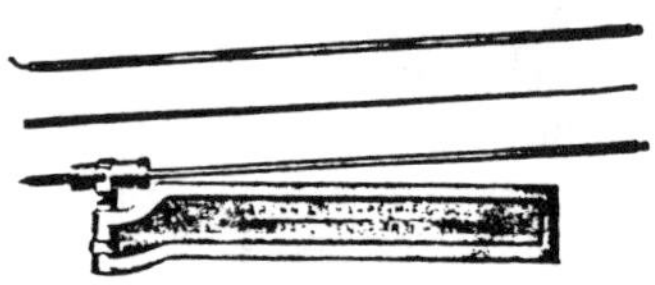

Price, $3 00.

B & S Indicator, No. 738.

English or Metric Measure.

Price, $5 75. Morocco Case, $0 50.

Patented July 24, 1906.

B & S Speed Indicator, No. 748.

Price, $5 50.

In Morocco Case, $6 00.

Patented March 8, 1910.

1915

Depth of Gear Tooth Micrometer.

No. 249.

ENGLISH MEASURE.

Range, 0 to 1".

Price, $4 50. Morocco Case, $0 50.

Patented January 29, 1907

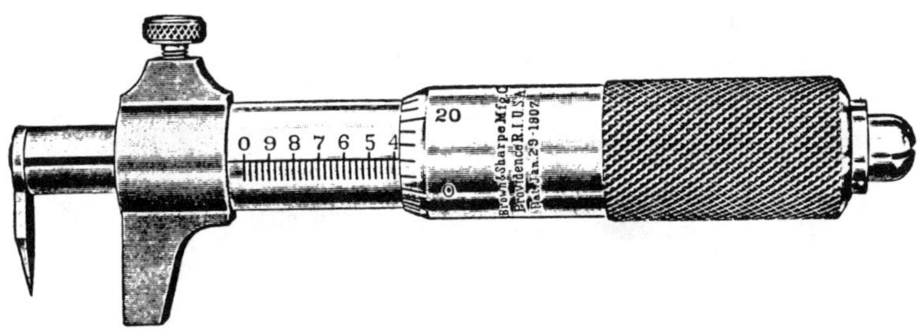

This Micrometer is designed for scribing a line on gear blanks to accurately indicate the extreme depth to cut the teeth. For this purpose it is a particularly economical tool in that it does away with the necessity of keeping a large number of separate gauges for the different pitches.

Tool Makers will also find it handy as a scratch gauge in scribing lines and measuring spacing within its range.

The scriber point is hardened, and a **Clamp Screw** is provided by means of which a setting may be preserved.

This Micrometer is made to measure all sizes less than one inch by thousandths of an inch.

1915

Inside Micrometer Caliper, No. 250.

Patented January 29, 1907.

ENGLISH OR METRIC MEASURE.

Range, .200″ to 1″ or 5 m/m to 25 m/m.

Price, $5 00. Morocco Case, $0 50.

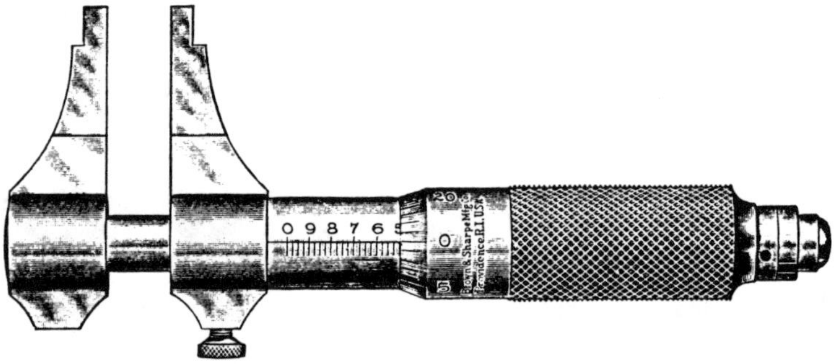

This Caliper is intended to meet the demand for a tool adapted to measuring small internal dimensions. It is graduated to read to thousandths of an inch.

The measuring screw is entirely enclosed, thus protecting it from dirt and injury.

The measuring surfaces are hardened and ground to a radius to insure accurate measurements and prevent cramping when measuring parallel surfaces.

The Caliper is provided with a **Clamp Screw** to preserve the setting.

Metric Measure. This Caliper is also made to measure sizes from 5 m/m to 25 m/m by hundredths of a millimetre.

1915

DARLING, BROWN & SHARPE

DARLING, BROWN & SHARPE Providence, RI

Formed May 24, 1866, by the merger of the two largest makers of machinist's steel rules, DARLING & SCHWARTZ, Bangor, ME, and the rule making part of J.R. BROWN & SHARPE, Providence, RI. The partners were Samuel Darling (1815-1896), Joseph R. Brown (1810-1876) and Lucien Sharpe (1830-1899). Darling continued to operate in Bangor until the spring of 1868 when, along with his stock, six workmen, grinder, and dividing machine, he moved to Providence. Schwartz was not associated with the new firm.

Although operated as a separate company, the new firm shared factory space with J.R. BROWN & SHARPE. Products are usually marked, D.B.& S., PROV., R.I.

The new firm made steel rules, Ames' universal squares (Nathan Ames patent of July 6, 1852, No. 9089), patent steel try squares, bevel protractors, hardened steel T squares and bevels, several types of gauges and a variety of steel and boxwood rules. All products had been previously made by either J.R. BROWN & SHARPE or DARLING & SCHWARTZ.

Some squares are marked "PAT EXT'D 1872". This would have been an extension of Darling's October 6, 1857 (No. 18327), patent, (reissued February 18, 1868) but the only patent extension granted Samuel Darling in 1872 was for an inkstand. He was, however, granted an extension for his square patent on September 21, 1871.

The Lucien Sharpe patent of December 12, 1879 (No. 222744), for a measuring rule was assigned to Darling, Brown & Sharpe. This now-familiar triangular draftsman's rule was the only Sharpe patented tool made by Darling, Brown & Sharpe.

In 1892, it was offering steel rules, patented August 3, 1880 (No. 230619), December 4, 1883 (No. 289512), and October 18, 1887 (No. 371341). Draughtsmen's protractors, patented August 2, 1887 (No. 367673), by Alton J. Shaw were offered in 1894. A second style of protractor was patented September 24, 1889 (No. 411741), by C.E.W. Dow, who assigned his patent to Darling, Brown & Sharpe.

In a letter written in December, 1895, the company noted that:

> "The prices for a number of our tools were reduced, the most notable being the price of rules. This list had been unchanged for over thirty years, and we felt that a revision was required. ...our chief success during the year had been the manufacture of a full line of tempered scales, which we are now commencing to place on the market. For a long time we have desired to do this, but we found that all the processes and machines made for this work produced results which were not up to our requirements, and we accordingly were compelled to devise methods and machines for this purpose. These have proved satisfactory, and we think it of importance to machinists that they can now obtain tempered rules which are as accurate as our standard rules."

During this period Samuel Darling, who apparently was in charge of tool development for the company, received the following tool related patents:

April 7, 1868 (No. 76410), Vise
July 12, 1870 (No. 105180), Improvement in Process and Apparatus for Hardening Files, etc. (assigned to Darling, Brown & Sharpe)
April 6, 1880 (No. 226162), Dividing Engine
May 25, 1880 (No. 227962), Dividing Engine
August 3, 1880 (No. 230619), Graduated Measuring Rule
December 7, 1880 (No. 235133), Micrometer-gauge for Sheet Metal
January 17, 1882 (No. 252476), Metallic Try-square

February 7, 1882 (No. 253266), Dividing Engine
October 30, 1883 (No. 287420), Micrometer
December 4, 1883 (No. 289512), Graduated Measure of Length
December 4, 1883 (No. 289513), Micrometer-gauge
July 19, 1887 (No. 366651), Bevel Protractor
July 19, 1887 (No. 366652), Try Square
October 18, 1887 (No. 371740), Center-square
October 18, 1887 (No. 371741), Measuring Rule
January 17, 1888 (No. 376430), Protractor
December 2, 1890 (No. 442020), Vernier-scale
March 15, 1892 (No. 470804), Key-seat Rule
June 14, 1892 (No. 477104), Caliper-rule
June 14, 1892 (No. 477105), Caliper-rule
December 31, 1895 (No. 552238), Screw Thread Gauge
Only the 1870 patent was assigned to the company.

On January 1, 1892, the Brown & Sharpe Mfg. Co. bought out Samuel Darling and, by late 1895, had merged the operation into the larger firm. An advertisment for the vernier caliper appearing in the *American Machinist* magazine for May, 1895, still shows Darling, Brown & Sharpe as the maker.The same ad, appearing in the December 1895 issue, now lists the Brown & Sharpe Mfg. Co.

For a detailed look at all the tools offered by DARLING, BROWN & SHARPE, a collection of catalogs, *Brown & Sharpe Mfg. Co. and Darling, Brown & Sharpe Catalogs for 1868, 1884 and 1891* is available from: The Astragal Press, P.O. Box 239, Mendham, NJ, 07945.

DARLING, BROWN & SHARPE'S GRADUATED SCALES.

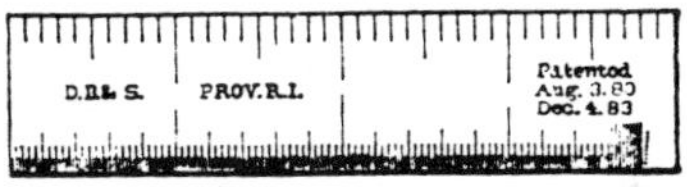

STANDARD STEEL RULES.

1 in.	2 in.	3 in.	4 in.	6 in.	9 in.	12 in.	18 in.	24 in.	36 in.	48 in.
20c.	30c.	40c.	50c.	75c.	$1.12	1.50	2.25	3.00	7.00	10.00

The rules in this list are divided seven ways in parts of inches as follows:

No. 1 Graduations.	No. 2 Graduations.	No. 4 Graduations.	No. 6 Graduations.	No. 7 Graduations.
1st cor. 10, 20, 50, 100	1st cor. 10, 20, 50, 100	1st cor. 64	1st cor. 32 whole length.	1st cor. 64
2d cor. 12, 24, 48	2d cor. 12, 24, 48	2d cor. 32	2d cor. 48 "	2d cor. 32
3d cor. 16, 32, 64	3d cor. 16, 32, 64	3d cor. 16	3d cor. 50 "	3d cor. 16
4th cor. 14, 28	4th cor. 8	4th cor. 8	4th cor. 64 "	4th cor. 100

12 in. Steel Rule of No. 5 graduation......$3.00 | 24 in. Steel Rule of No. 5 graduation......$6.00

No. 5 Graduations.

1st cor.............................16, 32, 64
2d cor.............................11, 14, 15, 17, 18, 19, 20, 21, 22, 23, 24, 25
3d cor.............................26, 27, 28, 29, 30, 31, 33, 34, 35, 36, 37, 38
4th cor.............................39, 40, 41, 42, 43, 44, 45, 46, 47, 48, 49, 50, 100

Rules longer than 12 in., No. 6 or 7 Graduations, are not kept in stock, but are made to order at special prices.

GEAR RULES.

No. 61, $3.00. 12 in. Steel Rule divided to 18, 20, 22, 24, 26, 28, 30 and 32 parts of an inch, whole length.
No. 78, $3.00. 12 in. Steel Rule divided to 6, 7, 8, 9, 10, 11, 12, 14, 16, 18, 20, 22, 24, 26, 28, 30, 32, 34, 36, 38 parts of an inch. One inch of each division.

The rules for No. 5 graduation can also be used for sizing gears.

SHRINK RULES.

$5.00, 24¼ in. Steel Rule Shrink on one side and Standard on the other. Divided on each side to 10, 20, 50, 100, 12, 24, 48, 16, 32, 64, parts of an inch.
$5.00, 24¼ in. Steel Rule Shrink on both sides, No. 1 graduation.
$3.00, 24¼ in. Boxwood Rule Shrink on both sides, No. 1 graduation.

STANDARD STEEL YARD MEASURES.

Divided on one side to inches and eighths, and on the other side to 1/16, ⅛, ¼, ⅜, ½, ⅝, ¾ and ⅞ of a yard. Size, 1 in. wide, ⅛ in. thick. Price, $3.00.

PATENT RULES WITH END GRADUATIONS.

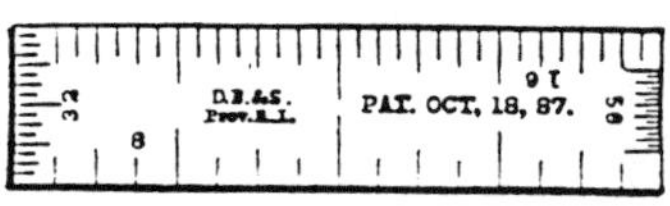

These Rules are made to specific widths.

2 in. rules are ½ in. wide | 4 in. rules are ¾ in. wide | 9 in. rules are 1 in. wide
3 in. " ⅝ in. " | 6 in. " 1 in. " | 12 in. " 1 in. "

The ends are graduated on one side to 48ths and 100ths, and on the other side, as shown in cut, to 56ths and 32ds. This last graduation extends entirely across the rule.

The No. 4 longitudinal graduation is used, and the prices are the same as those of Standard Steel Rules.
The 3 in. Rules are graduated to 200ths in place of 100ths on end.

SPECIAL SCALES.

2 Inches Equal 1 Foot.

This Scale is similar to our "Improved Scales for Draughtsmen," and is graduated on both sides alike, and figured from the same end. It is divided into 12ths of an inch, figured every 6th of an inch, and the first ⅙ on each side is divided into eight parts. Price, $1.25.

LADIES' WORK BASKET RULES.

This Rule is 6 in. long, 1 in. wide and 1/40 inch thick, made of steel and nickel-plated. It is light and serviceable, and is graduated in the manner found most convenient for ladies' use. The graduations on each side are figured from the same end. On one side both graduations are to 8ths of an inch; on the other side one graduation is to 8ths, the other to 16ths of an inch. Price, 75c.

STANDARD STEEL RULES,

French Measure.

1/20 Metre.	1/10 M.	1/5 M.	3/10 M.	½ M.	1 M.
45c.	85c.	$1.75	2.50	4.00	10.00

SQUARE STEEL RULES.

3 in.	4 in.	6 in.
45c.	60c.	90c.

Graduations.

8, 16, 32, 64 to the inch whole length. 16, 32, 64, 100 to the inch whole length.
16, 64, 50, 100 to the inch whole length.

1890

TRIANGULAR STEEL RULES.

3 in.	4 in.	6 in.	12 in.
50c.	70c.	$1.00	2.50

Graduations.

16, 64, 100 to the inch whole length. 16, 32, 64 to the inch whole length.

20, 50, 100,—12, 24, 48,—16, 32, 64 to the inch.

The 12 in. are divided only as follows: 8, 10, 12, 14, 16, 20, 24, 28, 48, 50, 64, 100 to the inch.

STANDARD STEEL STRAIGHT EDGES.

Of same width and thickness as Standard Rules.

Long	Wide	Thick	Price
6 in. long,	1 in. wide,	1/16 in. thick	$0.60
9 in. "	1 1/8 in. "	1/16 in. "	.90
12 in. "	1 1/4 in. "	1/16 in. "	1.20
18 in. "	1 1/2 in. "	1/8 in. "	1.80
24 in. long,	2 in. wide,	1/8 in. thick	$2.40
36 in. "	2 3/8 in. "	1/8 in. "	6.00
48 in. "	3 in. "	1/8 in. "	9.00

BEVELED STEEL STRAIGHT EDGES.

Long	Wide	Thick	Price
12 in. long,	1 3/8 in. wide,	3/16 in. thick	$2.00
18 in. "	1 3/4 in. "	3/16 in. "	3.00
24 in. "	2 in. "	3/16 in. "	4.00
36 in. long,	3 in. wide,	1/4 in. thick	$9.00
48 in. "	3 in. "	1/4 in. "	12.00

The beveled edges are 1/16 in. thick. Only one edge is beveled.

HARDENED STEEL STRAIGHT EDGES.

Long	Wide	Thick	Price
3 7/8 in. long,	7/8 in. wide,	1/16 in. thick	$0.60
5 1/2 in. "	1 1/8 in. "	1/12 in. "	1.00
7 in. "	1 1/2 in. "	1/12 in. "	1.25
10 3/4 in. "	1 5/8 in. "	1/12 in. "	2.00
14 in. "	2 in. "	1/12 in. "	3.00
17 in. long,	2 1/4 in. wide,	1/12 in. thick	$3.50
20 in. "	2 3/4 in. "	1/10 in. "	4.50
27 in. "	3 in. "	1/10 in. "	7.00
33 in. "	3 in. "	1/10 in. "	9.00
39 in. "	3 1/2 in. "	1/8 in. "	12.00

These Straight Edges are the tongues of the Hardened Cast Steel Try Squares, and are hardened on the edges only.

STEEL STRAIGHT EDGES,
For Draughtsmen.

Long	Wide	Thick	Price
15 in. long,	1 1/4 in. wide,	1/20 in. thick	$0.90
18 in. "	1 1/2 in. "	1/20 in. "	1.00
24 in. "	1 1/2 in. "	1/18 in. "	1.50
30 in. "	1 3/4 in. "	1/18 in. "	2.25
36 in. long,	2 in. wide,	1/16 in. thick	$3.00
42 in. "	2 1/4 in. "	1/16 in. "	4.00
48 in. "	2 1/2 in. "	1/14 in. "	6.00
60 in. "	2 3/4 in. "	1/12 in. "	8.00

KEY SEAT RULES.

With this rule parallel lines on shafts for key seats, mortises, etc., can be readily and accurately drawn. Its uses are so obvious to machinists as to need no further description.

4 in.	6 in.	8 in.
$2.50	3.00	3.75

STEEL CALIPER RULES.

This cut is a fac-simile of one side of these Rules; the other side is divided to 12ths, 24ths, 48ths, 8ths, 14ths and 28ths, on the outside and upon the slide to 32ds and 64ths of inches. When closed they are 3 in. long. The Caliper can be drawn out to measure 2 1/2 in. The thickness of the rule is 1/8 in.

These Rules are divided in four ways. A, divided on outside like cut, on slide to 32ds and 64ths. B, divided on outside like cut, on slide to 64ths and 100ths. C, divided on outside to 8ths, 16ths, 32ds and 64ths; on slide to 32ds and 64ths. D, divided on outside to 8ths, 16ths, 32ds and 64ths; on slide to 64ths and 100ths. Others are divided for Button Gauges, on outside to 16ths, 20ths, 32ds and 40ths, and on slide to 40ths and 80ths of inches.

Price.. $2.00

UNIVERSAL BEVELS.

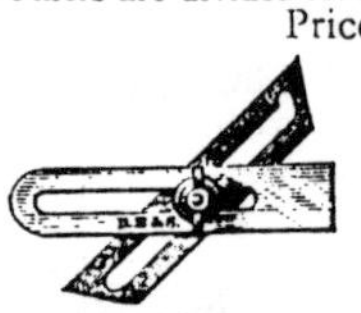

The head and tongue are each three inches long, five-eighths inch wide. Thickness of head one-eighth inch, and of tongue three sixty-fourths inch.

The head and tongue are each one and one-quarter inch long and one quarter inch wide. Thickness of head three thirty-seconds inch, and of tongue one-thirty-second inch.

Large Size............ $2.00. Small Size............ $1.50.

HARDENED CAST STEEL T SQUARE AND UNIVERSAL BEVEL.

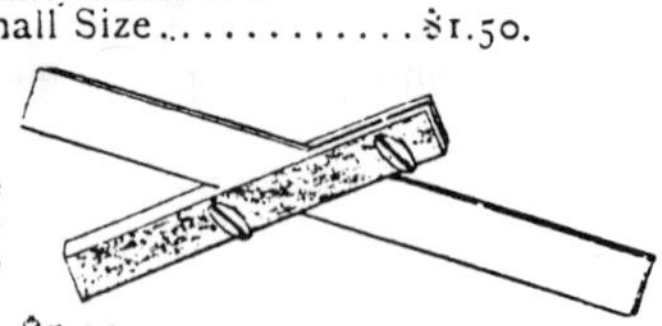

The head is five inches and the tongue eight inches long. Both parts are hardened, and ground straight and square. The tongue may be used at the extreme end of the beam. The wide side of the three cornered washer should always be placed next to the blade.

Price.............................. $5.00.

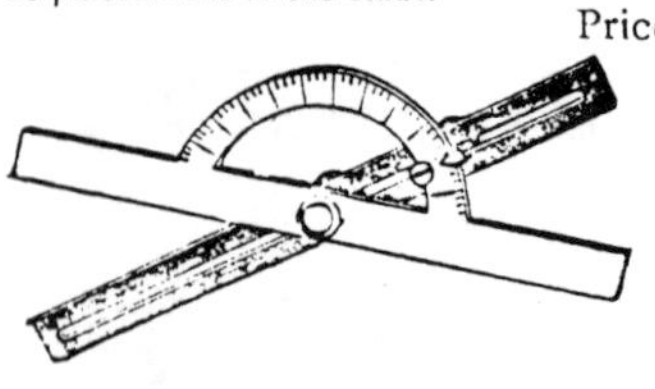

BEVEL PROTECTORS.

With sliding arm, and half circle divided to degrees.

With 6 inch sliding arm, $6.50.

" 10 " " " 7.00.

1890

COMBINED GAUGE AND CALIPER.

This tool combines Dividers, inside and outside Calipers, and a graduated double scratch gauge. It will caliper a round bar two inches in diameter. The beam is twelve inches long, one-half inch wide, three-eighths inch thick, and is graduated to 16ths, 32ds and 64ths of an inch.

Price........................$5.00

THIN STEEL SQUARES

For Machinists and Draughtsmen.

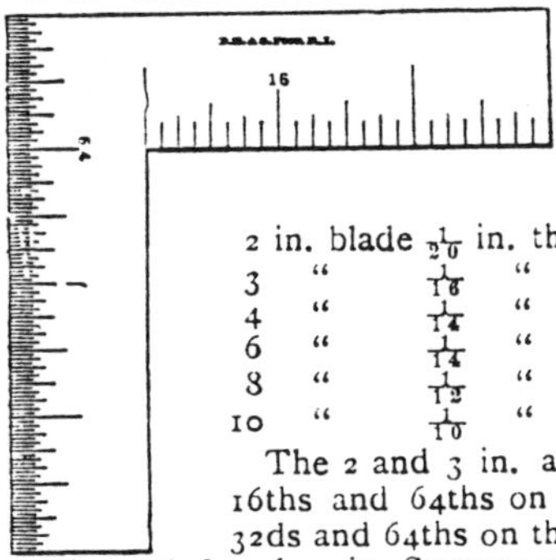

Full size of 2 in. Square.

2 in. blade	$\frac{1}{20}$ in. thick,	-	$1.00
3 "	$\frac{1}{16}$ "		1.50
4 "	$\frac{1}{14}$ "		2.00
6 "	$\frac{1}{14}$ "		3.00
8 "	$\frac{1}{12}$ "		4.00
10 "	$\frac{1}{10}$ "		5.00

The 2 and 3 in. are divided to 16ths and 64ths on one side, and 32ds and 64ths on the other.

The 4, 6, 8 and 10 in. Squares are divided on both sides to 16ths and 32ds of inches.

HARDENED CAST STEEL TRY SQUARES

For Machinists.

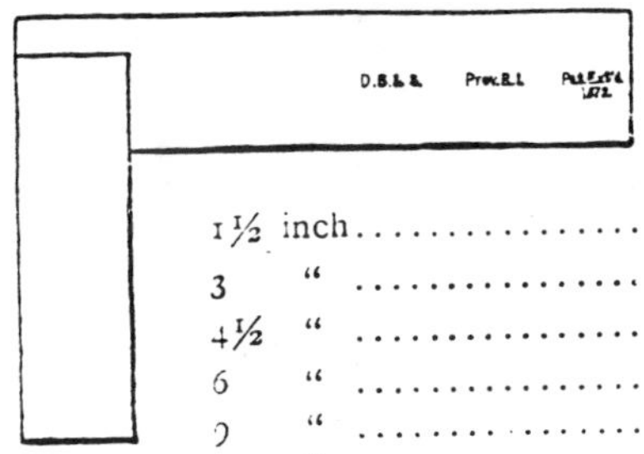

1½ inch	$2.00
3 "	3.00
4½ "	4.00
6 "	5.00
9 "	7.00
12 "	10.00
15 "	15.00
18 "	18.00

GRADUATED STEEL SQUARES

For Machinists.

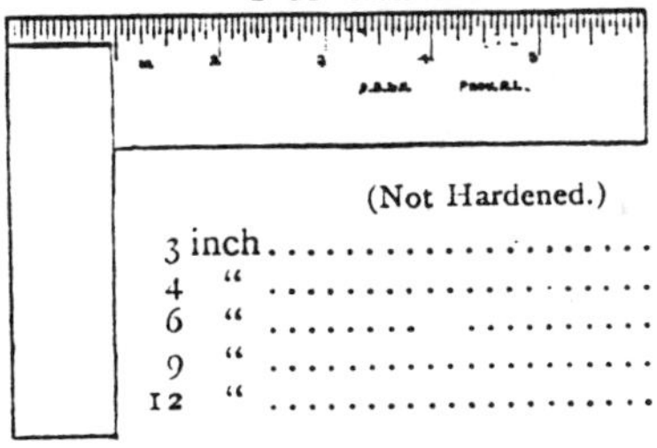

(Not Hardened.)

3 inch	$2.00
4 "	2.50
6 "	3.50
9 "	6.00
12 "	7.00

UNIVERSAL OR CENTRE SQUARES.

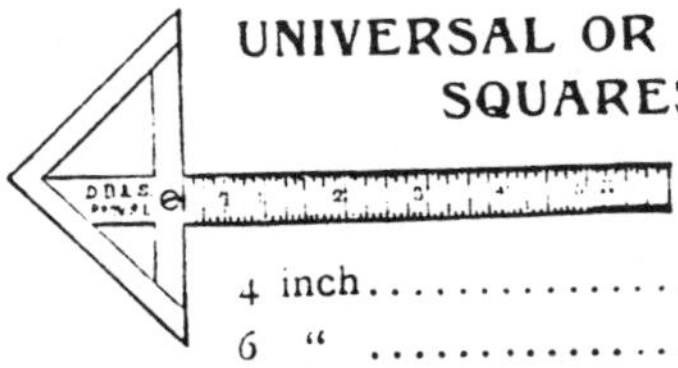

4 inch	$2.00
6 "	2.50
8 "	3.50
10 "	5.00
12 "	7.00

DRAUGHTSMEN'S T SQUARES.

Extreme length of head, 9 inches. Diameter of arc, 4½ inches. Divided to half degrees. Blade 1¼ inches wide, $\frac{3}{64}$ in. thick. The heads are made of bronze, and the blades of steel.

With 36 inch Blade...........$10.00.

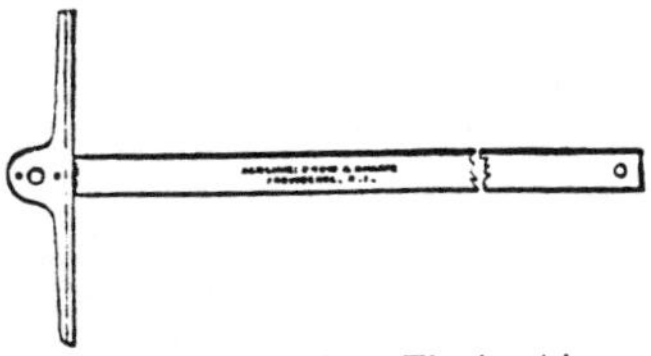

Extreme length of head 8 in. The head is made of cast iron and the blade of steel.

20 in. blade, 1 in. wide, $\frac{3}{64}$ in. thick.....$5.00
24 in. blade, 1 in. wide, $\frac{3}{64}$ in. thick..... 6.00
36 in. blade, 1 in. wide, $\frac{3}{64}$ in. thick..... 7.00
With 10 in. head and 36 in. blade, 1¼ in. wide, $\frac{3}{64}$ in. thick..................... 9.00

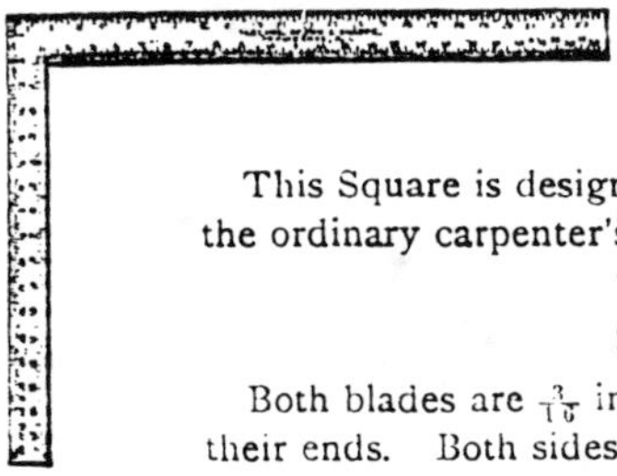

STEEL SQUARES FOR MILLWRIGHTS.

Price..................$10.00.

This Square is designed to meet the wants of parties desiring a more accurate tool than the ordinary carpenter's square.

Longest blade, 24 inches long, 2 inches wide.
Shortest " 18 inches " 1½ inches wide.

Both blades are $\frac{3}{16}$ in. thick at the corner where they unite, and taper down to $\frac{1}{16}$ in. at their ends. Both sides are divided to 8ths, 10ths, 32ds and 64ths of inches.

1890

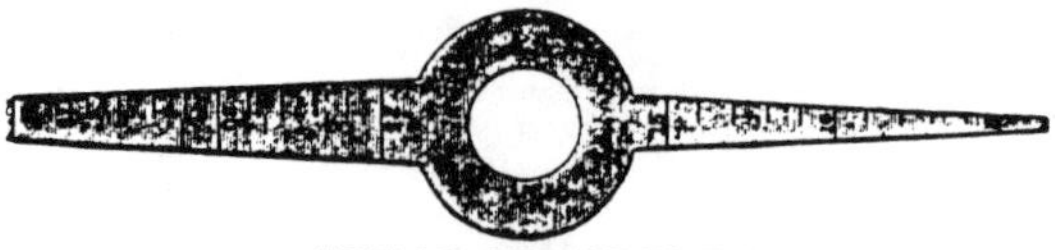

HOLE GAUGES.

This gauge will be found useful to underwriters and others, in determining the discharging capacity of sprinkling pipes, for fire extinguishers, in factories and other buildings.

Price.....................$2.00

AREAS OF CIRCLES from .05 in. to .25 in., by Hundredths of Inches.

Diameter.	Area, Inches.	Diameter.	Area, Inches.	Diameter.	Area, Inches.
.05	.0019635	.12	.011309	.19	.028352
.06	.0028274	.13	.013273	.20	.031416
.07	.0038484	.14	.015393	.21	.034636
.08	.0050265	.15	.017671	.22	.038013
.09	.0063617	.16	.020106	.23	.041547
.10	.0078539	.17	.022698	.24	.045239
.11	.0095033	.18	.025446	.25	.049087

NUT AND WASHER GAUGES,

For Measuring Diameter and Thickness, also Holes of Nuts, and Washers.

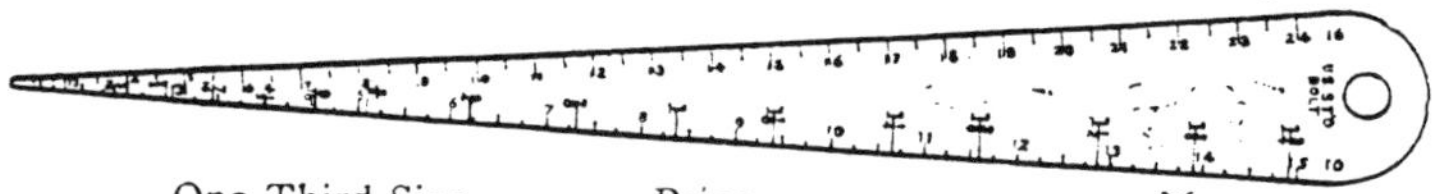

One Third Size. Price.................$6.00

The figures upon one edge are for 16ths and 32ds, and on the other for 10ths and 20ths of inches. Also U. S. Standard sizes for holes to tap for bolts. Opposite side is graduated the same as a steel rule to 32ds of inches.

PRINTERS' RULES.

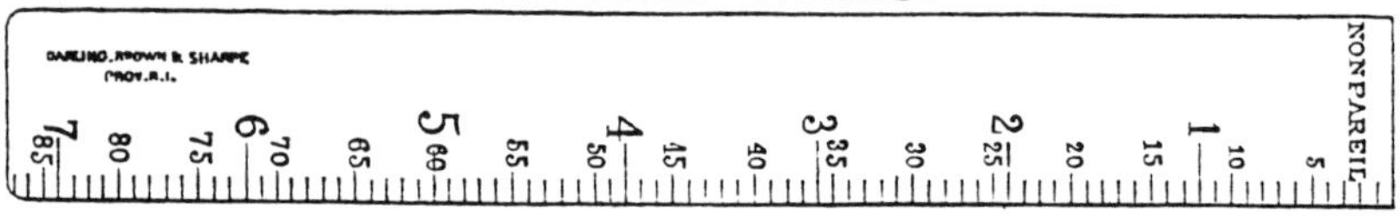

Two Thirds Size. Price.................................$1.25

This rule is made of German silver $\frac{1}{32}$ in. thick, one side is divided for Nonpareil and the other for Agate type. It is used by advertising agents in measuring space occupied by advertisements.

IMPROVED VERNIER CALIPERS.

Half Size.

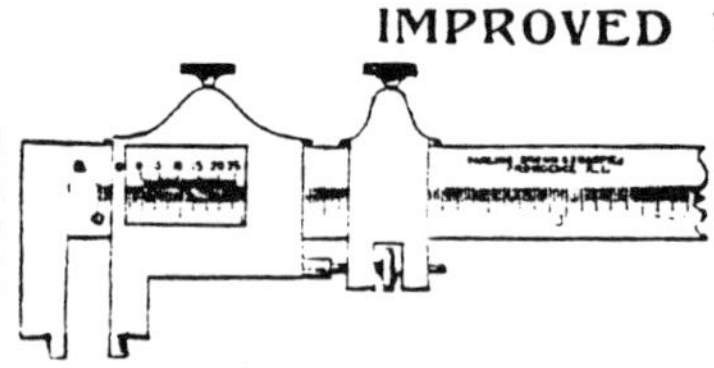

Front Side.

Half Size.

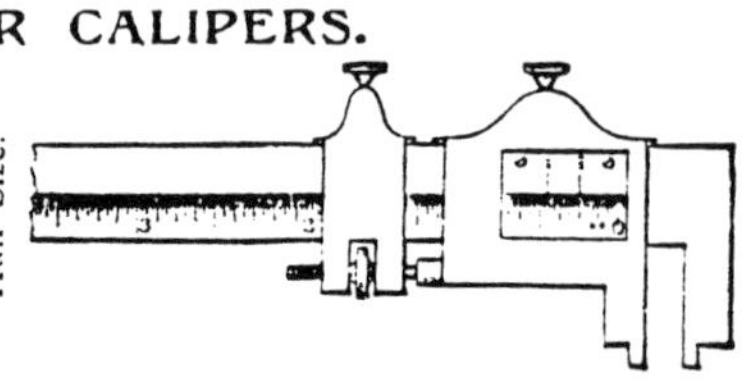

Back Side.

These Calipers are graduated as shown in cuts, also, in place of 64ths on the back side, some are divided to read with Vernier to 50ths of a millimeter. An explanation of the Vernier accompanies each instrument.

The jaws of the 6 inch, 12 inch and 24 inch Calipers are respectively $1\frac{1}{4}$ in. long, $\frac{1}{4}$ in. wide (when closed); $\frac{7}{8}$ in. and $2\frac{1}{4}$ in. long, $\frac{3}{16}$ in. wide; and $\frac{7}{8}$ in. long, $\frac{3}{16}$ in. wide. They are hardened and ground. The instruments are all $\frac{1}{8}$ in. thick, and are made of steel. They can be used for either inside or outside Calipers, and have points by which dividers can be set to transfer distances.

PRICE IN MOROCCO CASE.

6 inch....$15.00 12 inch....$20.00 12 inch, with long jaw.. .$25.00 24 inch....$25.00

A Standard is made for testing the accuracy of the adjustment of the Caliper. Price, $3.00

IMPROVED POCKET VERNIER CALIPERS.

Price, $10.00. In Morocco Case, $10.50.

The cut is a fac-simile of one side of our hardened cast steel Improved Vernier Caliper, a light, convenient and valuable instrument for machinists' and toolmakers' use in obtaining correct measurements. The side represented above is graduated upon the bar to inches and fiftieths of an inch, and by the aid of a Vernier is read to one-thousandths of an inch. The opposite side is graduated to inches and sixty-fourths of an inch. The outside of the jaws are of suitable form for taking inside measurements, and when the jaws are closed measure 250 one-thousandths of an inch in diameter. This caliper will measure one inch and eleven-sixteenths, outside diameters, when the jaws are opened full size.

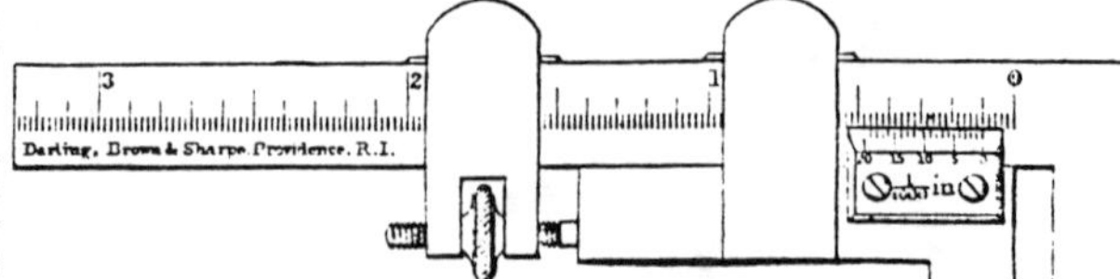

These Instruments can be graduated to millimeters (in the place of sixty-fourths of an inch), and provided with a Vernier to read to one-fiftieth of a millimeter.

OPEN STEEL TRIANGLES.

For. Draughtsmen.

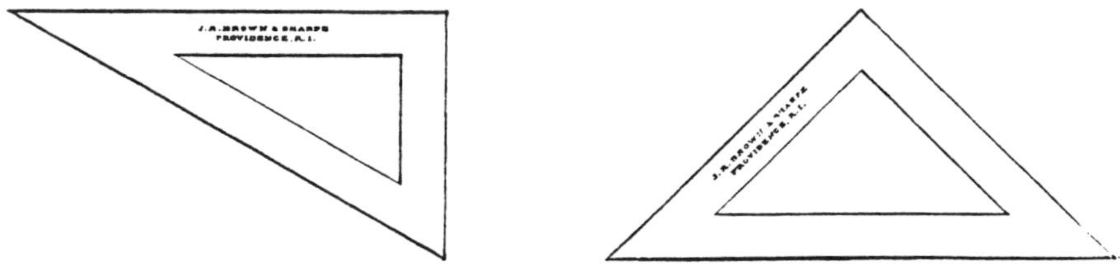

Angles, 45, 45 and 90 degrees. Large size, sides 8, 8 and 11¼ in., ¾ in. wide..... ....$4.00
Small size, sides 5, 5 and 7 1/16 in., ⅝ in. wide..................... 3.00

Angles, 30, 60 and 90 degrees. Large size, sides 6, 10⅜ and 12 in., ¾ in. wide.............. 4.00
Small size, sides 3½, 6 1/16 and 7 in., ⅝ in. wide.................. 3.00

Thickness, 1/16 inch.

A TREATISE ON ISOMETRICAL DRAWING.

By Prof. FARISH, transcribed from Volume First, Transactions of the Cambridge Philosophical Society, England. Illustrated by Wood Cuts and Diagrams.

Price.........................25 cents. Sent per mail, post paid, on receipt of price.

WILLIS' ODONTOGRAPH

Is an instrument invented by Prof. R. WILLIS, of Cambridge University, England, for describing the correct form for the teeth of wheels, and the templets and cutters used in making them. All wheels of the same pitch, of different sizes having their teeth drawn with this instrument, will run together correctly. No. 1 is used for drawing the teeth of small wheels by diametrical pitch when only a single arc is required. Price, $2.00, with drawing and directions for use.

No. 2 is for drawing the teeth of larger wheels by circular pitch, where it is necessary to have separate arcs for flanks and faces. Price, $3.00, with drawings and directions for use. Per pair, $4.50.

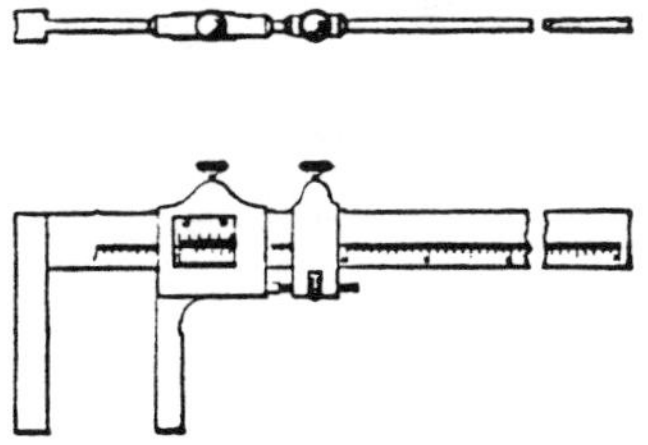

HEIGHT GAUGES.

This tool is used for obtaining the height of any projections from a plane surface, or the location of bushings in jigs, etc., the fixed jaw being thick enough to allow the instrument to stand upright. The blade is 10 inches long, and the Vernier on sliding jaw gives a reading in thousandths of an inch.

Price..............................$25.00

DEPTH GAUGES.

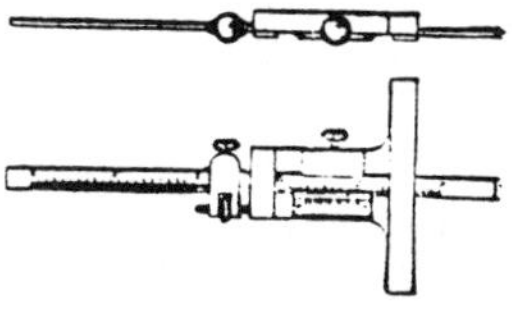

This gauge is of use for readily ascertaining the depth of any recess in dies, etc.; also, for determining distances from the top surface or from the edge of many pieces finished on Planing or Milling Machines. It is used on bench-work as well. Tool makers on Sewing Machine and Gun work most thoroughly appreciate this instrument.

The front is graduated as shown in cut, and reads by Vernier to thousandths of an inch. The back is divided to 64ths.

Price...............................$10.00

DRAUGHTSMEN'S PROTRACTORS.

About one-half size.

VERNIER READS TO TWO MINUTES.

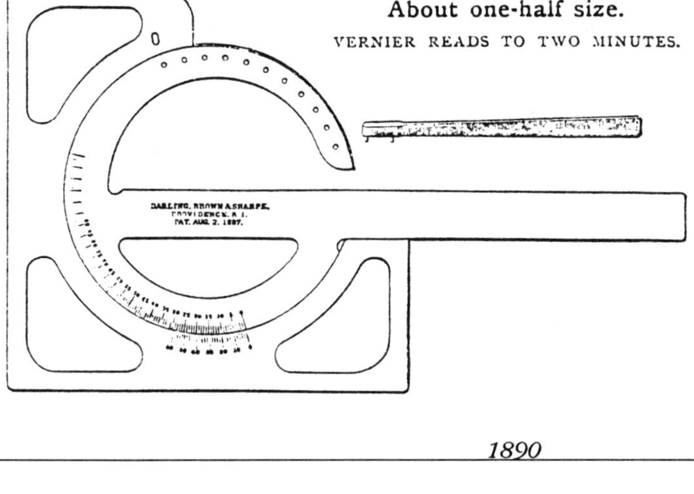

This protractor can be quickly set to any angle. It can be used either side up and on either of the two straight edges, and it is of advantage in dividing a circle, transferring angles, or laying off a given angle, without resetting, on either side of a line.

It forms a convenient extension to a **T** square, and frequently takes the place of 45° and 60° Triangle.

Price.....$6.50

In Morocco Case, $7.75

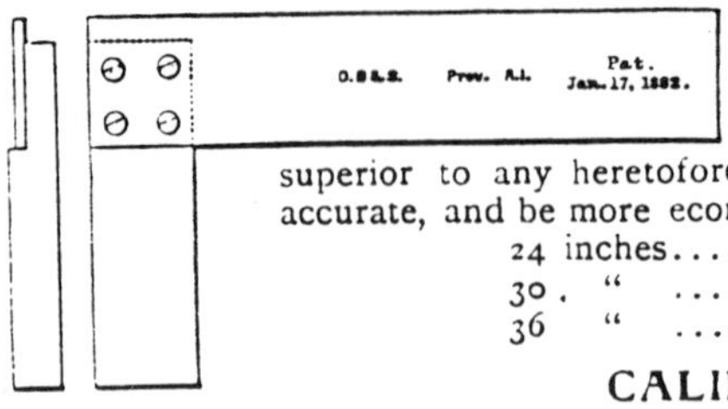

IMPROVED TRY SQUARES.

This method of making Large Try Squares is superior to any heretofore used, in that they can be made more permanent and accurate, and be more economically and readily repaired.

24 inches	$30.00
30 "	40.00
36 "	50.00

CALIPER SQUARES.

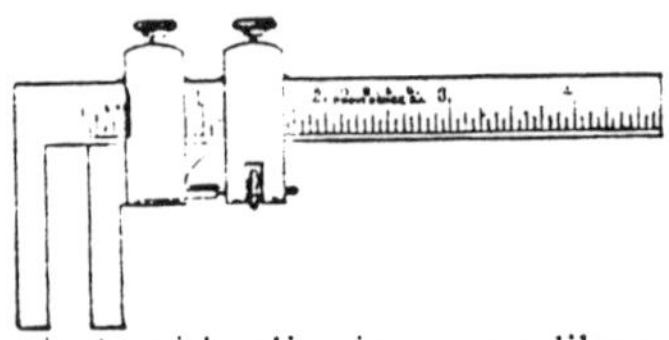

One side divided to 64ths, the other to 100ths of inches.

2 inch, without adjusting screw, jaws ¾ inch long	$ 2.25
4 " " " " " 1½ " "	3.50
6 " " " " " 2 " " hardened and ground	5.50
9 inch, without adjusting screw, jaws 3¼ in. long, hardened and ground	9.00
2 inch, with adjusting screw like cut, jaws ¾ inch long	3.50
4 " " " " " " " " 1½ " "	4.50
6 " " " " " " " " 2 " " hardened and ground	7.50
9 " " " " " " " " 3¼ " " " " "	11.00

The 6 inch Caliper Square is arranged for making inside measurements.

We make 4-inch Caliper Squares, with adjusting screw, with hardened jaws, used by United States Steamboat Inspectors for measuring boiler plate. The jaws are 1 inch long, and are cut out somewhat like those of Improved Pocket Vernier Caliper, page 61. Price, $6.00.

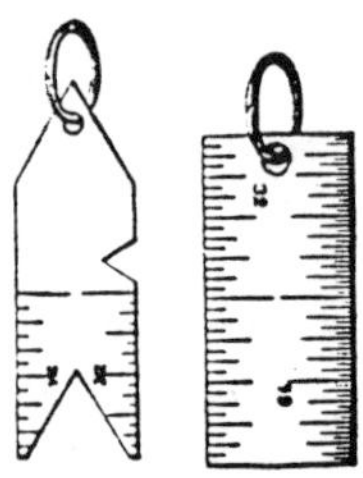

WATCH GUARD CHARMS.

These are one-inch steel rules and small centre gauges, furnished with split ring ready to attach. The prices are as follows :—

Centre Gauge or Steel Rule, plain, with ring,	$ 0.25 each.
Same, Nickel plated	.30 "
" Silver "	.40 "
" Gold "	.50 "

Either of two varieties of graduations on the rules will be sent (as ordered), as follows :—

One edge each to 16ths, 50ths, 64ths, and 100ths ; or, one edge each to 8ths, 16ths, 32ds, and 64ths.

METALLIC ·TRIANGULAR SCALES.

We desire to call the attention of Draughtsmen to our Patent Metallic Triangular Scales, which are of the exact size and shape of the common 12 inch Triangular Boxwood Scales. They are made from brass tubing with the ends closed, nickeled with a dull finish, and weigh less than 3½ ounces.

The liability of the wood scales to crack, warp or twist, the chipping of their edges, and their variation from standard measurement, are well known to all who have used them. These objections we have overcome in the new scales. The ends of these scales are covered with hardened steel plates which slightly raise the scales from the paper.

No. 64, $3.00. 12 inch divided to scales of $\frac{3}{16}$, $\frac{3}{32}$, ⅛, ¼, ⅜, ¾, ½, 1, 1½, and 3 inches to the foot, and 16ths of inches.

No. 73, $3.00. 12 inch, divided on one edge each to 10ths, 20ths, 30ths, 40ths, 50ths, and 60ths of inches, or to 20ths, 30ths, 40ths, 50ths, 60ths, and 80ths of inches.

In ordering No. 73, please state whether the divisions 10 to 60 or 20 to 80 are wanted.

TRIANGULAR BOXWOOD SCALES.

Nos. 63, 64 and 65, for Architects and Mechanical Draughtsmen.

No. 63, $5.00. 24 inch Triangular boxwood rule, divided to scales of ⅛, ¼, ⅜, ¾, ½, 1, 1½, 3, 2, and 4 inches to the foot and 16ths of an inch.

No. 64, $2.00. 12 inch Triangular boxwood rule, ditto, with $\frac{3}{16}$ and $\frac{3}{32}$, instead of 2 and 4 inches to the foot.

No. 65, $1.50. 6 inch Triangular boxwood rule, ditto.

No. 72, $5.00. 24 inch Triangular boxwood rule, on one edge to 10ths, 20ths, 30ths, 40ths, 50ths, 60ths of an inch.

No. 73, $2.00. 12 inch Triangular boxwood rule, ditto.

No. 75, $1.50. 6 inch Triangular boxwood rule, ditto.

Nos. 72 and 73 are divided either 10 to 60 or 20 to 80. Nos. 72, 73 and 75, for Railroad Engineers and Land Surveyors.

1890

JOBBERS' DRILL GAUGES.

For Gauging Twist Drills.

Size of Gauge in Decimals.

Price................$2.25.

Size.	Decimal.	Size.	Decimal.	Size.	Decimal.
1/16	.0625	7/32	.21875	3/8	.375
5/64	.078125	15/64	.23437	25/64	.39062
3/32	.09375	1/4	.25	13/32	.40625
7/64	.10937	17/64	.26562	27/64	.42187
1/8	.125	9/32	.28125	7/16	.4375
9/64	.14062	19/64	.29687	29/64	.45312
5/32	.15625	5/16	.3125	15/32	.46875
11/64	.17187	21/64	.32812	31/64	.48437
3/16	.1875	11/32	.34375	1/2	.50
13/64	.20312	23/64	.35937		

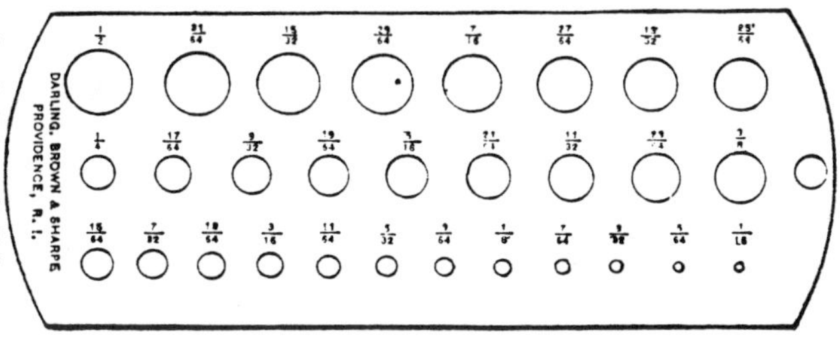

JEWELERS' WIRE GAUGES.

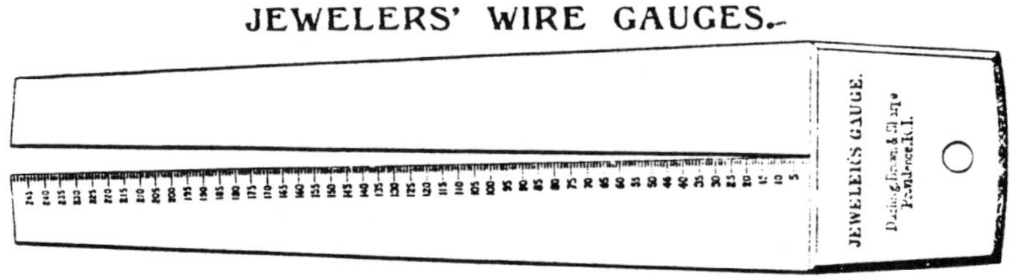

(This cut is one-half of actual size of Gauge.)

These Gauges are made with special reference to the wants of Manufacturing Jewelers. One edge of the angular slot is graduated into 250 parts, and figured to give the size in thousandths of inches. For example, a size of wire which passed down half-way into the slot, will stop opposite 125, is $\frac{125}{1000}$ of an inch in diameter. The angular slot has no sharp edge to injure the stock gauged.

Price..............................$5.00.

WIRE GAUGE AND CALIPER.

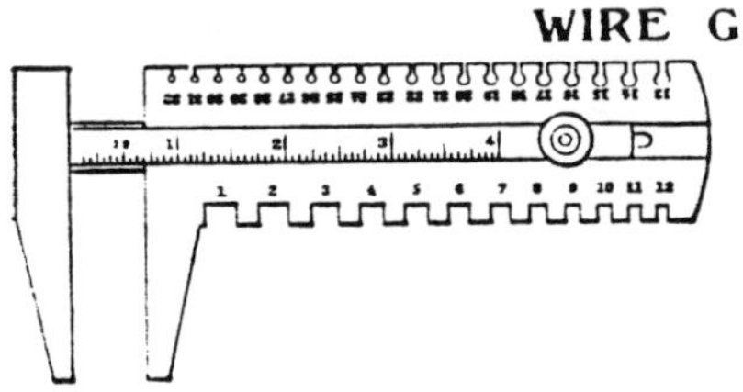

For store use in selecting Iron, Steel, and Sheet Stock, also for Iron and Steel Rollers' use.

The Wire Gauge Sizes are from Nos. 1 to 32, inclusive, Old Standard. The Caliper Tongue is graduated to inches and 32nds, and will measure 4-inch flat, round, or square stock. The Jaws and Wire Gauge are hardened.

Price............ .. ....$7.00

STEEL MUSIC WIRE GAUGES.

Sizes of the Numbers of Steel Music Wire Gauge.

No. of Gauge.	Size of each No. in decimal parts of an inch.	No. of Gauge.	Size of each No. in decimal parts of an inch.	No. of Gauge.	Size of each No. in decimal parts of an inch.
12	.0295	18	.0395	24	.0547
13	.0311	19	.0414	25	.0585
14	.0325	20	.043	26	.0626
15	.0343	21	.0461	27	.0663
16	.0359	22	.0481	28	.0719
17	.0378	23	.0506		

Price.....................$1.50.

(Full Size.)

These sizes are those used by the Washburn and Moen Manufacturing Company, of Worcester, Mass., manufacturers of Steel Music Wire.

ROLLING MILL GAUGES.

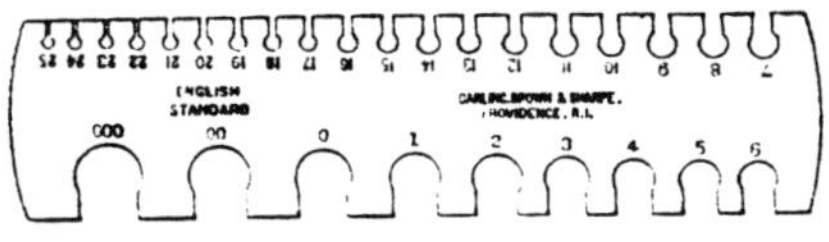

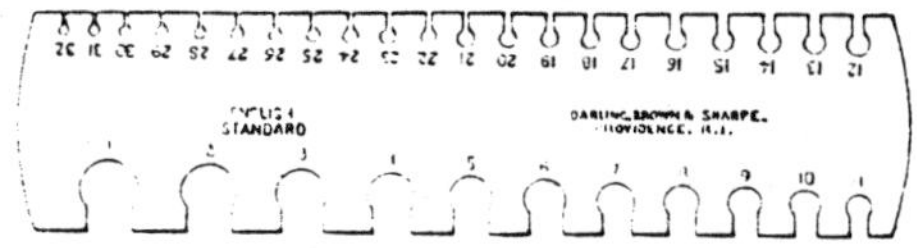

ANGULAR WIRE GAUGES.

000 to 9.......................................$6.00
7 to 17.......................................6.00
15 to 28.......................................6.00
25 to 40.......................................6.00
Divided with both old and new standard..............10.00

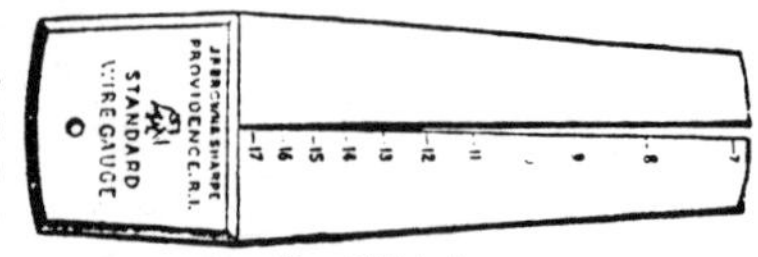

(One-Third Size.)

THE ENGLISH STANDARD WIRE GAUGE.
The same as "Stubs" or Birmingham Gauge.

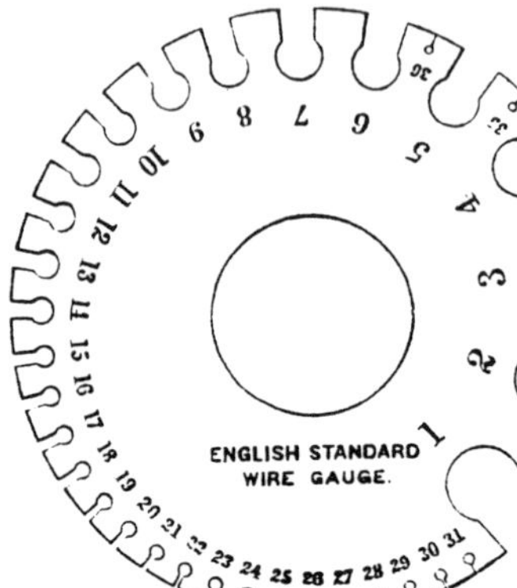

SIZES OF THE NUMBERS OF ENGLISH STANDARD WIRE GAUGE.

No. of Wire Gauge.	Size of each No. in decimal parts of an inch.	No. of Wire Gauge.	Size of each No. in decimal parts of an inch.	No. of Wire Gauge.	Size of each No. in decimal parts of an inch.
0000	.454	11	.120	25	.020
000	.425	12	.109	26	.018
00	.380	13	.095	27	.016
0	.340	14	.083	28	.014
1	.300	15	.072	29	.013
2	.284	16	.065	30	.012
3	.259	17	.058	31	.010
4	.238	18	.049	32	.009
5	.220	19	.042	33	.008
6	.203	20	.035	34	.007
7	.180	21	.032	35	.005
8	.165	22	.028	36	.004
9	.148	23	.025		
10	.134	24	.022		

Round Gauges, sizes 1 to 36............$2.00 | Round Gauges, sizes 6 to 36..............$1.50

POCKET SCREW AND WIRE GAUGES.

Front Side.

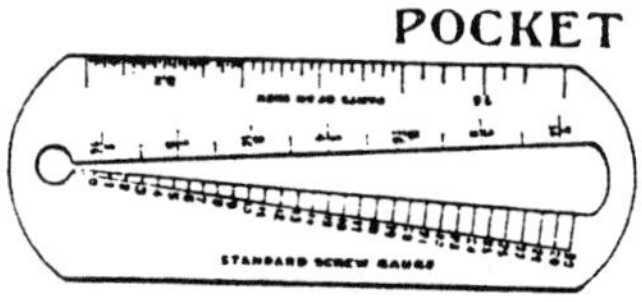

Cuts one-half size.

Price..............$2.50

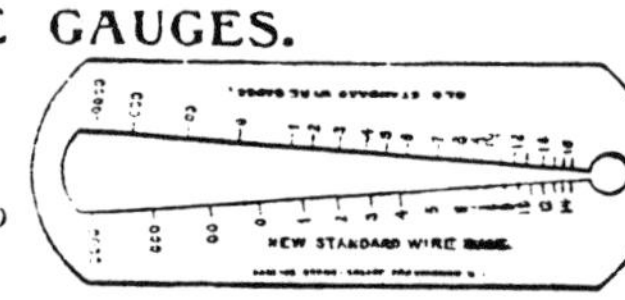

Back Side.

TWIST DRILL AND STEEL WIRE GAUGES.

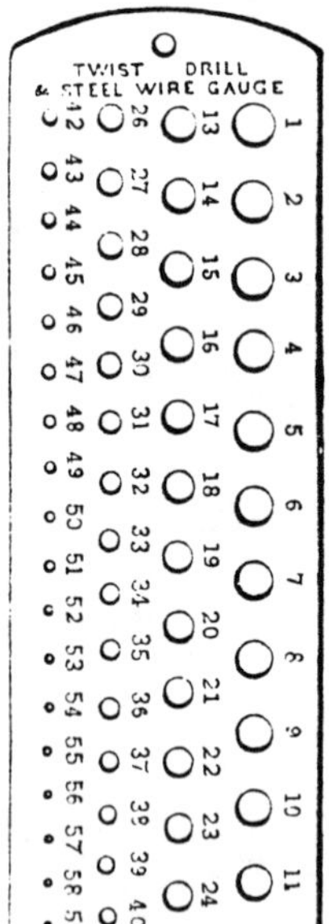

DIMENSIONS OF SIZE OF GAUGE IN DECIMAL PARTS OF AN INCH.

No.	Size of Number in Decimals.	No.	Size of Number in Decimals.	No.	Size of Number in Decimals.
1	.2280	21	.1590	41	.0960
2	.2210	22	.1570	42	.0935
3	.2130	23	.1540	43	.0890
4	.2090	24	.1520	44	.0860
5	.2055	25	.1495	45	.0820
6	.2040	26	.1470	46	.0810
7	.2010	27	.1440	47	.0785
8	.1990	28	.1405	48	.0760
9	.1960	29	.1360	49	.0730
10	.1935	30	.1285	50	.0700
11	.1910	31	.1200	51	.0670
12	.1890	32	.1160	52	.0635
13	.1850	33	.1130	53	.0595
14	.1820	34	.1110	54	.0550
15	.1800	35	.1100	55	.0520
16	.1770	36	.1065	56	.0465
17	.1730	37	.1040	57	.0430
18	.1695	38	.1015	58	.0420
19	.1660	39	.0995	59	.0410
20	.1610	40	.0980	60	.0400

Price............$1.25.

LARGE SCREW AND WIRE GAUGES.

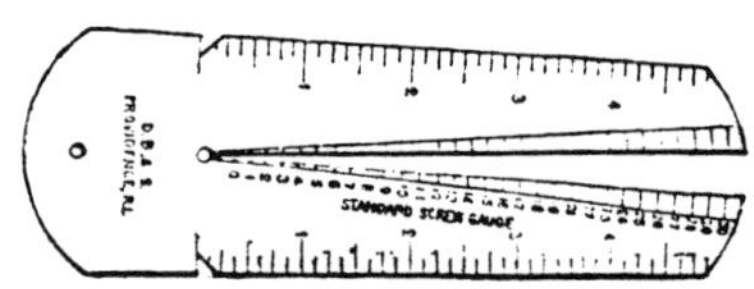

Front Side.

Cuts one-third size.

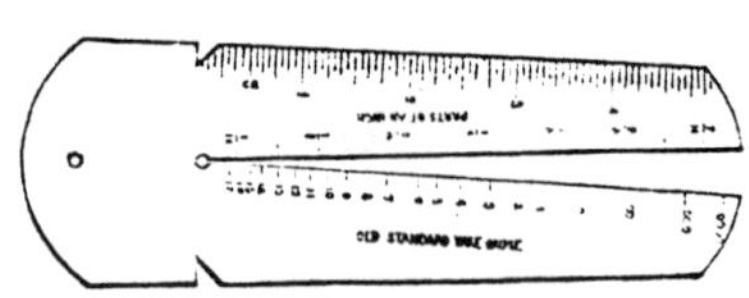

Back Side.

Price........ ....$3.50
Extra Thick.................... 4.50

1890

IMPROVED SCALES FOR DRAUGHTSMEN.

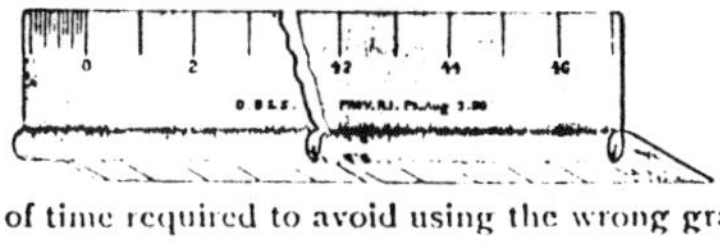

Though designed more particularly for draughtsmen, the form of these scales make them very convenient for many other purposes. Those we have in stock are made of steel, nickel plated; and a 12 inch scale weighs but 2½ oz. Each scale has one kind of graduation, and the same on both sides, or two kinds, one on each side. This relieves the draughtsman from the constant care and loss of time required to avoid using the wrong graduation, when there are many kinds on the scale.

List of Scales for Architects.

One Kind of Graduation.			Two Kinds of Graduation.
Twelve Inches Long.		Six Inches Long.	Twelve Inches Long.
3 inches = 1 foot.	¾ inches = 1 foot.	½ inch = 1 foot.	3 inches and 1½ inches = 1 foot.
2 " = 1 "	½ " = 1 "	¼ " = 1 "	1 " " ½ " = 1 "
1½ " = 1 "	¼ " = 1 "	3/16 " = 1 "	¾ " " 1½ " = 1 "
1 " = 1 "	⅛ " = 1 "	⅛ " = 1 "	⅜ " " 3/16 " = 1 "
		3/32 " = 1 "	¼ " " ½ " = 1 "
			¼ " " ⅛ " = 1 "

For Engineers.

One Kind of Graduation.		Two Kinds of Graduation.
Twelve Inches Long.	Decimal Foot.	Twelve Inches Long.
20ths of an inch.	1-200th of a foot.	10ths and 50ths of an inch.
40ths " "	1-250th " "	20ths " 30ths " "
50ths " "	1-400th " "	20ths " 50ths " "
60ths " "	1-500th " "	20ths " 80ths " "
80ths " "	1-800th " "	40ths " 30ths " "
100ths " "	1-1000th " "	40ths " 50ths " "
		40ths " 60ths " "
		40ths " 80ths " "
		40ths " 100ths " "

Miscellaneous.

12 inches long, graduated on one side 1/16 inch, other side, 1/32 inch.
12 " " " both sides, to 1/100 inch.
6 " " " on one side 1/16 inch, other side, 1/32 inch.
6 " " " both sides to 1/100 inch.
12 " " " to 32ds, ½ inch = 1 inch.
12 " " " on one side to 10ths, (100ths first three inches), other side to 12ths, (48ths first three inches).

12 inches................$1.25 6 inches................$1.00

DOUBLE CALIPERS.

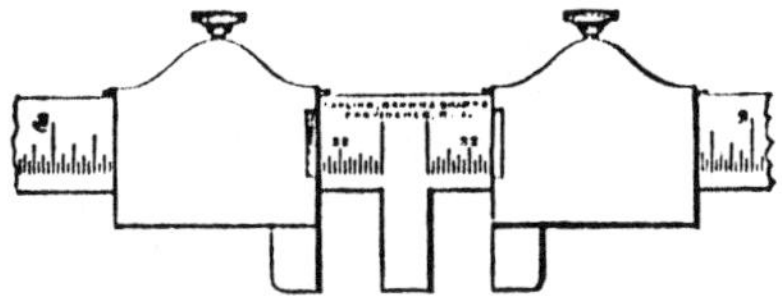

This Caliper is made with heavy bar and wood handle for use in rolling or forging iron and steel.

Extreme length, including handle, 12 inches, and will caliper 3½ x 1¼ inches.

Price.....................$15.00.

FULL SIZE.

CENTRE GAUGE.

And Gauge for Grinding and Setting Screw Tools.

The angles used in this gauge are 60 degrees. The four divisions upon the gauge of 14, 20, 24 and 32 parts to the inch, are very useful in measuring the number of threads to the inch of taps and screws. The following parts to the inch can be determined by them, viz.: 2, 3, 4, 5, 6, 7, 8, 10, 12, 14, 16, 20, 24 and 32.

The table on the gauge is used in determining the size of tap drills. The decimals represent the double depth of thread of the various pitches. To obtain the diameter of a tap drill, say 4 pitch, subtract the decimal .433, which is opposite the 4 on the table, from the size of tap, say 1 inch, and the result, .567 in this instance, will be the size of drill to be used.

Also on hand Centre Gauges of the Whitworth or English standard, 55 degrees.

Price......................................25 Cents.

DARLING, BROWN & SHARPE,

PROVIDENCE, R. I.

POCKET VERNIER CALIPER.

No. 680. Price, $10.00. In Morocco Case, $10.50.

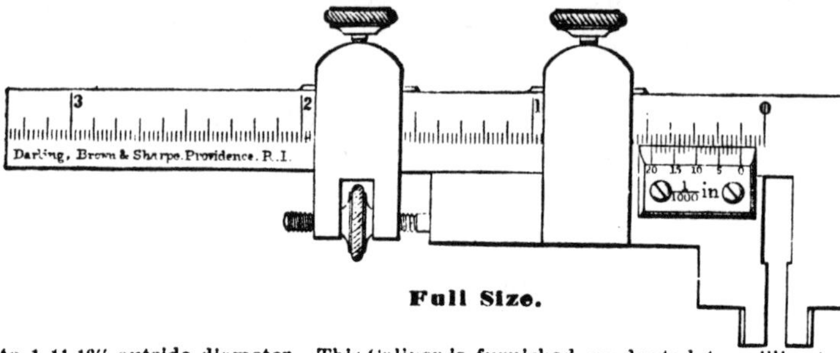

This Pocket Vernier Caliper is graduated on the front to read, by means of a vernier, to thousandths of an inch. It is graduated on the back to 64ths of an inch. The jaws are of steel, hardened and ground, are 3-4" long, 1-4" wide when closed, and take inside as well as outside measurements. The Caliper measures to 1 11-16" outside diameter. This Caliper is furnished graduated to millimeters in place of 64ths of an inch, with a vernier to read to 50ths of a millimeter.

ENGLAND—BUCK & HICKMAN 280 Whitechapel Road London, E
ENGLAND—CHAS. CHURCHILL & CO., Ltd., 21 Cross St., Finsbury, London. E. C.
GERMANY—SCHUCHARDT & SCHUTTE, 59 Spandauerstrasse, Berlin, C. (Small Tools).
GERMANY—G. DIECHMANN, Ansbacherstr 5 Berlin, W. 62.
FRANCE—FENWICK FRERES & CO., 21 Rue Martel Paris
FRANCE—F. G. KREUTZBERGER, 140 Rue de Neuilly Puteaux (Seine)
CHICAGO, ILL.—FRED. A. RICH, 23 So. Canal St.
NEW YORK CITY—F. G. KRETSCHMER, 136 Liberty St., Room 503.

May 1895

BROWN & SHARPE MFG. CO.

PROVIDENCE, R. I.

VERNIER CALIPERS.

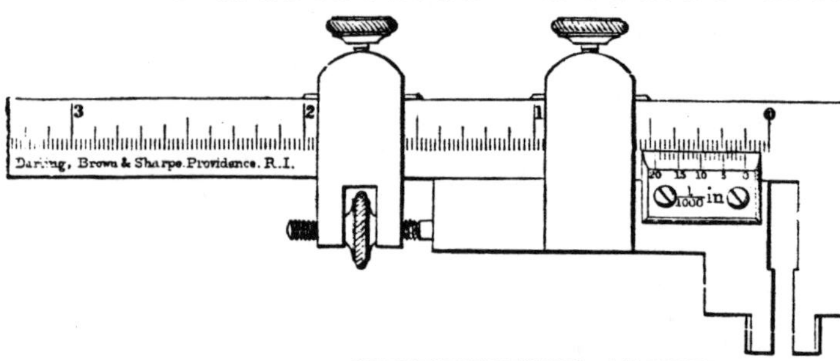

Accurate Tools make Accurate Measurements. Vernier Calipers are Accurate Tools. Vernier Calipers make Accurate Measurements.

Vernier Calipers have been a part of our product for forty years, and we believe we know how to attain accuracy. Four sizes in stock: 3 in., 6 in., 12 in. and 24 in. Larger sizes to order.

REPRESENTATIVES AT HOME:

CHICAGO, ILL.—FRED. A. RICH, 23 So. Canal St.
NEW YORK CITY—F. G. KRETSCHMER, 136 Liberty St., Room 507.
BUFFALO, N. Y.—R. HOFFELD & CO., 61 Carroll St.
CINCINNATI, OHIO—E. A. KINSEY & CO., 227 W. Fourth St.
PITTSBURGH, PA.—U. BAIRD MACHINERY CO., 74 Water St.

REPRESENTATIVES ABROAD:

ENGLAND—BUCK & HICKMAN, 280 Whitechapel Road, London, E.
ENGLAND—CHAS. CHURCHILL & CO., Ltd., 21 Cross St., Finsbury, London. E. C.
GERMANY—SCHUCHARDT & SCHUTTE, 59 Spandauerstrasse, Berlin C. (Small Tools).
GERMANY—GUSTAV DIECHMANN & SOHN, Zimmer-Strasse 87, Berlin, S. W. 12, Germany.
FRANCE—FENWICK FRERES & CO., 21 Rue Martel, Paris.
FRANCE—F. G. KREUTZBERGER, 140 Rue de Neuilly Puteaux (Seine).

December 1895

Advertisements from 1895, showing that the company's name change took place between May and December 1895.

SAWYER TOOL CO.

SAWYER TOOL CO. Athol, MA, later Fitchburg, MA

Founded by Burnside E. Sawyer (1861-?) in 1894, Sawyer, who was born in Templeton, MA, had been a long time employee of L.S. STARRETT, Athol, MA, starting soon after the company was formed in 1880.

In 1891 Montgomery & Co., a large New York tool dealer, offered Sawyer's Drawing Apparatus, No. 60, made under Burnside E. Sawyer's patent of September 9, 1890 (No. 435980), and a Protractor and Section Liner, No. 65, made under his patent of September 23, 1890 (No. 436842). It is uncertain whether these were made by Sawyer prior to founding the SAWYER TOOL CO., or by some other company under license from Sawyer. By 1907, the L.S. STARRETT CO. of Athol, MA, owned the patents but we do not know when it acquired them.

Sawyer's January 26, 1892 (No. 467664), patent for a Plotter was assigned to L.S. STARRETT, but no plotters appear to have been made under this patent.

The SAWYER TOOL CO.'s first product was a protractor miter introduced in *Iron Age* magazine in November, 1894. In 1895, it advertised a line of machinist's rules in *American Machinist* magazine. To introduce his company Sawyer offered to "send post-paid on receipt of $1.00, one each of 3, 6, and 12 inch rules, all guaranteed equal to any in the market."

Burnside E. Sawyer received the following patents after forming the SAWYER TOOL CO:

Screw Thread Gage, September 21, 1897 (No. 590486), used in an oddly shaped screw gauge made for a number of years.

Screwdriver, May 9, 1898 (No. 624875), (with William Arnott), used for the "Samson" screwdriver made until the company went out of business.

Surface Gauge, May 9, 1898 (No. 624876), used until replaced by another style about 1907.

In addition to the Sawyer patent tools the company made a combination bevel patented July 25, 1893 (No. 502199) by Carl G. Osteman and a variety of steel rules. The Osteman patent combination bevel is still in production by the L.S. STARRETT CO.

The 1896 catalog offered "rules, gages, squares, levels and kindred tools used by machinists." This included a case of rules which consisted of 12", 9", 6", 4", 3", 2" and 1" scales, with an additional 4" scale, very slim, and a center gauge. These early rules are extremely well made and are of higher quality than later offerings.

Sawyer advertised combination squares in 1898. Two styles were offered, one for machinists and a cheaper model for carpenters.

In 1898 the company moved to Fitchburg, MA, and incorporated with a capital of $50,000. Officers were K. Pardon Rawson, president and Burnside E. Sawyer, treasurer.

Catalog D, issued in 1900, offered: rules, gauges, squares, calipers and micrometers, protractors, bevels, straight edges, trammels, tap wrenches, hammers and hacksaws. Calipers and dividers of the J. STEVENS ARMS & TOOL CO., were also included in this catalog.

In 1901 the company name was changed to:

SAWYER TOOL MFG. CO. Fitchburg, MA, later, Ashburnham, MA

A major reorganization of the company took place soon after the name change, and was made public in the following announcement dated October 30, 1902.

> At the recent annual meeting of the stockholders of the Sawyer Tool Mfg. Co., Fitchburg, Mass., the following officers were chosen: President, Lyman H. Goodnow: vice-president, James F.D. Garfield: treasurer, Edmund D. Garfield: directors: L.H. Goodnow, J.F.D. Garfield, Emory E. Ellis.

> Mr. Ellis succeeds B.E. Sawyer on the Board of Directors, as Mr. Sawyer has sold out his entire interest in the company. Mr. Ellis was re-elected manager.

New keyseat rule blocks were introduced in 1902. Identical units were later made, and are still made, by the L.S. STARRETT CO.

Catalog F, issued in 1903, offered steel rules, combination squares, straight-edges, surface and pitch gauges, beam trammels, micrometers, screw-drivers and bits, hammers, nail-sets, etc.

In 1903, the firm announced that J.H. Cook, "who for over 20 years has been connected with the L.S. Starrett Co., Athol, Mass., in the sales department, has resigned to accept a similar position with the Sawyer Tool Mfg. Co, of Fitchburg, Mass." The announcement also noted that Cook "had much to do with the development of a number of fine tools which are on the market today." His only known contribution to tool development at Sawyer is his caliper patent of August 22, 1905 (No. 797559), which was assigned to Sawyer Tool Mfg. Co. Sawyer made these calipers as its "Beeswax Joint" calipers for many years.

Catalog G, issued in 1904, offered steel rules, combination squares, screw pitch gauges, machinist's hammers, straight edges, try squares, planer gauges, trammels, center gauges, combination bevels, hack saws, Samson screwdrivers, surface gauges, depth gauges, inside micrometers, nail sets, and center punches.

A 1905 article in *American Machinist* magazine describes the company's newly introduced precision bench levels as the first available "fit to do accurate work." The article describes tests made with the 6" level which would indicate an accuracy of about .001" per foot. The levels were offered in 6, 9, 12, 18 and 24 inch lengths.

Catalog H, issued in 1905, offered steel rules, one inch rule with handle, center gauges, straight edges, depth gauges, bevel protractors, combination squares, try squares, double adjustable squares, toolmaker's squares, clamps, combination bevels, angle indicators, key seat rule blocks, beam trammels, micrometer surface gauges, beam calipers, tap wrenches, inside micrometers, screw pitch gauges, hammers, bench levels, Sampson screwdrivers, nail sets, scribers, and friction joint calipers with Cook's patent beeswax joint.

Rules, squares, depth gauges, a new style of surface gauge, hack saws, thread gauges, and an automatic two-stroke center punch were offered in its 1908 catalog. The new style surface gauge, introduced in 1907, must have been well accepted as it is much less scarce today than the earlier styles.

In 1908, the company was sold to Carl A. Hubbell who had moved his company, the T.R. ALMOND MFG. CO., from Brooklyn, NY, to Ashburnham, MA, in 1907. Emory E. Ellis left the SAWYER TOOL MFG. CO. when Hubbell became the owner and formed the UNION CALIPER CO. in Fitchburg, MA.

Hubbell moved the SAWYER TOOL MFG. CO. to Ashburnham, MA, in November, 1912. The new plant had 84,000 square feet of floor space. About 1916, the company was absorbed into the T.R. ALMOND MFG. CO. and ceased to exist.

For a detailed look at all the Sawyer tools offered in 1904, a reprint of the SAWYER TOOL MFG. CO., Catalog G, is available from: Martin J. Donnelly Antique Tools, P.O. Box 281, Bath, NY, 14810-0281.

MONTGOMERY & CO., 105 FULTON STREET, NEW YORK.

SAWYER'S DRAWING APPARATUS, No. 60.

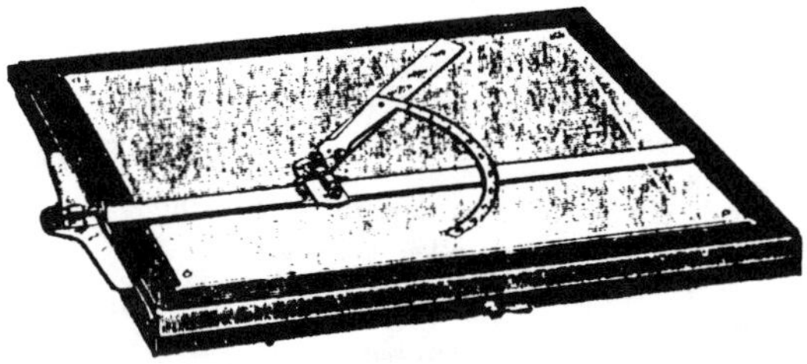

The great trouble with Drawing Boards and Tables is that the changes of atmosphere cause them to shrink and swell, so that a draughtsman is uncertain at any time whether his board or table is true or not. Then the wood being soft becomes worn and indented, so that it presents a rough and uneven surface for the use of a T Square. The apparatus here described overcomes these difficulties, and presents many advantages that make it not only a convenience, but an actual necessity to a practical draughtsman. Two styles are made, one a plain surface board, the other a frame for holding light inside boards. In both are adjustable metal edges of T shaped iron, the edge of the board being grooved and cut away so that the wood does not touch the iron except at the points where it is fastened to it. If the wood shrinks or swells it will not spring the iron. By a screw-headed eccentric pin in the corner, where the metal edges meet, the same can be instantly squared with each other. The boards are rabbeted on the edge, forming a groove between the wood and metal edge. Into this groove the guide plate of the attachment to T square drops, and by a spring tension draws the T square to the face of metal edge, and by an extra turn of the nut, the T square can be locked at any part of the board. This leaves both hands free to use triangles, pencil, etc. For ordinary use there is no necessity of locking the square.

	With Plain Boards. No. 60 A.		With Frames. No. 60 B.
20 x 25	One Metal Edge, $24.00	Two Metal Edges, $27.00	Two Metal Edges, $26.00
23 x 31	" " " 28.00	" " " 30.00	" " " 29.00
28 x 41			" " " 31.00

	Plain Surface Boards. No. 61.		Frames. No. 62
16 x 20	One Metal Edge, $1.90	Two Metal Edges, $3.00	
20 x 25	" " " 4.00	" " " 6.00	Two Metal Edges, $9.50
23 x 31	" " " 6.00	" " " 8.00	" " " 11.00
28 x 41			" " " 12.00

INSIDE BOARDS FOR FRAMES. No. 63.

(Made of Three Layer Stock, best quality.)

36 x 48	each, $3.00	Half doz., $14.00	23 x 31	each, $1.40	Half doz., $7.50
27 x 41	" 2.00	" " 9.00	20 x 25	" .90	" " 4.50

T SQUARE WITH ATTACHMENTS. No. 64.

20 inch Blade, Nickeled	$3.50	36 inch Blade, Nickeled	$5.75
25 " " "	4.25	41 " " "	6.50
31 " " "	5.00	48 " " "	7.50

PROTRACTOR AND SECTION LINER, No. 65.

The same principle is employed in a Protractor and Section Liner that is used on the blade of T Square. This instrument is the most rapid working and accurate tool of its kind on the market. It is a head with a swivel, spring-tempered blade, that can be locked rigidly at any angle; in the blade are two holes that are accurately in line with the axis. In the arm of a graduated quadrant are two pins that exactly correspond to the holes in the Protractor Blade. By applying this quadrant and reading on the edge of the T square, the blade can be swung around to any degree or part of degree. After blade is locked the quadrant can be removed; you can then use the extreme length of blade on both sides. The accurately graduated quadrant being of a 14 inch circle magnifies the degrees to such an extent that it is an extremely close working tool. The Section Liner is back of the Protractor Blade, and consists of a milled wheel, to the back of which is fastened a ratchet wheel. Hung upon the same stud that the milled and ratchet wheel revolve upon, is a lever upon which is a pawl that engages with the ratchet wheel. The whole is mounted on a boss that slides on a vertical stud. Turning the thumb nut on the end of vertical stud forces the milled wheel on to the paper, then by pressing the finger button on the end of lever, the Protractor Blade is caused to move along more or less, being regulated by a stop screw, or by the number of times the button is pressed. For rapid uniform hatching and tinting it is unrivalled. The bearings, screws, etc., are case hardened. The blades are spring tempered, and accurately ground, and the workmanship throughout is first-class. When the Protractor Blade, which is 15 inches long, or more, is set at right angles to the T square as an auxiliary it cannot be excelled for making the construction lines to a drawing, and you are not obliged to remove it when using the T square. Its narrowness permits its use flush to either edge of the board.

15 inch Blade, Nickeled .... $14.00 | 18 inch Blade .... $1.00 extra.

Protractor No. 65 A, without Section Liner $10.00

FRAME ON ADJUSTABLE IRON STAND, No. 66.

We mount the smaller-sized Frames on Iron Stands. These for home and office use make the Apparatus the best thing out. The T Square and Angle Protractor remaining in position with the board at an incline, makes it very convenient for the draughtsman.

The Frame is made to use a number of light inside boards. By having a frame with a true edge or edges, a drawing can be quickly removed, and another board substituted and the original drawing replaced without changing its position. A number of drawings can thus be worked along at the same time with the same frame. They are nicely finished with cherry surface. We make boards and frames with one metal edge on end.

Adjustable Iron Stand, Mounted on Castors $8.00

Mounting Frame 1.00

Protractor Miter.

Sawyer Tool Company, Athol, Mass., for whom Peoli & Willey, 97 Chambers street, New York, are selling agents for domestic and export trade, are introducing a protractor miter for use in combination with a steel square, as herewith shown. The device consists of a frame with a circular edge accurately graduated in degrees, by notches, and a swinging arm pivoted at a point central to the graduations. At the end of the arm are two latches that snap into the graduations on the circle and determine the angle or miter, it is stated, positively. At the opposite end of the protractor the parts are rabbeted to form bearings by which, in connection with the pivotal stud, the tool may be located on any part of the square, as shown in Fig. 2. It is preferable to use the protractor at the center of the square for ordinary work, but correct angles can be laid out with the tool on any part of the square if two of the three bearing points are kept in contact. It is further explained that when the protractor is set at any angle, the angles of the two square tongues form a miter; as, for instance, if the arm is set at an angle of 30°, or one-third miter, to one tongue of the square, it will form an angle of 60°, or two-thirds miter, at the other tongue of the square. This feature will be found useful in cutting in rafters, &c. By using the second catch 180 divisions of a quarter circle, or 720 divisions of a full circle, can be made. The tool is finished in japan and nickel plate and is designed for use by carpenters, pattern makers, cabinet makers and in wood working establishments, for laying out different pitches, cutting in rafters, getting out house finish, making correct patterns for duplicate work, projecting and taking off angles from drawings and for laying out winding stairs, &c. The manufacturers claim that the tool can be attached rigidly to a square in an instant and as quickly detached; that the settings are positive; that it is light, strong, simple and durable, and that it is warranted accurate. Useful tables accompany each tool.

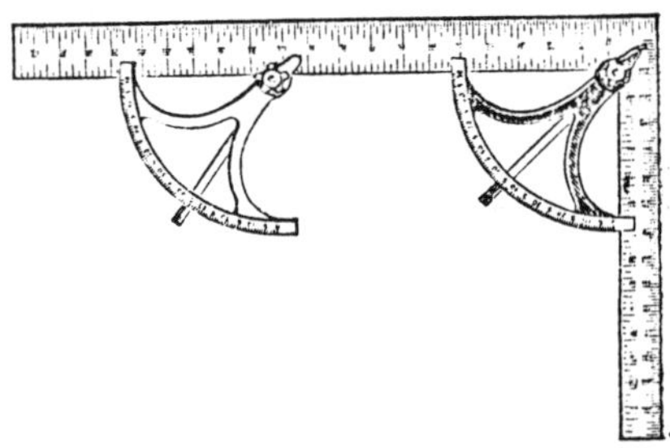

Fig. 2.—Protractor Miter Attached to a Square.

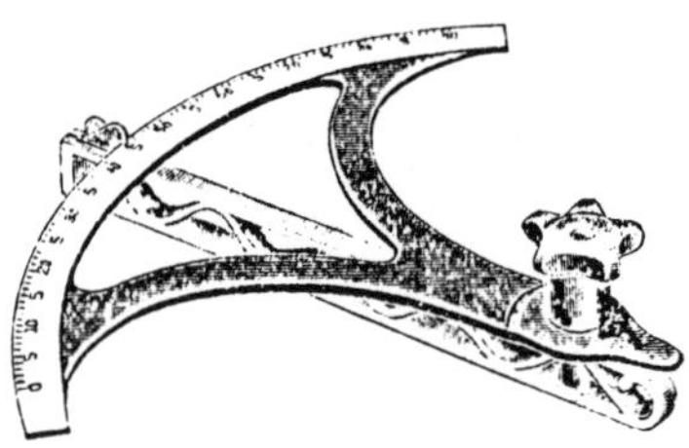

Fig. 1.—Protractor Miter.

1894

MACHINISTS, ATTENTION.

To introduce, will send post-paid on receipt of $1.60, one each of 3, 6 and 12 inch rules, all guaranteed equal to any in the market. Offer good until April 1. No stamps.

SAWYER TOOL CO., Athol, Mass.

1895

Sawyer's Improved Mechanics' Tools.

The Sawyer Tool Company, Athol, Mass., have brought out several new tools for mechanics' use, as shown in the accompanying engravings. Fig. 1 is their No. 37 hard-

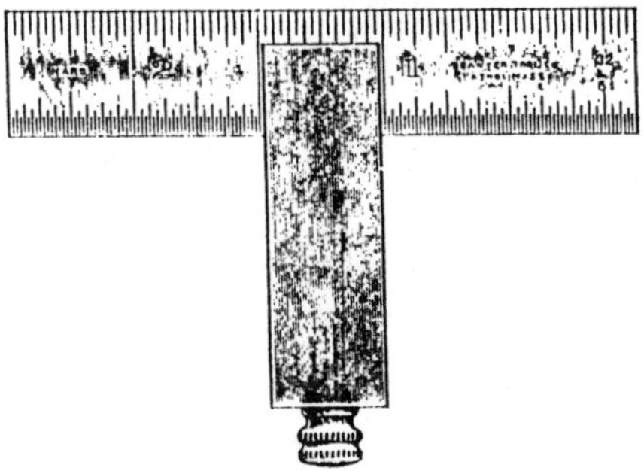

Fig. 1—No. 37 Tool Makers' Square.

ened tool makers' square with sliding graduated blade. The stock and blade of the square are hardened to prevent wear and getting out of true. When desired they furnish a narrow blade ⅛ inch wide, 2 inches long. Fig. 2 represents No. 41 Sawyer's improved try square, the blade of which is tempered throughout and firmly seated in fine gray iron stock by a powerful device with special refer-

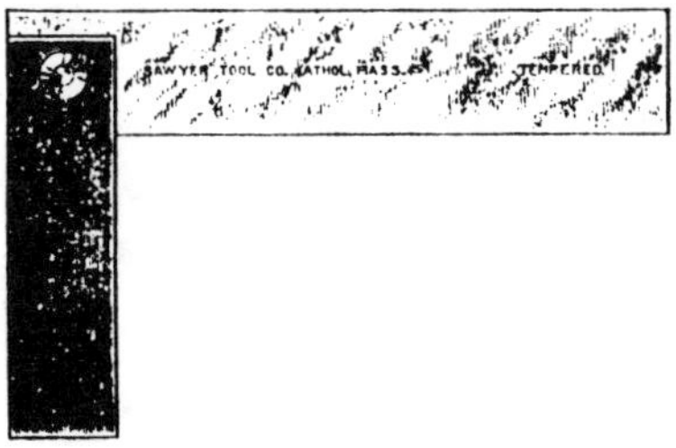

Fig. 2.—No. 41 Improved Try Square.

ence to accuracy. They are made in 4, 6, 8, 10 and 12 inch sizes. Fig. 3 is No. 85 screw pitch gauge. The form of the blades allows the points to be very narrow, making it specially useful for determining the pitch of internal

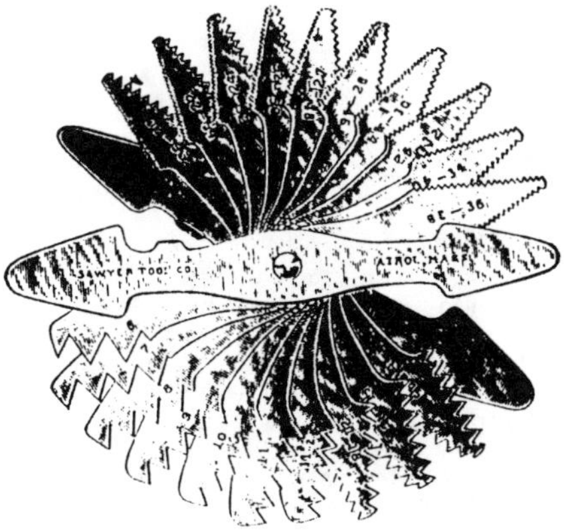

Fig. 3.—No 85 Screw Pitch Gauge.

threads. It contains the screw pitches in general use, from 6 to 56, inclusive, and the desired pitches are quickly and easily found.

1897

1897

Combination Squares.

We claim for these Squares simplicity, elegance of design and finish combined with the best material and workmanship. They are made especially for machinists, who wish a tool that will wear and remain accurate. These blades are absolutely hard. The very best value for the money. Graduated in both English and Metric measure.

Price:

inch, with center head	$2.00, without	$1.60
" " " "	2.25, "	1.90
12 " " " "	2.50, "	2.20

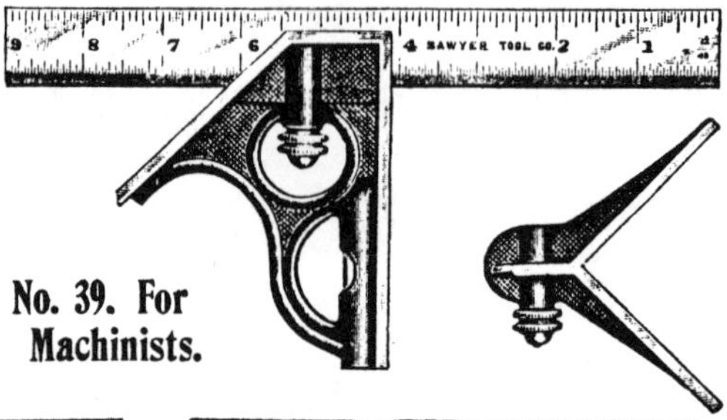

No. 39. For Machinists.

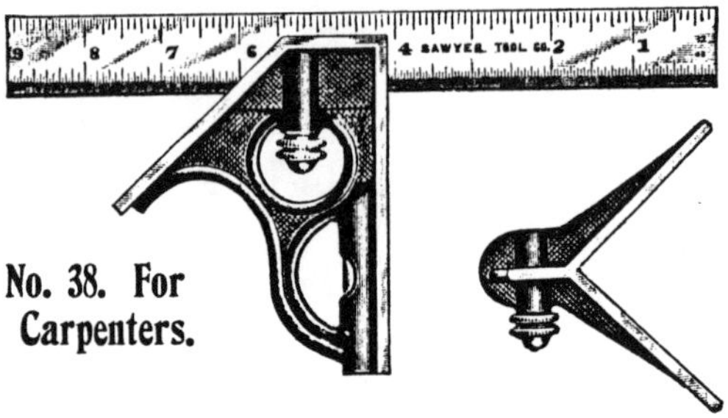

No. 38. For Carpenters.

Made especially for Carpenters and Woodworkers. The blades are tempered, although they are not so hard as our No. 39. Made in three styles of graduations No. 38 has 12ths, 48ths, 32ds and 8ths. No. 38a has No. 4 graduation. No. 38b has Metric graduation.

Price:

6 in. with center head	$1.40, without,	$0.95
9 " " " "	1.65, "	1.20
12 " " " "	1.90, "	1.45

New Catalogue Just Out.

Sawyer Tool Co.,
Athol, Mass., U. S. A.

1898

Samson Screw Driver.

Sawyer Tool Company, Fitchburg, Mass., are making the Samson screw driver here illustrated. The features of this tool are referred to by the manufacturers as follows: It supports the blade up to the actual straining point; it supports the head of the screw so that it will not wabble; a wood screw can be started by striking the end of handle with the palm of the hand and screwed to place without touching the screw after once in position; the blade being parallel it bears as much at the bot-

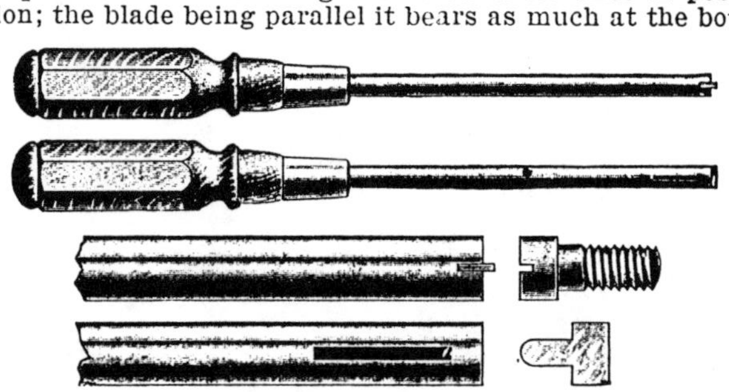

Samson Screw Driver.

tom of the slot as it does at the outside, thus preventing a heavy burr on each side of the slot as is caused by tapered blades; as the blades and shanks of drivers run in regular sizes corresponding to the average sizes of wood and machine screws, the blade does not wear and mar the screw as it would with a tapered blade; a broken blade can be quickly replaced by a new one, extra blades being furnished with each tool The blades can be supplied in ¼, 5-16, ⅜ and ½ inch diameters. The engravings show both screw driver complete and section of blade in two different positions, and illustrate how the detachable blade with parallel sides is inserted or replaced and easily removed if necessary.

1899

Surface Gauge No. 100.

The accompanying cuts relate to a surface gauge put on the market by the Sawyer Tool Company, Fitchburg, Mass. In the gauge, it is explained, an exceptionally easy and fine adjustment is provided, being positive and operative in whatever position the spindle happens to be. The large cut shows the gauge, while the smaller ones represent the different positions in which the tool can be used. One of them shows where it can be readily adapted to a lathe spindle, for use in laying out work on a face plate of the lathe or for internal boring. Another one represents its use on a planer slot by swinging the spindle one-half around. Another position shows it as it would be used in laying out parallel lines where the center cannot be obtained with dividers. The other cuts represent still different positions.

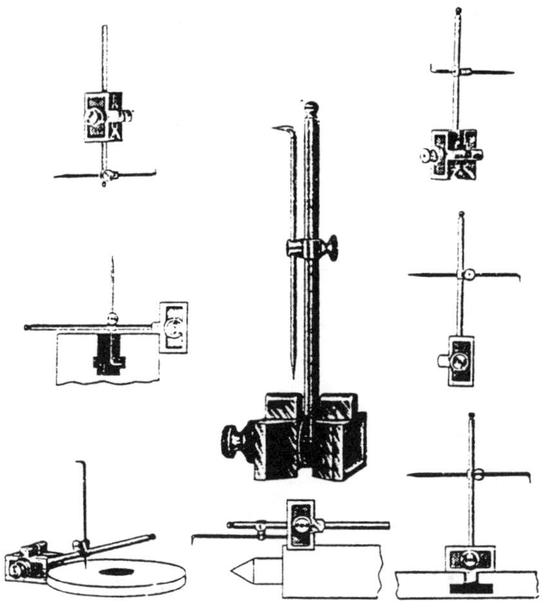

Surface Gauge No. 100.

1899

Bevel Protractors.

The Sawyer Tool Company, Fitchburg, Mass, are putting on the market protractors herewith shown. The stock of the tool in Fig. 1 is referred to as well proportioned, with good width of face. The polished rib at the back is alluded to as strengthening and improving the appearance of the tool. One of the best features of the tool, the manufacturers state, is a level accurately adjusted and permanently attached to the turret, forming an excellent inclinometer. The fact of not having to remove the blade when using the level, it is remarked, will many times be found convenient. The blade of the tool is hardened. The set shown in Fig. 2 comprises the company's No. 38 combination square with hardened blade and the No. 36 bevel protractor, shown in Fig. 1. The same blade is used independently with either square, protractor or center head. The blade being used in all of these combinations, attention is called to the importance of having the blade hardened so that it will stand the wear and usage caused by the variety of work for which it is designed.

Fig. 1.—Bevel Protractor No. 36.

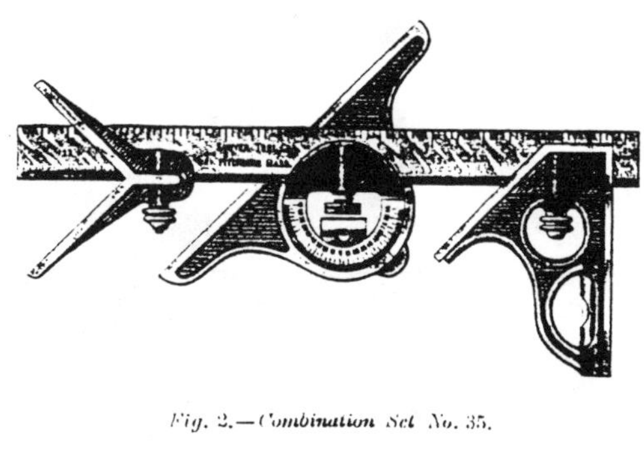

Fig. 2.—Combination Set No. 35.

1900

Double Squares, No. 44.

WITH TEMPERED BLADES.

Send for enlarged catalog which will be out this month showing

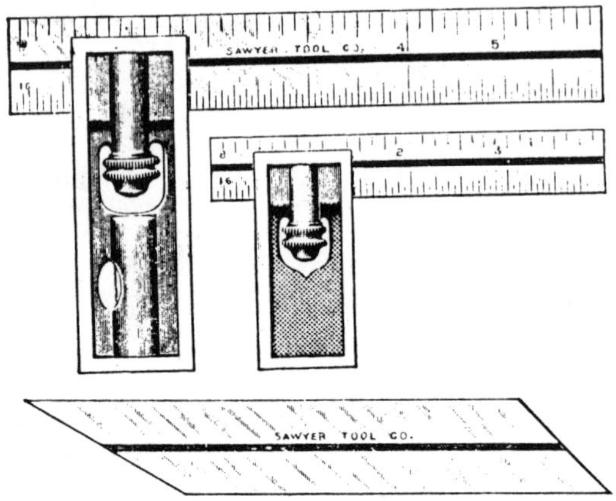

Made in four and six inch sizes only.

HIGH GRADE MACHINISTS TOOLS

MADE BY THE

SAWYER TOOL CO., FITCHBURG, MASS., U. S. A.

NEW TOOLMAKERS' SURFACE GUAGE No. 99

ALL STEEL,

HARD BASE,

GREAT SCOPE.

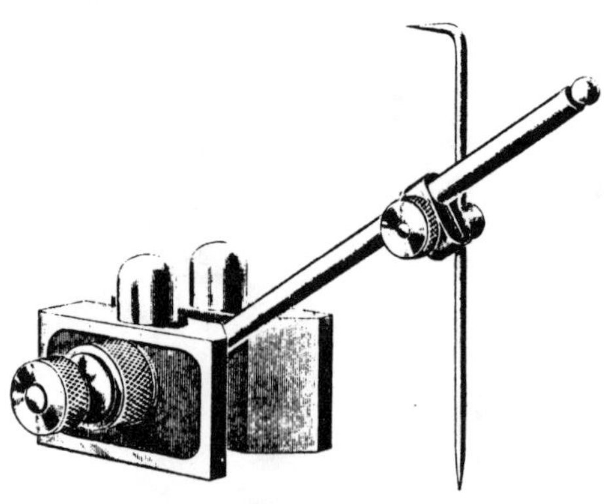

BEST GUAGE FOR THE PURPOSE MADE.

Send for Enlarged Catalogue, just out, of

High-Grade Machinists' Tools

Made by

SAWYER TOOL CO., FITCHBURG, ... MASS., U.S. A.

1900

Improved Hardened Center Gauge No. 20 B.

The gauge illustrated in the accompanying cut is put on the market by the Sawyer Tool Company, Fitchburg,

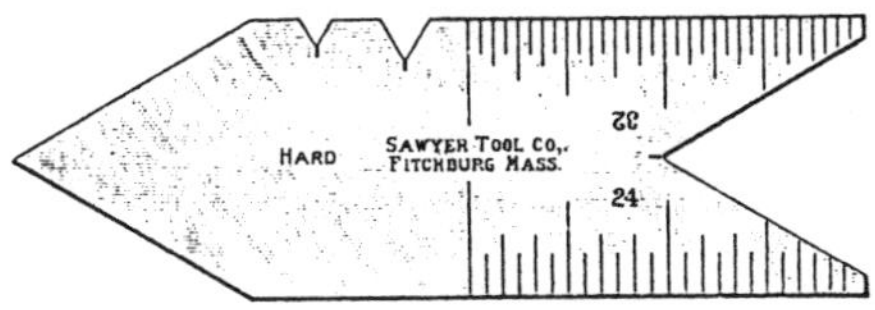

Improved Hardened Center Guage No. 20 B.

Mass. The gauge has small slits at the bottom of the angular notches to provide for the sharp edge or burr which is naturally caused by grinding an angular or threading tool. The slits also allow an exit for dirt that may accumulate.

1900

April, 1901. MACHINERY.

New Combination Square No. 38B.

FINEST TOOL OF THE KIND MADE WITH

HARD BLADES.

ACCURATE.

Like our Protractor No. 36, it has a polished back ⅛ in. wide which fits the hand perfectly. Blade, 1½ in. wide. Stock, 5⅝x⅞ in. wide

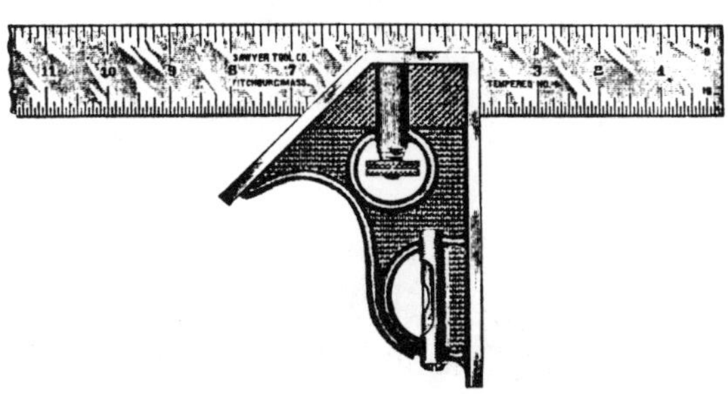

The blade alone is worth price asked for the Square. Made with two lengths of Blades. Price, 18 in. $3.75. 24 in. 4.50.

Sent PREPAID on receipt of list price and FULLY WARRANTED.

DID YOU KNOW that we furnished the BEST QUALITY of HARD BLADES in ALL our SQUARES? WE OUGHT TO as we have been longest at it now making HARD GRADUATED BLADES from 2½ to 24 inches, in both ENGLISH and METRIC graduations.

SEND FOR CATALOG D OF HIGH GRADE TOOLS ***Sawyer Tool Co.,***

FITCHBURG, MASS., U. S. A.

1901

Improved Screw Pitch Gauge

Sawyer Tool Company, Fitchburg, Mass., have just placed on the market an improved patented screw pitch gauge, as here illustrated. The blades tapering almost to a point allow them to be used in the smallest possible hole that a gauge can be used in, yet at the same time are not weakened or liable to bend. The finer pitches are cut upon each edge of the blade, which reduces the bulk of the tool nearly one-half. The convenient form in which it is made enables the desired pitches to be found readily and renders it particularly

Screw Pitch Gauge.

a pocket tool. It is furnished with 39 pitches, as follows: 4, 4½, 5, 5½, 6, 7, 8, 9, 10, 11, 11½, 12, 13, 14, 15, 16, 18, 20, 22, 24, 26, 27, 28, 30, 32, 34, 36, 38, 40, 42, 44, 46, 48, 50, 52, 54, 56, 58, 60.

1901

Improved Depth Gauge No. 57.

The Sawyer Tool Mfg. Company, Fitchburg, Mass., are putting on the market the improved depth gauge shown herewith. It is referred to as combining in one

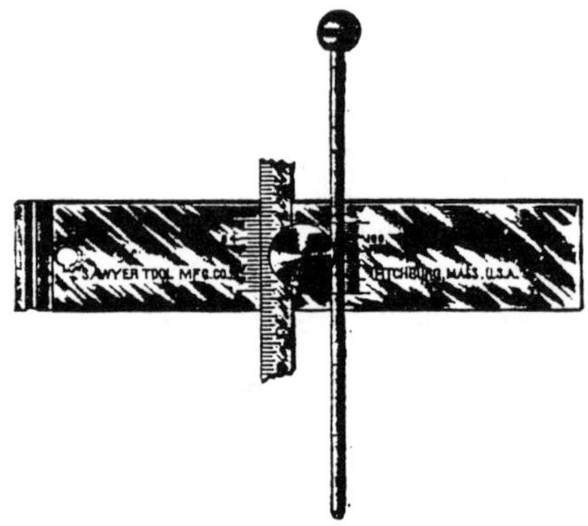

Improved Depth Gauge No. 57.

tool the advantages of the rule and the round rod, and as intended to meet the most exacting requirements. The stock is graduated in 64ths and 100ths inch and measurements may be read at a glance by comparing the ¼-inch divisions on the rod with the graduations on the stock. The rod is held against either graduation by the beveled head of the bolt, a slight turn of the nut being sufficient to clamp the rod in place. The same bolt holds the rule. By transferring the bolt to the hole near the end of the stock the gauge may be used in many places where otherwise its use would be impossible. The tool is made in two sizes of stock, 4 and 6 inch, and in two sizes of rod and rule, 4 and 6 inch.

1902

Key Seat Rule Blocks No. 27.

The Sawyer Tool Mfg. Company, Fitchburg, Mass., are offering key seat rule blocks, as shown in the accompanying cut. As the name indicates, the blocks are de-

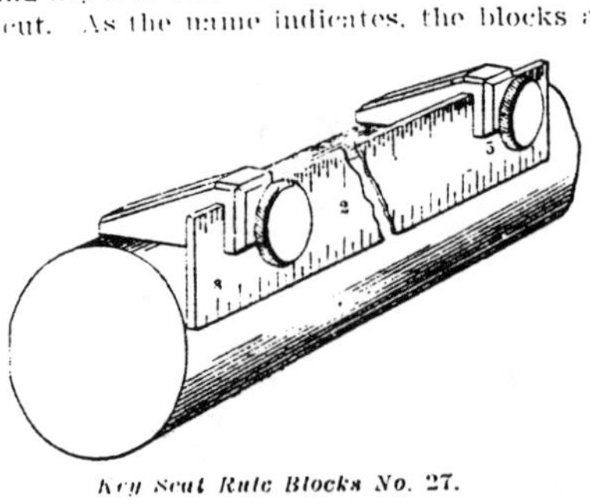

Key Seat Rule Blocks No. 27.

signed to convert an ordinary rule into a key seat rule. They are used for key seating and various other purposes, such as setting calipers and dividers and various other tools. The method of attaching the blocks is shown in the illustration.

1902

Inside Micrometer.

The Sawyer Tool Mfg. Company, Fitchburg, Mass., are manufacturing the inside micrometer shown full size in

Inside Micrometer, with Six Rods.

the accompanying illustration. While the micrometer is said to be the smallest manufactured for inside use, and compactness has been a chief consideration in designing it, other features of an inside micrometer, it is stated, have not been omitted. The screw has a movement of ½ inch, and has adjustment for wear, as has also the hardened anvil at the end of the thimble. The six rods that accompany the instrument are ground to standard sizes, and screw into the body of the tool, assuring firmness, accuracy and freedom from slipping.

Steel Clamps.

The steel clamp shown in the accompanying illustration and made by the Sawyer Tool Mfg. Company, Fitchburg, Mass., has as a feature the reinforcement of the jaws where the outer screws penetrate them, which does

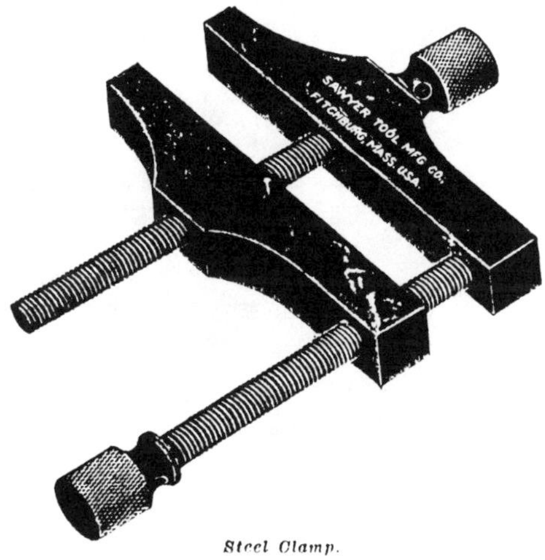

Steel Clamp.

away with the tendency of this form of clamp to break when the usual great strain is put upon them. The screws have a ball and socket bearing upon the jaws, for the purpose of eliminating unnecessary strain upon the threads.

Cross Level.

The Sawyer Tool Mfg. Company, Fitchburg, Mass., are making the cross level here illustrated. It is de-

Cross Level.

signed to level both ways on a surface, is made of iron enameled, with polished ends. Its size is 2½ x 3 x 5 inches.

Angle Indicator.

The Angle Indicator shown in the accompanying illustration is designed to be used in connection with any rule, bevel or a universal bevel. It is furnished either with the half button shown in the cut, or with a bevel protrac-

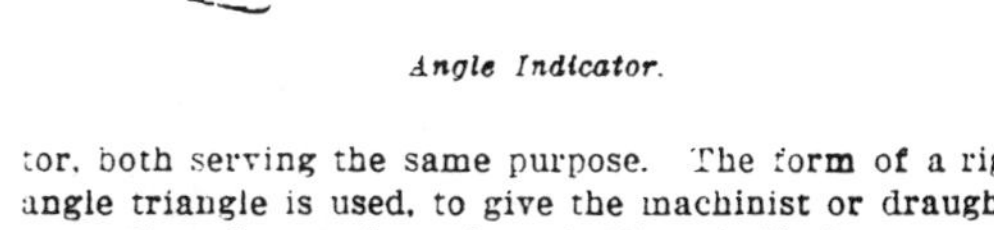

Angle Indicator.

tor, both serving the same purpose. The form of a right angle triangle is used, to give the machinist or draughtsman the often-used angles of 45 and 90 degrees. By use of this tool any angle within its scope may be quickly and accurately transferred. It is manufactured by the Sawyer Tool Mfg. Company, Fitchburg, Mass.

Adjustable Bench Level.

The Sawyer Tool Mfg. Company, Fitchburg, Mass., have added to their line of fine mechanics' tools the Adjustable Bench Level here shown. The frame is made of gray iron, accurately ground and neatly enameled. It has a longitudinal groove in base, permitting use upon shafting and kindred round bodies. The vials are made from straight glass tubing, ground barrel shape on the inside by a patented process, thus insuring accuracy

Adjustable Bench Level.

whether the level is upright or tilted sideways. Both level and plumb vials are covered by revolving nickel plated shields for protection, when not in use, from dust and liability to break. The sizes made are 6, 9, 12, 18 and 24 inches.

One-Inch Rule with Holder No. 2.

The 1-inch rule illustrated herewith has for its principal use measuring in places where a longer rule cannot be used, as in holes and cavities. The rule may be swung

One-Inch Rule with Holder No. 2.

at any angle with the holder and will retain its position. It measures 5 inches over all. The rules are furnished in Nos. 1, 2, 4, 7, 8, 15 and 16 graduations, and are offered by Sawyer Tool Mfg. Company, Fitchburg, Mass.

Scriber.

The Sawyer Tool Mfg. Company, Fitchburg, Mass., are manufacturing the Scriber here illustrated, a feature of which is that the finely tempered crucible steel mark-

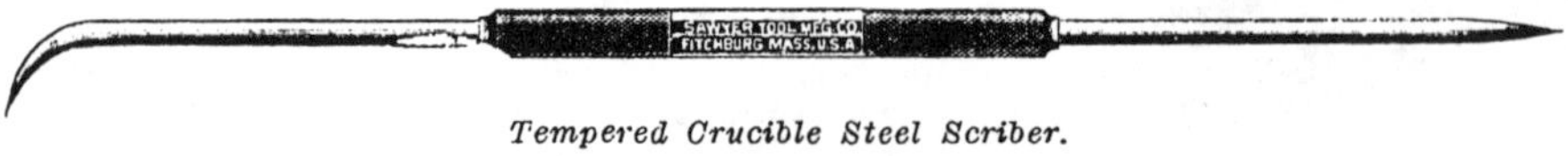

Tempered Crucible Steel Scriber.

ing points are held in the body of the tool by a taper fit, instead of screwing them in. The body or center portion is knurled, and of sufficient size to be securely held in use. The length, over all, is 8 inches.

1904

The Sawyer Surface Gauge.

The surface gauge shown in the illustration has several features in which it differs from accepted types at present on the market. As will be seen, the tool has a heavy base of hardened steel, slotted at one side of its upper surface to take the rocker arm. This arm is pivoted at the rear instead of in the center, the purpose being to secure less backward and forward motion of the scriber when it is necessary to use the adjustment, the longer leverage securing this result. The adjustment screw works against a spring underneath the rocker arm in the slot. The scriber arm is held in a new yoke which is so designed that it will slide easily on the spindle

Sawyer Surface Gauge.

much the same as if the spindle were ground. The auxiliary base replaces the usual plugs. It is held in place when its use is desired by a thumbscrew working in a groove at either side of the base. The auxiliary base is reversible, one edge being straight, the other milled at an angle so that it may be used on a shaft or other round piece of any size, thus accomplishing, according to the claim, what cannot be done with plugs when the round piece is of small diameter. The straight edge of the auxiliary base is beveled slightly at its upper edge, to allow for the presence of a burr on the edge of the

Sawyer Outside Calipers

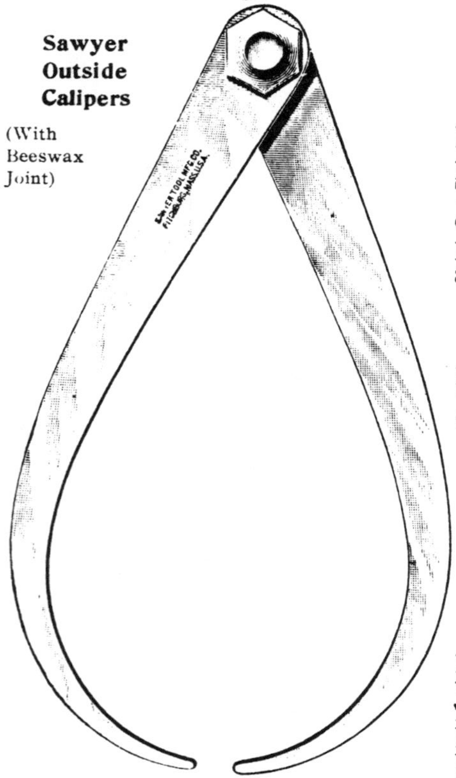

Sawyer Outside Calipers

(With Beeswax Joint)

$1.15 Value for 60 Cents

SPECIAL LIMITED OFFER

Every machinist will be interested in our very liberal business proposition we are making to patrons who lend us their co-operation. To place this proposition before appreciative purchasers, we send one 4-in. and one 6-in. Caliper for 60 cents. The regular catalog price for these two tools is $1.15. You have opportunity to make a saving and hear something to your advantage.

AN IMPROVED TOOL

Applying beeswax to a caliper joint in a manner that the beeswax is retained for an indefinite period, whether the tool is constantly or only occasionally used, has been successfully solved by this invention. The pockets for holding the beeswax are made in such a manner that they in no way impair the utility of the joint, but rather add to it. These Caliper joints must be tried to appreciate their real value. They have that firmness of feeling so much desired by every mechanic. Great care is used in their construction. Points are all hardened and of the proper shape. Calipers are of a neat design and proper proportion, nicely finished. The joints are adjustable for any desired tension. Dealers will find these Calipers a profitable line to handle. Mechanics will find them the best Calipers they have ever used.

SAWYER TOOL MFG. CO., 1 Winter St., Fitchburg, Mass.

July.

Sawyer Tool Mfg. Co.,
Fitchburg, Mass.

Enclosed find 60 cent for one 4-in. and one 6-in Caliper. Both to be sen postpaid with your specia proposition.

Name ....

Address..............................

....

Name of Shop . ...

1907

STANDARD TOOL COMPANY

STANDARD TOOL COMPANY Athol, MA

Officers were: George T. Johnson, President; D. Appleton Newton, General Manager and Treasurer; and Stephen H. Bellows, Superintendent. These men held the same positions at the ATHOL MACHINE CO. The Standard Tool Co. was operated as part of the ATHOL MACHINE CO. and was never an independent company. There is, in fact, some evidence from the Starrett patent litigation, that the Standard Tool Co. was set up by the ATHOL MACHINE CO. as part of its attempt to evade the Starrett patents. (Starrett sued the ATHOL MACHINE CO. and part of their defense was that the squares which were the subject of the suit were not made by them but by the Standard Tool Co. See Appendix III for more details.)

One of its first tools appears to have been a quick-adjusting monkey wrench patented June 8, 1880 (No. 228437), by Stephen H. Bellows. It is no coincidence that this patent is very similar to a wrench patent granted Laroy S. Starrett on May 8, 1877 (No. 190636). Starrett, of course, was working for the ATHOL MACHINE CO. in 1877, so it would appear that Standard appropriated his wrench patent much as it attempted to evade his patent for the combination square. The wrench will be found with either STANDARD TOOL CO. or ATHOL MACHINE CO. markings.

The company's first machinist's tool was the Chaplin try and center square, originally patented by Orril R. Chaplin on May 8, 1866 (No. 54503). The squares made by the Standard Tool Co. are marked "Chaplin's Patent, Reissue May 4, 1880 (No. 9183). This reissue was assigned by Chaplin to Daniel A. Newton and Stephen H. Bellows and contained many of the features of Laroy S. Starrett's patent for his combination square. It would appear that the officers of the company set out to find, and buy, a patent which would allow them to prevent Starrett from making his combination squares. This, of course, set the stage for the later patent litigation with Starrett.

In 1882, the company announced that it was "beginning to manufacture a new combination tool, consisting of pipe-tongs, gas-fitters's plyers, and the ordinary flat plyer." This tool was still offered in its 1900 catalog.

It introduced a new combined level, plumb, surface and scribe gauge in 1883. This tool, offered in 6", 8" and 10" sizes, contains features patented by Stephen H. Bellows on September 16, 1873 (No. 142761) and by David B. Woolson on October 7, 1879 (No. 220323). This gauge proved very popular and was offered until the firm went out of business.

William A. White's caliper patent of April 1, 1884 (No. 296259) was one-half assigned to D. Appleton Newton. Stephen H. Bellows' patent of April 15, 1884 (No. 297056) for a caliper rule was assigned to Standard Tool Co. Tools were made under both patents for many years.

A universal bevel was introduced in 1887, and in 1888, a line of spring calipers based on the Stephen H. Bellow's patent of June 3, 1884 (No. 299722).

The STANDARD TOOL CO.'s screw pitch and center gauge combined, bevel protractor, pocket center gauge, steel caliper rule, and universal bevel are illustrated in the 1884 Charles H. Besly & Co. catalog. All are listed with the brand name of "Bay State". This may have been a brand name used by the Standard Tool Co., but no tools with that name or any other reference to the name have been located to date.

By 1889, the company was offering Chaplin try and center squares, steel rules, caliper rules, universel bevels, caliper gauges, surface gauges, bevel protractors, depth gauges, screw pitch and center gauges, steel squares, spring calipers, and straight edges.

A protractor based on J.P.B. Wells' patent of July 22, 1890 (No. 432779) and a beam micrometer made under Bellow's patent of July 28, 1891 (No. 456875) were introduced in 1891.

George T. Johnson, president, died in 1893 and was succeeded by his son William G. Johnson. A business report for that year lays stress on the good business the company was doing in monkey wrenches and bicycle wrenches. No mention was made of machinist's tools except to claim that it made the best and that its business had reached record levels.

In 1897, it was advertising Bellows' combination square and protractor. This tool had been patented by Bellows on March 11, 1884 (No. 295100), but Laroy S. Starrett had sued, claiming infringement on his patent. The case was decided in Starrett's favor in July, 1885. Starrett's patent, as reissued in 1880, expired in 1897 so Standard Tool Co. was free to make the Bellow's design. It also began offering a combination square almost identical to the Starrett design.

Bellows received a patent for a spirit level on May 11, 1897 (No. 582517). These are shown in a c1897 catalog in a machinist's version in 6", 9" and 12" sizes and a plumber's model in 6" and 12". The plumber's model "has a graduated gate on one end that can be adjusted to give the exact pitch of pipes, etc." This was the feature covered by Bellow's May 11, 1897 patent. This same catalog also offered a complete line of C.F. RICHARDSON & SON's levels.

In July, 1899, *Iron Age* magazine announced that:

> "Standard Tool Company are making an iron level ... in many respects it is similar to the ordinary cast iron level, the level glass, however, being inserted in a cast iron shell before being placed in the level, the shell resting on a central point. Two screws pass through the frame and fit on to a projection at each end of the receptacle, making an adjustable level of it, as a turn of the screws will readily adjust the glass."

They noted that the levels were supplied in sizes from 6 inches to 24 inches.

The level was also made in a 4" pocket size which lacked the adjusting feature. In 1900, the company began offering a line of micrometers, including an odd "shifting" design patented by J.P.B. Wells on January 9, 1900 (No. 641173). In 1901, the company begin making nearly exact copies of the L.S. STARRETT CO.s "Fay" and "Yankee" calipers. This was probably after the expiration of the Starrett owned Fay patent of June 2, 1885 (No. 319215).

A newly designed "Never Slip" surface gauge, made under the Stephen H. Bellows patent of April 23, 1901 (No. 672634), was likely the last new tool to be introduced. Although the patent was assigned to the ATHOL MACHINE CO., these rare gauges are marked with the STANDARD TOOL CO. name. In 1902, the ATHOL MACHINE CO., and therefore the STANDARD TOOL CO., was sold to Frank S. Ewing, Dresser T. Bates and Stephen E. French of Orange, MA. They, in turn sold out to Laroy S. Starrett in 1905.

The STANDARD TOOL CO. ceased operation in 1905 although Starrett kept the ATHOL MACHINE CO. in operation. It must have been very satisfying to Starrett to win out so completely over his old enemy.

For a detailed look at all the STANDARD TOOL CO. tools offered in 1900, a reprint of its 1900 catalog is available from: Martin J. Donnelly Antique Tools, P.O. Box 281, Bath, NY, 14810-0281.

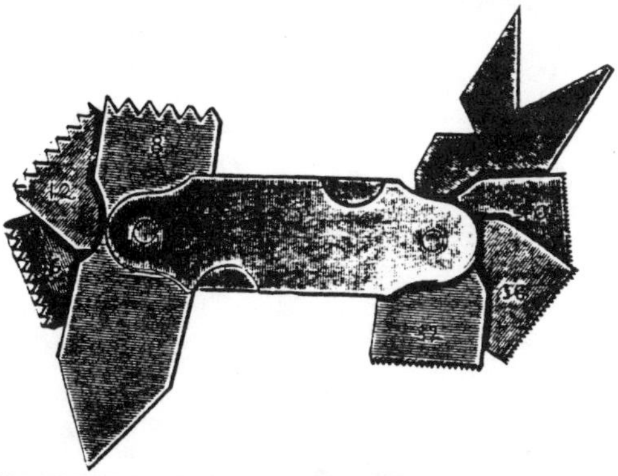

No. C 294

BAY STATE SCREW PITCH AND CENTER GAUGE COMBINED.

Twenty Pitches.

Price.................... $1.00

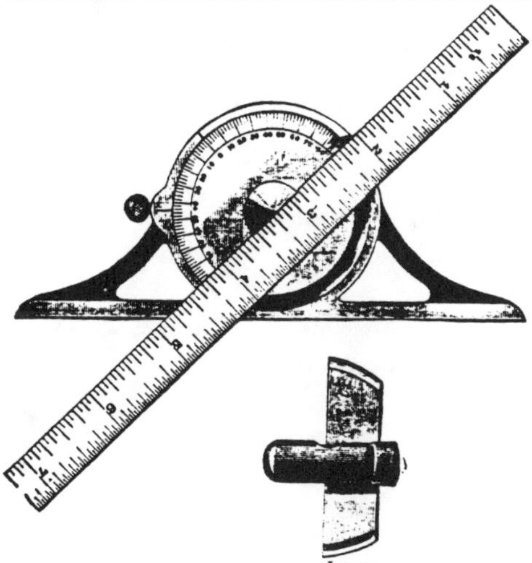

LEVEL.

No. C 295

BAY STATE BEVEL PROTRACTOR.

(WITH LEVEL ATTACHMENT.)

The Blade, a graduated Steel Rule, is firmly held in a groove in the revolving disc by a conical-headed screw, drawn to its seat by a thumb nut at the back. The Blade can be removed and the Level inserted in its place, the Level being secured in its proper position in relation to the base by a suitably adjusted stop.

Price, 9 inch.............. $2.50
" 12 "............... 3.00

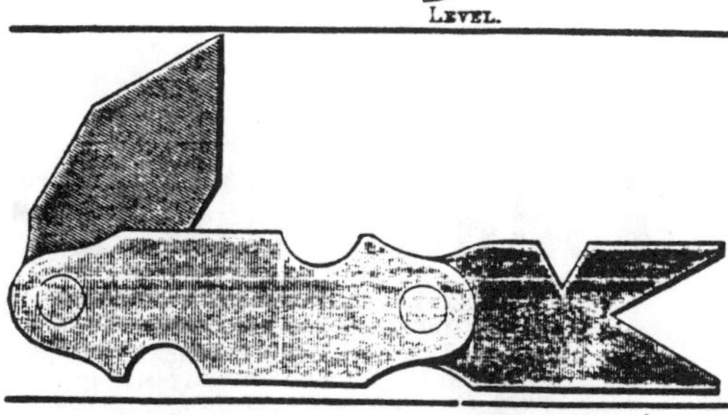

No. C 296

BAY STATE POCKET CENTER GAUGE.

[*Full Size.*]

Angle 60 degrees. Price, 40c.

This Gauge being made with folding blades, can be used in many places where the use of an ordinary Center Gauge would be inconvenient.

No. C 297 BAY STATE STEEL CALIPER RULE.

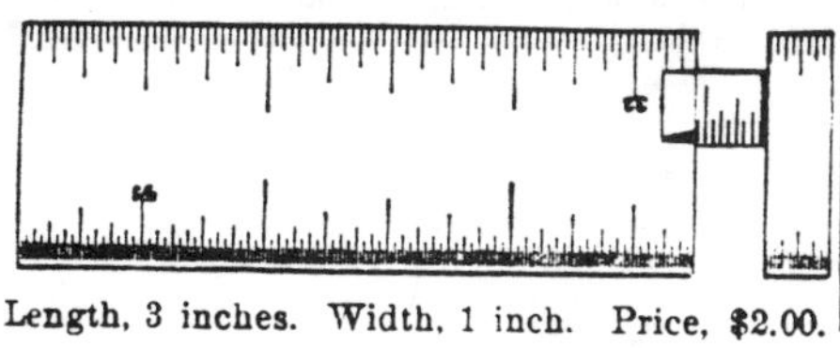

Length, 3 inches. Width, 1 inch. Price, $2.00.

BAY STATE UNIVERSAL BEVEL.

No. C 298

Length, 3 in. Width, 5-8 in. Price, $1.50

This page from the Charles H. Besly & Co. catalog shows the "Bay State" trade name used for Standard Tool Co. tools — 1884

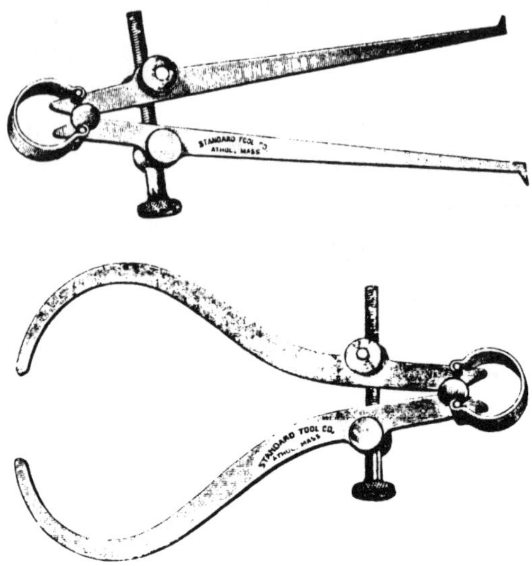

STANDARD SPRING CALIPERS.

American Machinist, 1888

STANDARD TOOL CO.,

ATHOL, MASS.,

MANUFACTURERS OF

The Celebrated Chaplin Try and Center Square, Standard Steel Rules, Steel Caliper Rules, Universal Bevels, Caliper Gauges, Improved Surface Gauges, Bevel Protractors, Depth Gauges, Screw Pitch and Center Gauges, Hardened Steel Squares, Graduated Steel Squares, Spring Calipers, Pliers, Straight Edges, &c., &c.

WRITE FOR ILLUSTRATED CATALOGUE AND PRICE LIST OF FULL LINE.

1889

Standard Try Square.

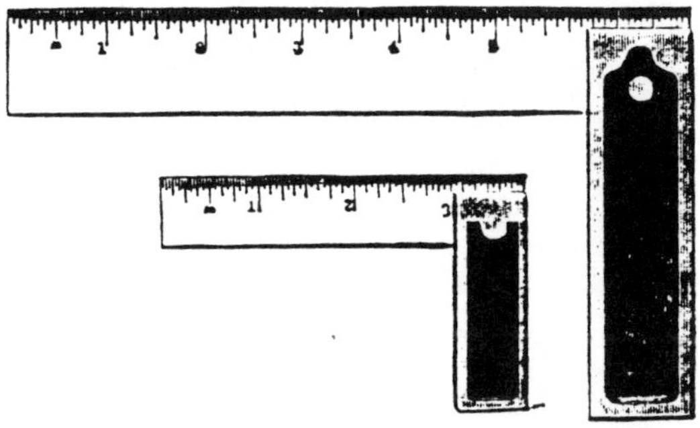

Graduated steel blade—not hardened—recessed iron handle. Light and convenient, and warranted accurate.

Inches	3	4	6	8	10
Each	75c.	$1 00	1 25	1 75	2 25

Chaplin Try and Centre Square.

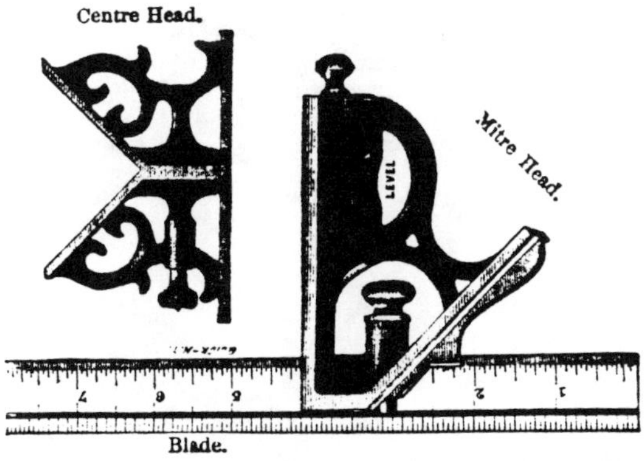

A set complete comprises a mitre head, centre head and blade, as shown in cut—one blade fitting both heads; but as the blade, with either head, forms a complete tool in itself, the several parts will be furnished separately, when desired.

The set complete will be sent in all cases, unless otherwise specially ordered.

Set Complete.

Size, inches,	6	9	12
Each	$2 00	2 50	3 00

Price of Separate Parts.

Size, inches,	6	9	12
Blades, each,	65c.	80c.	$1 00
Mitre head "	60c.	80c.	1 00
Centre head "	85c.	$1 00	1 25

Graduated Steel Square.

Not Hardened.

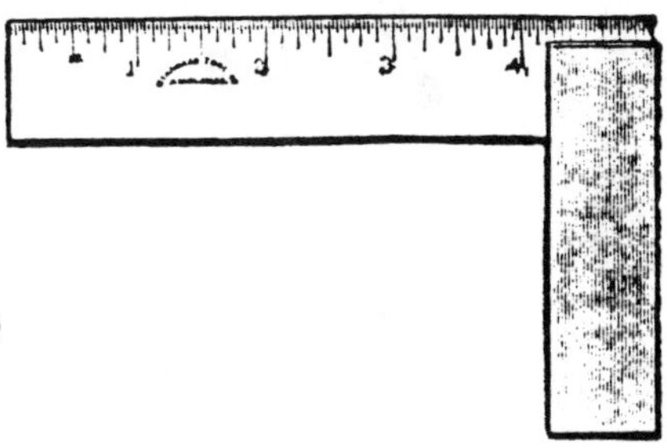

Size, inches,	3	4	6
List, each,	$1 25	1 50	2 00

1889

ROUND BASE SURFACE GAUGES.

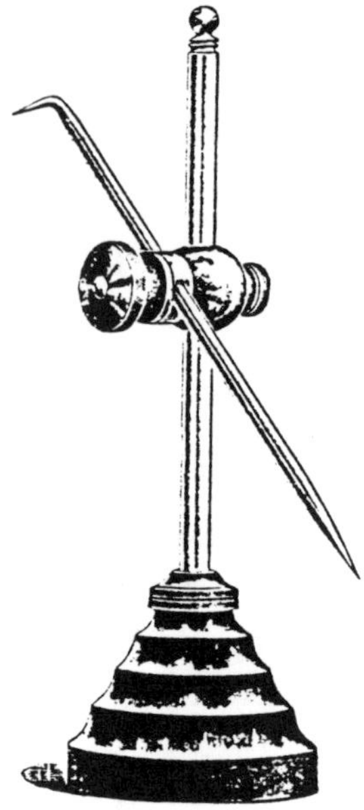

This cut represents our improved Surface Gauge. The base is turned and polished, and is lapped true on the bottom. The spindle proper is firmly fastened into the base, and has fitted over it a steel tube, on which is mounted the device for holding the pointer. A fine adjustment is made by forcing the tube up or down on the spindle by means of a screw with a knurled head, tapped into the top of the spindle, and when adjusted can be securely locked in position by a slight turn of the set screw opposite the pointer, while the pointer can be turned in either direction without loosening any of the parts.

6 inch.................................$2.25
9 " 2.50
12 " 3.50

IMPROVED SURFACE GAUGES.

This tool is provided with an adjustable attachment for the quick and accurate adjustment of work on a line with the bed of a planer, or parallel to its edge; also with a spirit level, making a very convenient level and plumb.

Inches..................	6	8	10
Each.....................	$2.00	2.50	3.00

OPEN FACE CALIPER GAUGES.

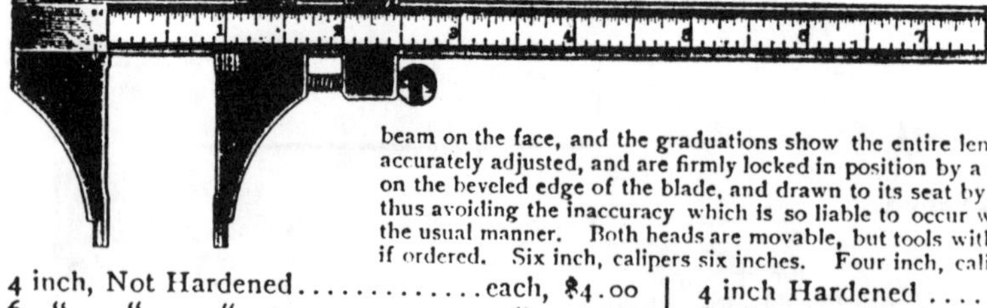

Graduated in 50ths and 64ths and verniers to 100ths, and finer if desirable.

The advantage of this tool over all similar tools consists in having heads dove-tailed, so that the heads are flush with the beam on the face, and the graduations show the entire length. The heads are quickly and accurately adjusted, and are firmly locked in position by a screw, with beveled head bearing on the beveled edge of the blade, and drawn to its seat by a thumb-nut on the reverse side, thus avoiding the inaccuracy which is so liable to occur when fastened with a set-screw in the usual manner. Both heads are movable, but tools with one fast head will be furnished if ordered. Six inch, calipers six inches. Four inch, calipers four inches.

4 inch, Not Hardened............each, $4.00	4 inch Hardened................each, $5.00
6 " " " .... " 5.00	6 " " .... " 6.00

Morocco Cases for above, $1.00 each.

1890

STANDARD TOOL CO.

CHAPLIN'S TRY AND CENTER SQUARE.

(*SET COMPLETE.*)

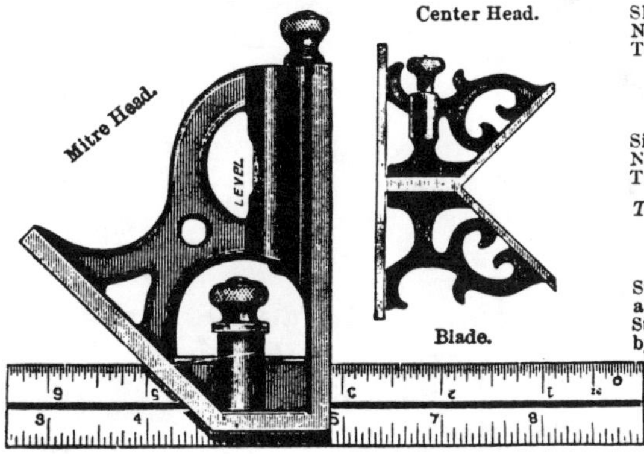

SIX INCH,	- - Price,	$2.00
NINE "	- - "	2.50
TWELVE "	- - "	3.00

PRICE OF SEPARATE PARTS.

	Blade.	Mitre Hd.	Center Hd.
Six In.,	$0.65	$0 60	$0.85
Nine "	0.80	0 80	1.20
Twelve "	1.00	1 00	1.50

The SET COMPLETE will be sent in all cases, unless otherwise specially ordered.

It consists of a Mitre Head or Stock provided with a Spirit Level, a T or Center Head and a sliding Steel Blade, firmly held to its seat by a clamp bolt.

The Blades are graduated on each of the four edges, the entire length, on one side in Sixteenths and Thirty-Seconds, and on the other in Tenths and Twelfths.

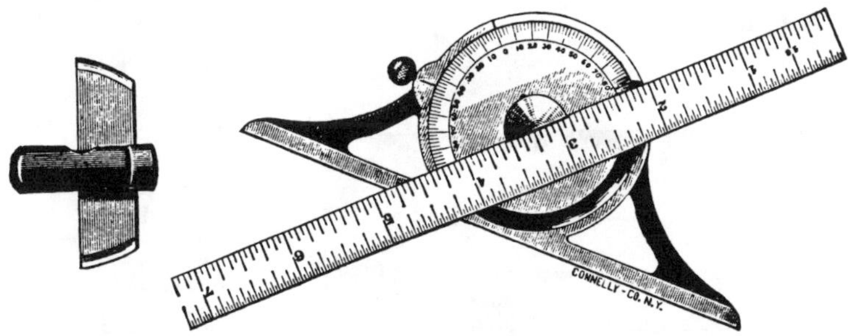

STANDARD TOOL CO.'S BEVEL PROTRACTOR,

(*WITH LEVEL ATTACHMENT.*)

NINE INCH, . - Price, $2.50 | TWELVE INCH, - - Price, $3.00

The uses and advantages of this Tool can readily be seen. The Blade, a graduated Steel Rule, is firmly held in a groove in the revolving disk, at any required angle with the base, by a conical-headed screw, drawn to its seat by a thumb nut at the back. The Blade is further held in position at right angles with the base, and at an angle of 45 degrees, by a suitably adjusted stop in the disk, making it a fixed Try Square or Set Mitre. The Blade can be removed and the Level inserted in its place by simply loosening the thumb nut—the Level being secured in its proper position in relation to the base by suitably adjusted stop, thereby making the tool a perfect adjustable Plumb and Level.

STANDARD TOOL CO.'S CALIPER GAUGE,

With Trammel Points and Adjusting Screw.

SIX INCH,	- Price,	$3.00
NINE "	- "	3.50
TWELVE "	- "	4.00

The Blades in these Gauges are graduated the same as our Standard Steel Rules—in Eighths, Sixteenths, Thirty-Seconds and Sixty-Fourths—and when removed from the sliding heads, may be used for the same purposes.

STANDARD TOOL CO.

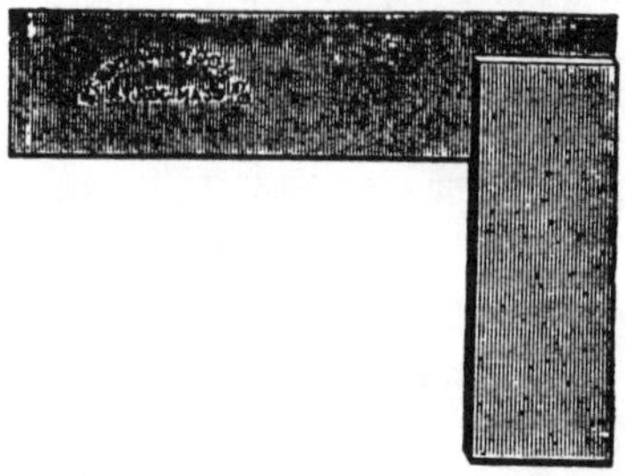

Hardened Steel Square.

Price, 1½ inch		$1 50
" 3 "		2 25
" 4½ "		3 00
" 6 "		3 75

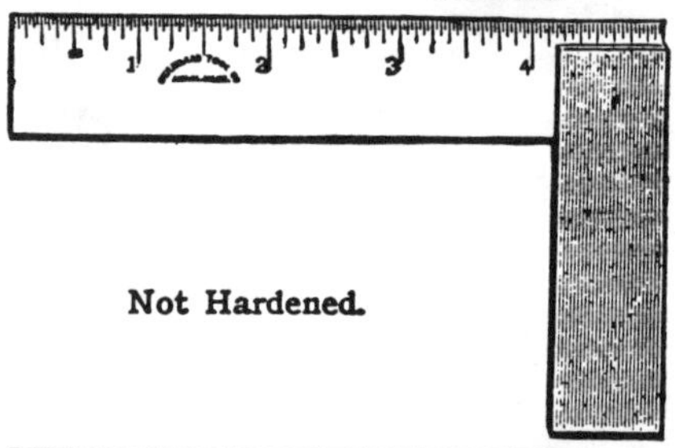

Not Hardened.

Graduated Steel Square.

Price, 1½ inch		$1 00
" 3 "		1 25
" 4 "		1 50
" 6 "		2 00

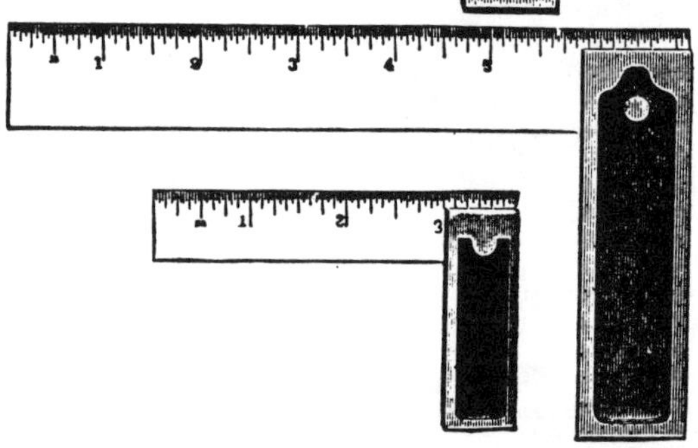

Standard Try Square.

Graduated Steel Blade—not hardened—Recessed Iron Handle. Light and convenient, and warranted accurate.

Price, 3 inch		$ 75
" 4 "		1 00
" 6 "		1 25
" 8 "		1 75
" 10 "		2 25

(Nine-tenths full size.)

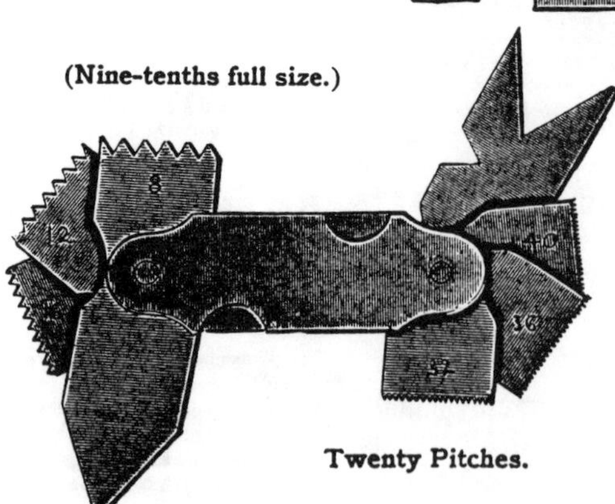

Twenty Pitches.

Standard Screw Pitch and Center Gauge Combined.

A first-class tool in every respect, and warranted accurate. Angle of Center Gauge, 60 deg.

Price $1 00

"Standard Tool Co." Lock Slide Caliper Rule.

The slide in this rule is graduated in 50ths and 64ths and verniers to 100ths, is quickly and accurately adjusted, and when adjusted can be firmly locked in position by a thumb nut on reverse side.

Price of 3 inch......... $2 00
" " 4 " 2 50

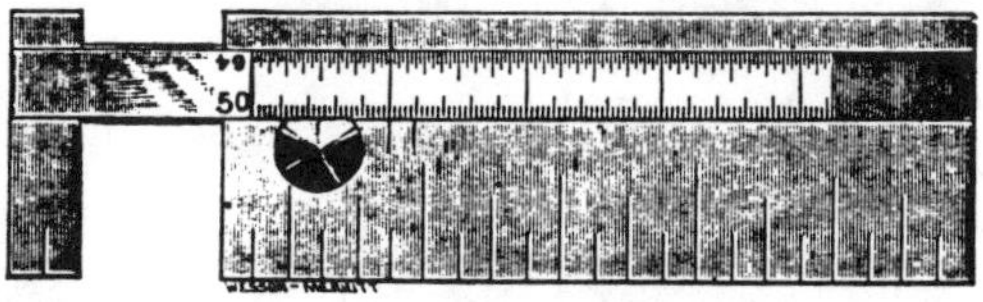

"STANDARD TOOL CO." STEEL RULES.

Graduated on one side in eighths and thirty-seconds, and on the other in sixteenths and sixty-fourths.

One inch.	$0 20
Two "	25
Three "	35
Four "	50
Six "	65
Nine "	1 00
Twelve inch...........	1 25
Eighteen inch.........	2 00
Twenty-four inch.....	2 50

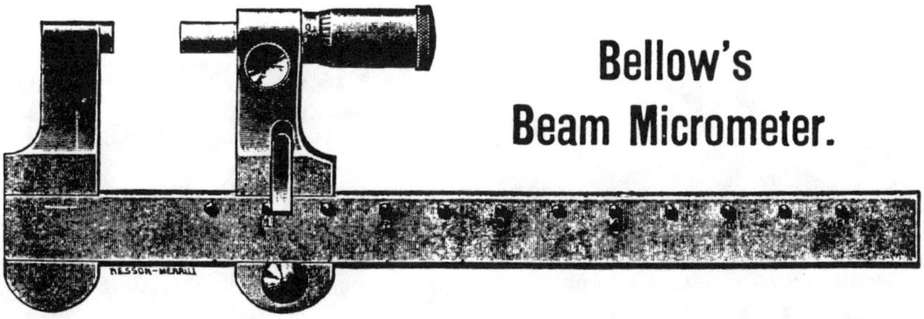

Bellow's Beam Micrometer.

Price each, 4 inch ....$12 00 In Morocco Case, $13 00
" " 6 " .... 15 00 " " " 16 00

Standard Depth Gauge.

Indicator has a movement of two inches, can be used either in center or end of beam, and is quickly adjusted to any depth,—the screw being used only for very fine adjustments.

1891

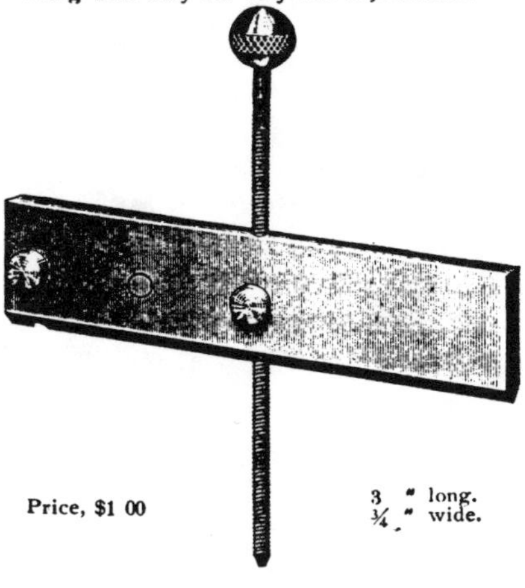

Price, $1 00

3 " long.
¼ " wide.

BOSS SPRING CALIPERS.
Made From
Tool Steel
8 inch mailed till April 1, for 50 cts.
STANDARD TOOL CO.
ATHOL, MASS., U. S. A.
MANUFACTURERS OF
MECHANICS' FINE TOOLS.
CATALOG FREE.

MAILED FOR 80 CENTS
TILL APRIL 1ST.
STANDARD TOOL CO.
Athol, Mass.
MANUFACTURERS
Mechanics'
FINE TOOLS.
CATALOG FREE.
6-INCH FIRM-JOINT TRANSFER CALIPERS.

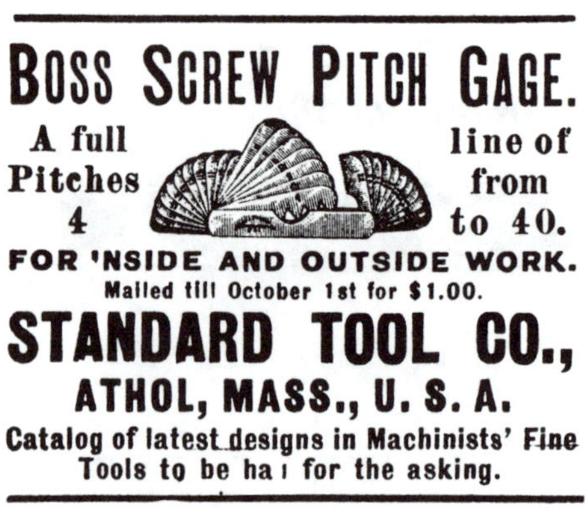
BOSS SCREW PITCH GAGE.
A full line of Pitches from 4 to 40.
FOR 'NSIDE AND OUTSIDE WORK.
Mailed till October 1st for $1.00.
STANDARD TOOL CO.,
ATHOL, MASS., U. S. A.
Catalog of latest designs in Machinists' Fine Tools to be ha i for the asking.

3-INCH UNIVERSAL BEVEL.
MAILED TILL JULY 1ST FOR $1.00.
CATALOG FREE.
STANDARD TOOL CO.,
ATHOL, MASS., U. S. A.
MANUFACTURERS OF
MECHANICS FINE TOOLS.

1895

Graduated Depth Gauge.

Athol Machine Company, Athol, Mass., are just introducing a graduated depth gauge, as here illustrated.

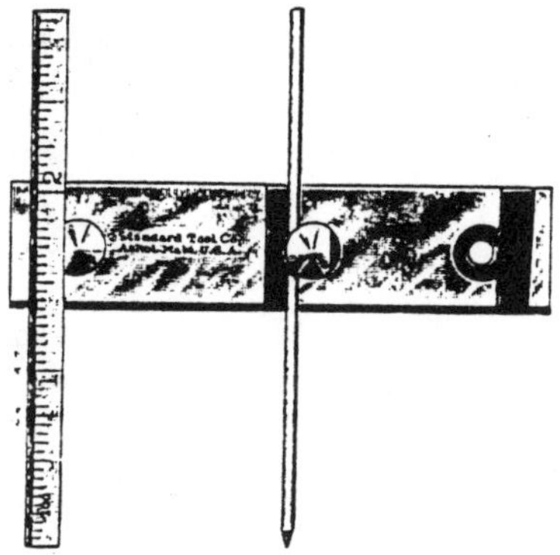

Depth Gauge with Graduations.

The blade on the left is 3 inches long and 3-16 inch wide, is drawn to its seat with a screw having a beveled edge head and locked with a thumb nut on the reverse side. It is graduated in sixty-fourths on one side and one-hundredths on the other. Both edges of the beam and blade are accurately ground, so that the tool can be used as a square when desired. A wire is sent with the gauge for use in holes too small to admit the blade, and is held by a groove under the round edge of screw head. The third slot is cut at an angle of 1 degree for use in die work and for getting the draft on patterns.

1897

Bellows' Combination Square and Protractor.

The cut shows a new combination square, level and protractor made by the Standard Tool Company, Athol, Mass. It requires no explanation. The central disk is graduated, and the blade may be set at any desired angle and clamped by a

COMBINATION SQUARE, LEVEL AND PROTRACTOR.

screw at the back. The blade also clamps into the left hand end, making a perfect square. The V in the right hand end is a right angle, and may be applied to the corner of a planer platen or any similar place. The little level shown below also clamps to the disk at the center, and may be adjusted to any angle.

▲ ▲ ▲

1897

Adjustable Cast Iron Level.

Standard Tool Company, Athol, Mass., are making an iron level, as here illustrated, which is offered as possessing many of the advantages of higher cost levels. In many respects it is similar to the ordinary cast iron level, the level glass, however, being inserted in a cast iron shell before being placed in the level, the shell resting on a central point. Two screws pass through the frame of the level and fit on to a projection at each end

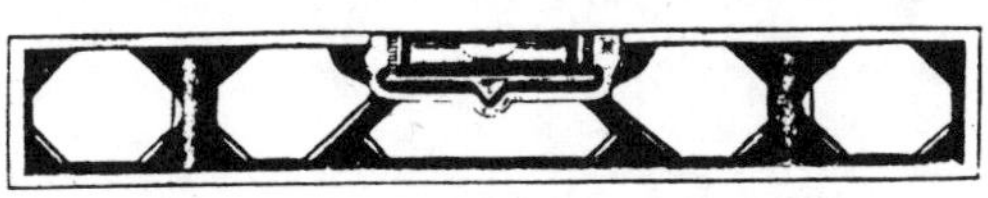

Adjustable Cast Iron Level.

of the iron bottle receptacle, holding it securely in place, and at the same time making an adjustable level of it, as a right turn of the screws will readily adjust the glass should it by any accident get out of place. If the glass is broken a new one can be inserted and adjusted by any one. The levels can now be supplied in sizes from 6 to 24 inches inclusive.

1899

553
Never-Slip Surface Gage.

PRICE.

6 inch,	- -	$2.25
9 "	- - -	2.50
12 "	- - -	3.50

FIG. 2.

The Surface Gage here shown is very simple in construction, compact and rigid. The base has two lugs on it, one of which is drilled to receive a bolt and a large knurled headed screw is tapped into the other.

In operation the screw is loosened sufficiently to allow the clutch shown in Fig. to be opened and the adjustment made when it is again closed, holding the spindle rigidly and positively in position

Fine adjustments are made by use of the small nnt shown on outer end of the adjusting arm and the nut on the opposite end connects with an adjustable attachment for the quick and accurate adjustment of work on a line with the bed of the planer or parallel to its edge.

1900

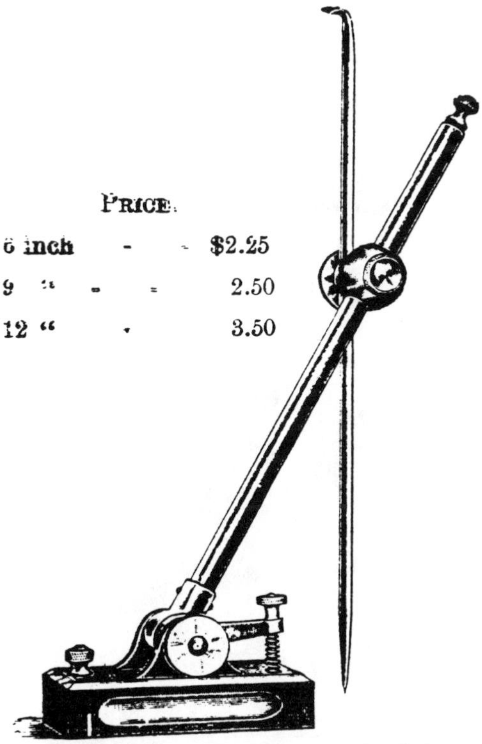

PRICE.

6 inch	$2.25
9 "	2.50
12 "	3.50

The Surface Gauge here shown is very simple in construction, compact and rigid. The base has two lugs on it, one of which is drilled to receive a bolt and a large knurled headed screw is tapped into the other. The spindle holder has a taper hole through it into which fits the adjusting arm and a turn of the screw forces it into the spindle holder, holding the spindle rigidly between the two ears, and it is not necessary to loosen it in use as all movable parts swing on friction joints which cannot become loose by turning either the spindle or the pointer.

Fine adjustments are made by use of the small nut shown or outer end of the adjusting arm and the nut on the opposite end connects with an adjustable attachment for the quick and accurate adjustment of work on a line with the bed of the planer or parallel to its edge.

Universal Steel Bevel Protractor.

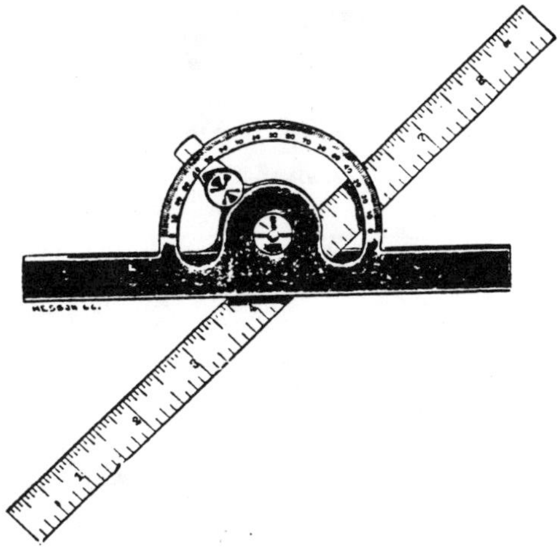

9 inch,	$5 00
12 "	6 00

Cases, extra, - - $1 25

1900

DON'T BUY A MICROMETER CALIPER

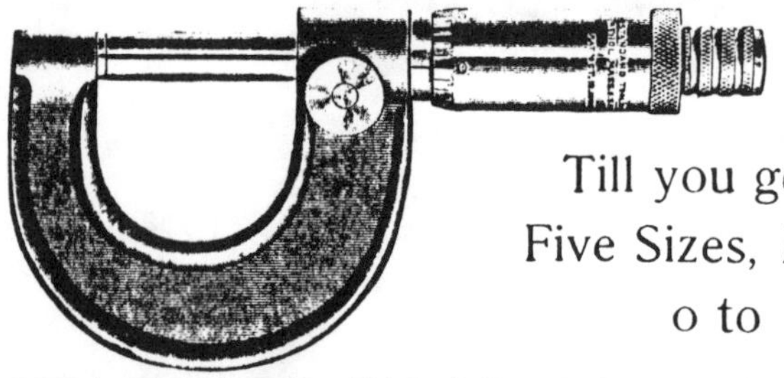

Till you get our catalog.
Five Sizes, Measuring From
0 to 12 inches.

STANDARD TOOL CO., Athol, Mass., U. S. A.

MANUFACTURERS OF MECHANICS' FINE TOOLS.

Agents for Great Britain, C. W. BURTON, GRIFFITHS & Co., Ludgate Square, Ludgate Hill, London.

STANDARD HAND VISE.

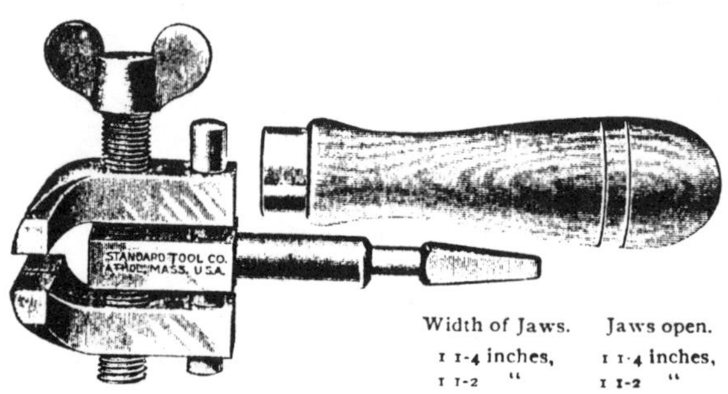

Drop Forged and tempered steel jaws, Black and Bright finished.

CATALOG
87 PAGES
FREE.

Width of Jaws.	Jaws open.	Weight.	Price, Blk.	Bright.
1 1-4 inches,	1 1-4 inches,	14 ozs.	$1.25	$1.50
1 1-2 "	1 1-2 "	18 "	1.50	1.75

STANDARD TOOL CO., Athol, Mass., U. S. A.

Manufacturers of Mechanics' Fine Tools.

Agents for Great Britain, C. W. BURTON, GRIFFITHS & Co., Ludgate Square, Ludgate Hill, London.

1900

MICROMETER CALIPER.

No. 2	0 to 3 inches.
No. 3	3 to 6 "
No. 4	6 to 9 "
No. 5	9 to 12 "

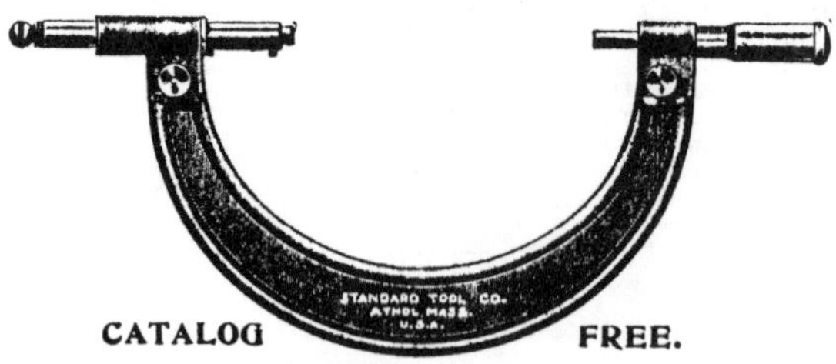

CATALOG FREE.

We show herewith a micrometer Caliper which while moderate in price and adapted to a range of work, for which expensive tools have heretofore been required, possesses all the accuracy of any now in use.

Standard Tool Company, Athol, Mass., U. S. A.

Manufacturers of Mechanics' Fine Tools.

Agents for Great Britain, C. W. BURTON, GRIFFITHS & Co., Ludgate Square, Ludgate Hill, London.

1901

L.S. STARRETT CO.

STARRETT, L.S. Athol, MA

Laroy S. Starrett (1836-1922) was born in China, Maine, April 25, 1836. He moved to Newburyport, MA, in 1853, and by 1863 was renting a 600 acre farm. On May 23, 1865, Starrett was granted a patent for a meat chopper which he began to produce and sell.

A report on the fair of the New England Agricultural Society, held at New Haven, CT, in August, 1868, notes the following:

> L.S. Starret & Co., of Newburyport, Mass., exhibited "Starret's Patent Meat and Vegetable Chopper", an apparatus in which a vertically moving chopping knife is worked within a suitable bowl or vessel by means of a crank and appropriate intermediate mechanism.

Later in 1868, Starrett moved to Athol, MA, where the ATHOL MACHINE CO. was to take over the production and sales of his meat chopper. The ATHOL MACHINE CO. was formed June 3, 1868, for the "special purpose of making Starrett's food chopper and other articles of Mr. Starrett's invention". The company was capitalized at $25,000 and Starrett was one of the directors but not an officer.

Starrett also obtained a patent for a washing machine on June 27, 1865 (No. 48458). This appears to be simply a variation of the food chopper. Some examples are known, but it is not clear if they were made by Starrett, ATHOL MACHINE CO., or some other company. Most likely the washing machines were made by the ATHOL MACHINE CO. A food press, which Starrett patented April 15, 1873, was made by the ATHOL MACHINE CO.

Starrett's first tool patents were granted May 8, 1877 (No. 190636), and February 25, 1879 (No. 212760), covering a quick adjusting wrench. No examples of this wrench are known. However, both the ATHOL MACHINE CO. and its subsidiary, the STANDARD TOOL CO., made wrenches with a very similar design patented by S.M. Bellows on June 8, 1880 (No. 228437).

In 1876, Starrett invented the combination square, the form of which has remained unchanged to this day. Probably due to his unhappiness with the situation at the ATHOL MACHINE CO., where "business conditions were not to his liking" he began making the squares on his own.

The first combination squares were made for him at the Richardson Machine Shop in Athol, MA, beginning in 1877. Starrett, who was forced out of ATHOL MACHINE CO in 1878, received his patent on May 6, 1879 (No. 215024) (reissued October 19, 1880, No. 9419).

In 1880, his former associates at the ATHOL MACHINE CO. threatened legal action against Starrett and anyone else who was involved in making the combination square. Starrett wrote that Richardson "stopped work in spite of all I could say or do to assure them the Athol Machine Co. was wrong and infringing my patent and that I was able to protect them and myself. But they lacked the courage to go ahead." Starrett then offered to buy out the part of the business used to make the squares. Richardson accepted and L.S. STARRETT was founded. See Appendix II for Starrett's own story of the combination square.

The Athol Transcript for January 25, 1881 noted:

> "The wooden building on River Street lately used as a grist mill has been leased by Mr. L.S. Starrett, who will occupy it in the Spring for the manufacture of his patent Try Square, having purchased the stock and machinery used therein of C.F. Richardson who has been making the goods for him."

The Chronicle for March 31, 1881 noted:

> "Owing to the increased demand for Starrett's improved try and mitre squares, the proprietor has been obliged to remove from Richardson's Machine Shop to larger and more convenient quarters in Fisher and Co.'s Block on River Street, which has recently been raised. He will turn out from 1800 to 2000 squares per month."

On June, 29, 1880, Starrett patented a center head for his combination square (No. 229283), and on December 4, 1883, patented a protractor head (No. 289463), thus completing the combination square set as we know it today.

Articles in the October and December, 1881 issues of *Machinery* magazine praised his combination square and a new surface gauge for which he received a patent on March 21, 1882 (No. 255123).

Starrett was now in litigation with the ATHOL MACHINE CO., which had bought the rights to the Chaplin try and center square, patented May 8, 1866 (No. 54503), by Orril R. Chaplin. It persuaded the patent office to reissue the patent on May 4, 1880 (No. 9183), and then sued Starrett for patent infringement. Starrett was not a man to be intimidated and printed the following notice in his 1882 price list.

CAUTION

> Since the success of my Patent Squares has become known, unprincipled parties have had the foolhardiness to make and put on the market a tool, under the Pretext of Chaplin's Patent, which not only infringes on mine, but is not accurate. Having commenced suit against said parties, I shall soon put a stop to their manufacture, and shall hold all persons, selling or using them responsible.
>
> Respectfully
> L.S. STARRETT

Starrett won the patent suit filed against him by the ATHOL MACHINE CO., but was not able to stop it from making the Chaplin square through its subsidiary, the STANDARD TOOL CO. See Appendix III for details of the litigation.

In 1882, Starrett began offering a line of flexible steel rules made on the graduating machine he had designed and built for graduating the blades for the combination square. Rules quickly became a major part of the business and, by 1909, over 400 styles and sizes were offered.

Starrett introduced, in 1885, a universal bevel and a screw thread gauge, for which he was granted a patent on May 19, 1885 (No. 318051).

In 1886, he began offering a line of lock joint calipers with fine adjustment, also made as lock joint transfer calipers (both lines made under the Frank G. Lilja patent of January 6, 1885, No. 310292), a 12" vernier caliper "with the top edge unobstructed" and an inclinometer. The inclinometer was covered by the L.S. Starrett patent of August 7, 1883 (No. 282583).

Starrett also expanded in September, 1886, by leasing the 40' x 80' wooden mill on River Street opposite his existing shop. The lease was to run for five years but Starrett bought the building in 1888 and added a third story. The building was thereafter called the "Wooden Shop" of the Starrett Company. In May, 1890, Starrett bought more land on River Street south of the river. In 1893, he used this land to erect the southern wooden shop and the shop spanning the river, both new shops were 40'x 80' and of three stories.

His January 1, 1887, catalog offered: Starrett's combination squares, improved bevel protractors, Starrett's patent double squares, universal bevels,, Starrett's patent inclinometer #10, caliper squares, Starrett's improved caliper squares, improved screw pitch gauges, graduated scratch gauges, steel rules, flexible rules, narrow rules, centre gauges, shrink rules, steel straight edges, lock-joint transfer calipers (Lilja's patent), Starrett's patent surface gauges, iron levels (made by C.F. Richardson, Athol, MA), Cook's improved trammel points (their position in the catalog would indicate that they were not then made by Starrett, but this is uncertain. Starrett eventually did make the Cook trammel points), Woodman's improved speed indicator (not made by Starrett).

Later in 1887, Starrett bought the stock, machinery and patents of CHARLES P. FAY & CO. of Springfield, MA. Fay had worked for Starrett in 1883 but had left to start his own business about 1884. With the purchase of the Fay business, Starrett acquired the "FAY" and "YANKEE" lines of calipers and dividers which are still made under those names. Charles P. Fay joined Starrett as part of the buyout, staying until 1896 when, at his father's death, he became part owner and vice president of J. STEVENS ARMS & TOOL CO., Chicopee Falls, MA. Starrett introduced, in 1888, a micrometer caliper square which he patented March 5, 1889 (No. 399167), and combination dividers for which he received a patent on September 24, 1889 (No. 411537).

He introduced, in 1889, a new micrometer with Starrett's patented (July 29, 1890, No. 433311) "speeder" attachment (a smaller diameter knob at the rear of the thimble) and later in the year an improved micrometer which incorporated a split bushing within the thimble to take up wear on the screw threads and a knurled locking ring for clamping the spindle.

He introduced, in 1893, a tool maker's surface gauge patented March 17, 1896 (No. 556703) by Charles P. Fay and Laroy S. Starrett. This patent covered the #55, #56 and #57 surface gauges. The latter two are still in production.

In 1893, Starrett bought half interest in the F.J. Gay Co., a maker of milling and forming cutters in Providence, RI. The Gay company was moved to Athol and consolidated with L.S. Starrett.

In a letter written to *American Machinist* magazine in December, 1893, Starrett noted the following:

> "Up to the first of June we were behind orders, having more than we could do, but since that time orders have dwindled down, until I have let many of my help go; nevertheless, as I had contracted to have built a three-story building, 160x40 feet, and an annex brick building, 70x44 feet for engine house, etc., I have carried them through to completion, taking advantage of the dull times to prepare for good times when they come.
>
> These buildings are heavily timbered and finished in A1 shape with all modern improvements, and are now about ready for the reception of machinery. A 110 horsepower wheel, and a 100 horsepower engine have been put in to furnish the additonal power.
>
> I have during the past year interested myself in the manufacture of milling cutters, forming tools, slitting saws, etc., with Mr. F.J. Gay, formerly superintendent of this branch of business for the Brown & Sharpe Mfg. Co., and now with advanced ideas of his own, we shall endeavor to lead in this line of business.
>
> I have also during the year added to my line of fine tools an improved speed indicator, adjustable jaw cut nipper, a line of firm joint calipers, new surface gauges, flat steel squares, etc., all of which will be shown in my 1894 catalogue of 86 pages."

Starrett's excursion into the cutter business was short lived, however. In 1899 he sold it to Francis J. Gay and Edgar J. Ward who then formed Gay & Ward to continue making cutting tools. Gay & Ward was incorporated as the Union Twist Drill Co. in 1905, with Laroy S. Starrett as a major stockholder.

Catalog #13, issued July, 1895, contains the #52 surface gauge, patented March 21, 1882(No. 255123), the #53 micrometer surface gauge, the #55 surface gauge (patented March 17, 1896, No. 556703 by L.S. Starrett and Charles P. Fay), and the #56 toolmaker's surface gauge which is also found marked with the March 17, 1896 patent. The #57 surface gauge is not yet listed. The catalog has several pages of iron levels, all the product of C.F. RICHARDSON, Athol, MA. Other new tools are: "Reliable" try squares, patent July 23, 1895 (No. 24503, design) by L.S. Starrett; lock joint calipers, patented May 21, 1895 (No. 539759) by L.S. Starrett and Charles P.

Fay (these replaced the Lilja patent calipers); and adjustable jaw cut-nippers, patented September 22, 1896 (No. 568242) by Charles P. Fay. The two styles of trammel points, offered as Cook's Trammel Points in catalogs from 1887 to 1894, were now treated as Starrett products.

The company name on the 1895 catalog is still shown as L.S. STARRETT. However, many of the newer tool cuts clearly show a marking of THE L.S. STARRETT CO. or L.S. STARRETT CO. Starrett advertisements, beginning about 1895, also used the L.S. STARRETT CO. label. This would indicate that the assumption that tools marked L.S. STARRETT were made before 1900 and those marked L.S. STARRETT CO. were made beginning in 1900 is in error by about five years.

Catalog #15 was issued in 1897 under the name of the L.S. STARRETT CO. New tools in this catalog were: nail sets, center punches, scribers, an offering of 30 different styles of micrometers instead of six as before, a six-inch beam micrometer patented by M.E. Smith on May 28, 1889 (No. 404057), Starrett's patent (February 9, 1897, No. 576626) hack saw frames, Starrett's pocket companion tool patented by Starrett on October 19, 1897 (No. 592176), and a complete line of new levels consisting of: bench levels, #130, #132, #95 and #96; pocket levels, #135; and engineer's and plumber's levels #133. The new levels replaced the C.F. RICHARDSON products offered in the earlier catalogs, leaving transits as the only Richardson product still offered by Starrett.

Starrett, who had operated as a sole proprietor, incorporated January 1, 1900 as:

STARRETT CO., L.S. Athol, MA

On April 5, 1900, Starrett filed for the now familiar trademark of "The representation of a micrometer-gauge, square, and calipers or dividers interlocked." He claimed use since April, 1897. The trademark was granted on May 1, 1900,

Catalog number 16 was issued in January, 1900, and offered the following new tools: new style combination square No. 17; carpenter's square; center square; pocket slide calipers; micrometer gauge No. 24, 12 to 36 in.; micrometers with finger ring (L.S. Starrett design patent of December 18, 1900, No. 33734); micrometers with ratchet stop (L.S. Starrett patent of April 17, 1900, No. 647910); attachment for converting a 2 in. into a 1 in. micrometer; micrometer No. 127, 0 to 12 in.; inside micrometer, 32 to 207 in.; thickness gauge, No. 172; inspector's gauge (L.S. Starrett patent of November 13, 1900, No. 661562); Little Giant jack screws; complete line of hack saws; universal surface gauge No. 57; screw adjusting calipers (L.S. Starrett patent of April 16, 1901, No. 672424); double calipers; carpenter's dividers, No. 92; stair gauge fixtures, No. 111; trammels, No. 59; and "Time Saver" drill and tap gauges.

The "Time Saver" drill and tap gauges had been introduced in 1898 by GEORGE SCHRADE, New York, NY, who had been granted a copyright for the gauge the same year. Starrett bought the product from Schrade soon after.

Catalog number 17, issued in 1903, included the following new tools: Draughtsmen's scales, patented May 13, 1902 (No. 700156), by R.J. Simpson; #118 spacing center punch, patented November 4, 1902 (No. 713021), by L.S. Starrett; #550 and #551 screwdrivers, patented May 26, 1903 (No. 729004) and June 9, 1903 (No. 730804), by L.S. Starrett; #119 nail set and holder, patented June 7, 1904 (No. 762258), by L.S. Starrett.

In 1903, Starrett eliminated a large competitor when he bought the tool business of the J. STEVENS ARMS & TOOL CO., of Chicopee Falls, MA. The transaction was made public by the following announcement in the *Iron Age* magazine of November 19, 1903.

> The L.S. STARRETT Co., Athol, Mass., have bought the entire fine Machinist's Tool business of the J. STEVENS ARMS & TOOL CO., Chicopee Falls, Mass., including manufactured stock, stock in process, special tools for manufacturing and good will. The firearms part of the Stevens business having grown so rapidly that the company found it necessary either to erect more buildings, curtail

> their product, or dispose of some department, and the L.S. STARRETT CO. having made a satisfactory offer for the Tools, the sale of this portion of the company's business was closed on the 13th inst., the transfer to take place on the 20th.

A 1904 supplement to the Starrett number 17 catalog, lists the #174 tap wrench, #195 double-lip countersink and #199 cut-nipper which were previously made by Stevens. The offering of only three tools out of the scores made by Stevens would indicate that the purchase of the Stevens tool line was to get the tools off the market rather than to make additions to the Starrett line.

The 1904 supplement also introduced the #16 protractor (L.S. Starrett patent of December 27, 1904, No. 778808), #19 protractor, #298 key seat clamps, #367 desk rule, #257 new universal surface gauge, #70 pocket scriber, #553 pocket screwdrivers, #552 jewelers' screwdrivers (L.S. Starrett patent of March 20, 1906, No. 815758), #555 jewelers' screwdrivers, #162 pin vises, #281, #282, #283 wire gauges, #447 height gauge attachment, and the line of #226 micrometers.

In 1905, Starrett quietly bought controlling interest in his old adversary, the ATHOL MACHINE CO. He announced that the company would not be merged with the L.S. STARRETT CO., but would be operated separately and would continue to make vises, meat choppers and mechanic's tools. The tools produced by the ATHOL MACHINE CO. from that time on tended to be lower priced versions of the Starrett tools. Also in March, 1905, the following announcement appeared in the trade press:

> The L.S. Starrett Company, Athol, Mass., announces that it has purchased the Leveling Instrument and Transit business of the Richardson-Oliver Company, Athol, Mass., formerly C.F. Richardson & Son, who for many years made a line of leveling instruments adapted for the use of architects, carpenters, stone masons and others. The Starrett Company will hereafter produce and market these Tools in connection with its present extensive line.

Starrett continued to expand the works, adding a brick office building in 1901 and a large extension to the factory in 1906. A new boiler house was built in 1908, and by 1909 the total floor space was about four and one-half acres. Four 500 HP boilers supplied steam engines with a capacity of 1800 HP, this in addition to the water power generated by the company's dam on the Millers River. Starrett employed 600 men in 1909 and claimed to have the largest plant in the world devoted exclusively to the manufacture of small tools for mechanics. The company offered, in 1909: "over 80 different micrometers, over 200 styles of calipers and dividers, and over 400 styles and sizes of steel rules".

The claim to be the largest would certainly have been true, at least as far as the United States was concerned. Its only real competitor, the Brown & Sharpe Mfg. Co., was much larger but it was primarily a machine tool builder and small tools were only a fraction of its business.

New tools in the 1909 catalog were: improved bevel protractor, planer and shaper gauge, telescope inside gauges (L.S. Starrett and J.A. Adell patent of November 24, 1908, No. 905110), toolmaker's buttons, quick adjusting micrometers, toolmaker's calipers and dividers, automatic adjustable-stroke center punch (J.A. Adell and L.S. Starrett patent of February 12, 1907, No. 843655), level sight attachments, center gauge attachment (J.L. Baltzer and L.S. Starrett patent of February 27, 1906, No. 813997), and improved calipers.

The quick adjusting micrometer was a unique design patented January 2, 1906 (No. 809272) by L.S. Starrett and J.A. Adell. The design allowed the nut to be disengaged from the screw threads by depressing a plunger. The tool could then be opened or closed rapidly. Even though offered as late as 1938, these micrometers are very rare today and were probably not a commercial success.

The company announced that it had "Discontinued use of Union label" in 1914. No explanation was given.

About 1913, the L.S. Starrett Co., bought the ADELL BROTHERS MFG. CO., Orange, MA. John A. Adell, one of the owners, had shared four joint patents with L.S. Starrett from 1906 to 1908, so he had probably worked at the L.S. STARRETT CO. at that time. The tools produced by ADELL BROTHERS MFG. CO., thread gauges patented August 29, 1911 (No. 1002019), and machinist's clamps formed of sheet metal, were made by the L.S. STARRETT CO. after the buyout. The clamps were listed in the Starrett Number 20 catalog as their #445 "Paramount Indestructible Pressed Steel Clamps". Adell apparently rejoined Starrett as he shared another joint patent with him in 1921.

Catalog number 20, issued in 1914, offered the following new tools which were aimed at crafts other than the machinist: #240 expansion pliers, patented January 3, 1911 (No. 980361), by L.S. Starrett; #443 socket wrench set; #429 carpenters' scratch gauge; #439 builders' combination square, patented January 3, 1911 (No. 980362), by L.S. Starrett.

Catalog number 21 was issued in 1916, and included the following new tools: quick-reading steel rules, metric shrink rules, blacksmith's steel rules, metric and English folding pocket rules, drop forged combination squares with metric and English graduations, improved bevel protractors, #453 diemaker's square, #123 pocket vernier calipers, #650 low cost "Yankee" micrometer calipers, vernier height gauges, adjustable hacksaw frames, #437 cutting nippers, #276 cutting pliers (L.S. Starrett patent of January 2, 1917, No. 1210558), and drive- pin punches.

Catalog number 22, issued in 1920, included the following new tools: #359 universal bevel protractor (L.S. Starrett and J.A. Adell patent of July 12, 1921, No. 1384105), #24-A micrometer caliper gauge for tire molds, #494-D toolmaker's buttons (A.H. Starrett patent of January 11, 1921, No. 1365203), #272-M metric radius gauges, #570 screwdriver, and an accurate set of V-blocks.

In 1920, Starrett bought out the REISNER MFG. CO. of Hagerstown, MD, and began making the dial indicator product line which William H. Reisner had been making since 1905. Reisner may have been hired at that time since his indicator patent of June 13, 1922 (No. 1419306), was assigned to the L.S. STARRETT CO.

Laroy S. Starrett, who had kept firm control of the company throughout his life, died in 1922 at the age of 86. He was succeeded by his son-in-law, Frank A. Ball (1859-1943).

Catalog number 23, issued in 1924, included the following new tools: #1000 and #1020 stainless steel rules; #418 hook rules, patented August 24, 1926 (No. 1597058), by J.W. Crawford; #423 toolmaker's rules with holder, patented December 12, 1922 (No. 1438617), by A.H. Starrett; #493 protractor and depth gauge; #55 beveled edge squares; #452 cylinder gauge, patented April 19, 1927 (No. 1625401), by R.J.O. Simpson; #458 dial bench gauge; #170 dial sheet gauge; #48 spring depth gauge; #436 line of micrometers; #175 inspector's micrometer caliper gauge; #466 angle gauge, patented June 5, 1928 (No. 1672695), by R.J.O. Simpson; #86 combination hand vise and #200 hand vise, patented November 13, 1923 (No. 1473946), by A.W. Amsden and W.D. Starrett; #62 rule holder; #129 bench block, patented January 16, 1923 (No. 1442415), by A.H. Starrett; #22 drill point gauge, patented December 19, 1922 (No. 1439282), by A.W. Amsden.

The #175 inspectors' micrometer caliper gauge had previously been made by the RIEHLE BROS. TESTING MACHINE CO., Philadelphia, PA, since 1892. It was made under patents issued to John D. Sloane on July 7, 1891 (No. 455644) and March 27, 1894 (No. 517073), but these were long since expired when Starrett took over production.

The number 24 catalog, issued in 1927, offered fewer new tools, including: #456 gear tooth vernier calipers; #438 dovetail vernier calipers; #236 depth and angle gauge, patented November 9, 1926 (No. 1606192), by J.G. Sisson; #806 feeler gauge holder, patented January 28, 1930 (No. 1744856), by A.W. Amsden.

Catalog number 25, issued as a fiftieth anniversary catalog in 1930, included very few new tools, probably due to the poor business conditions which had been felt in the heavy manufacturing industries since the mid-1920's. Those offered included: #362 Draftsman's protractor (a copy of the Brown & Sharpe tool made since the 1890's); #251 steel beam trammels; #459 cutter clearance gauge; #199 master precision level.

New tools in the number 26 catalog, dated 1938, included: #711 "Last Word" indicators, #564 "Universal Junior" indictors, patented March 27, 1934 (No. 1953028); and #829 small hole gauges, patented January 17, 1928 (No. 1656180), by Gustav A. Eisele. The "Last Word" indictors had been made by H.A. LOWE, Cleveland, OH, from 1914 until selling out to Starrett.

For a detailed look at all the Starrett tools offered in 1895, a reprint of the L.S. STARRETT, 1895 catalog is available from: The Astragal Press, P.O. Box 239, Mendham, NJ, 07945. Later catalogs, especially the number 26 issued in 1938 and used until at least 1946, are easily located.

L. S. STARRETT'S FINE MECHANICAL TOOL MANUFACTORY.

1891

LE ROY S. STARRETT.

Meat Cutter.

No. 47,875. Patented May 23, 1865.

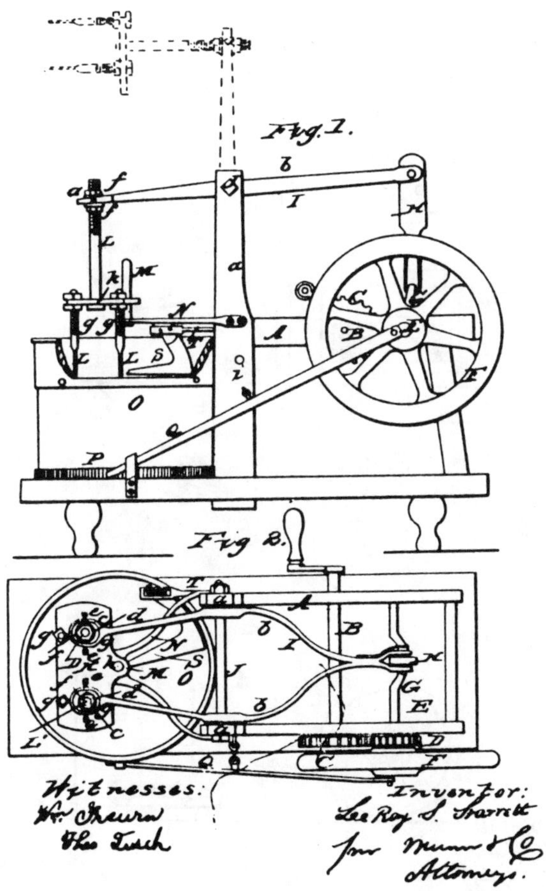

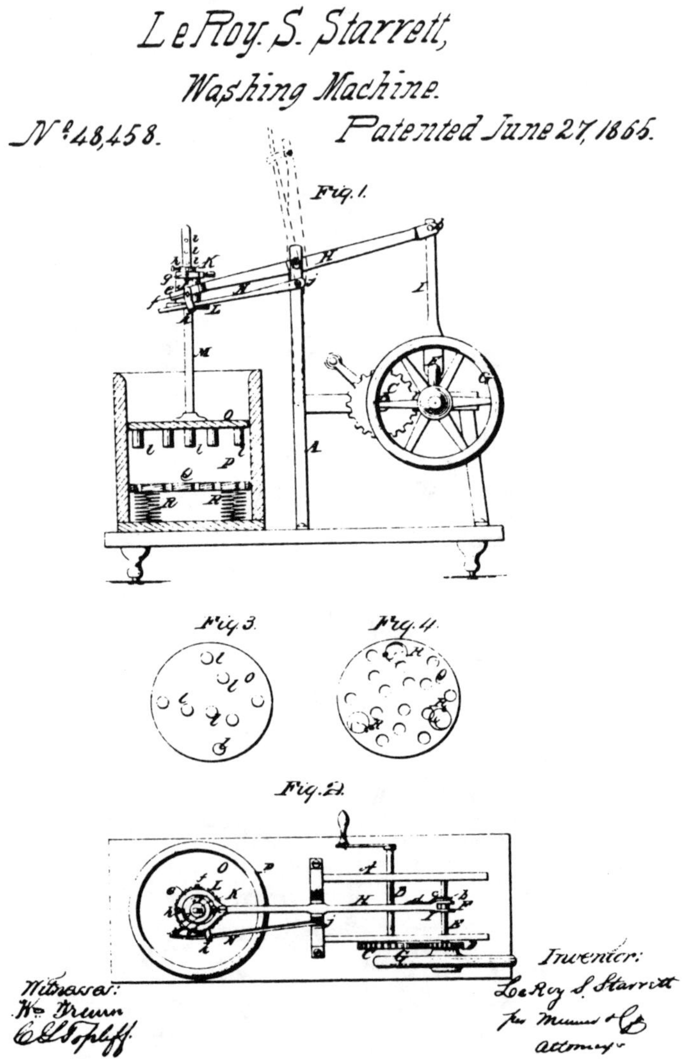
LeRoy. S. Starrett,
Washing Machine.
No 48,458.
Patented June 27, 1865.
Fig. 1.
Fig. 3.
Fig. 4.
Fig. 2.
Witnesses:
Inventor:
LeRoy S. Starrett
per Munn & Co
attorneys

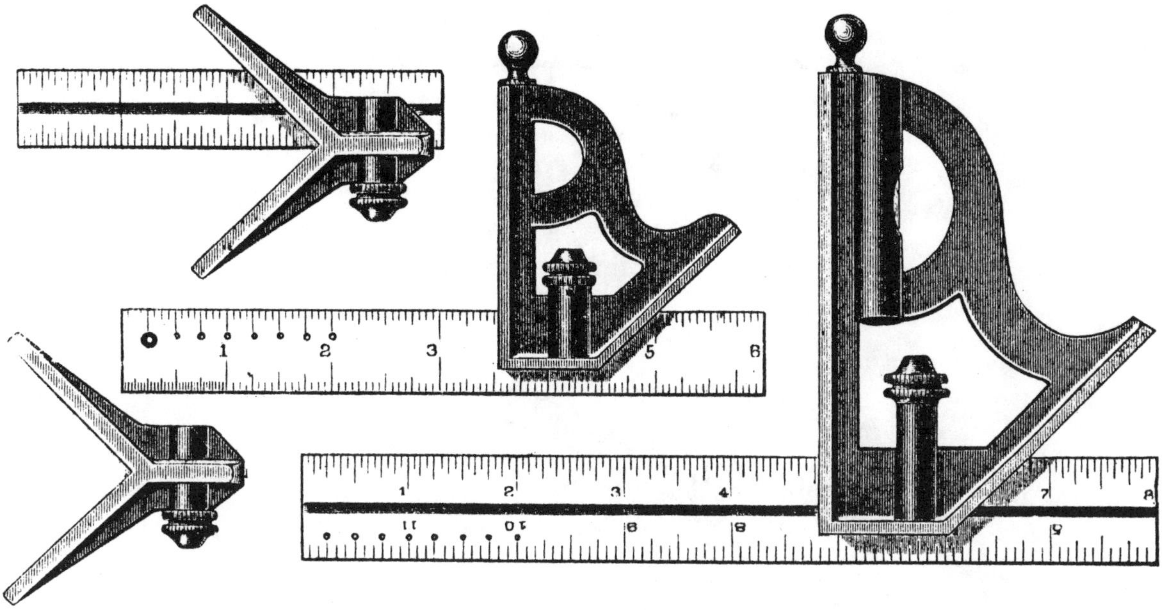

STARRETT'S COMBINATION TOOL.—See page 7.

1881

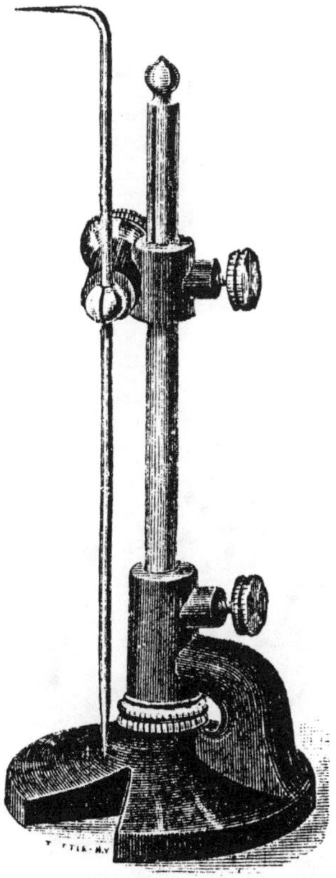

NEW SURFACE GAUGE.

1881

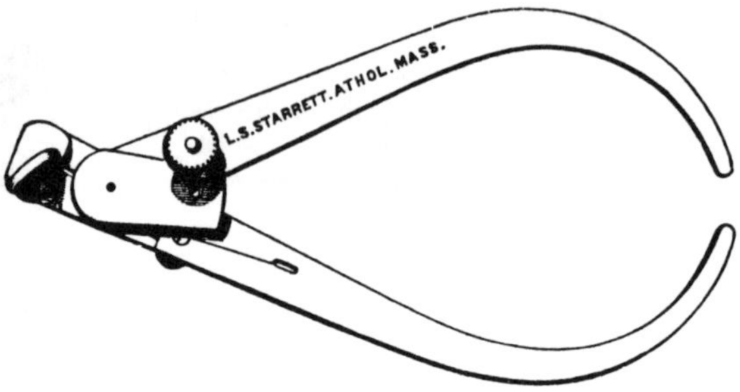

LOCK JOINT CALIPER.

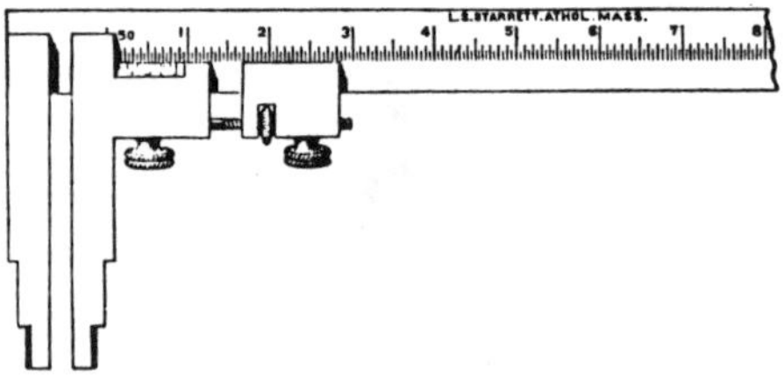

VERNIER CALIPER.

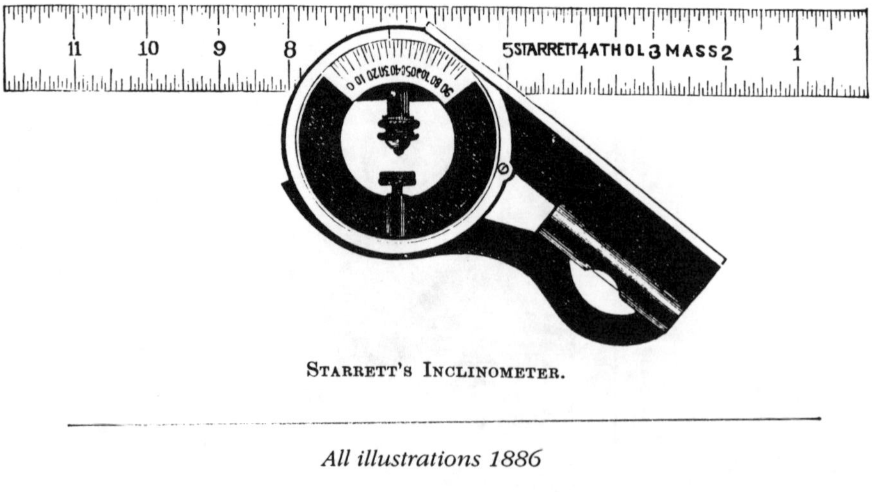

STARRETT'S INCLINOMETER.

All illustrations 1886

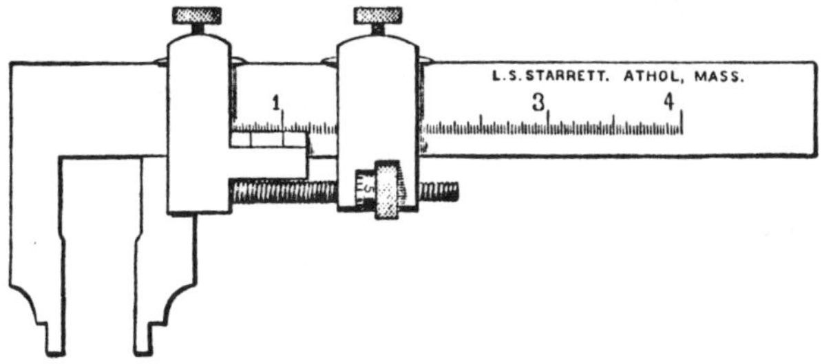

MICROMETER CALIPER SQUARE.

1888

COOK'S NEW EXTENSION BEAM.

TRAMMEL POINTS.

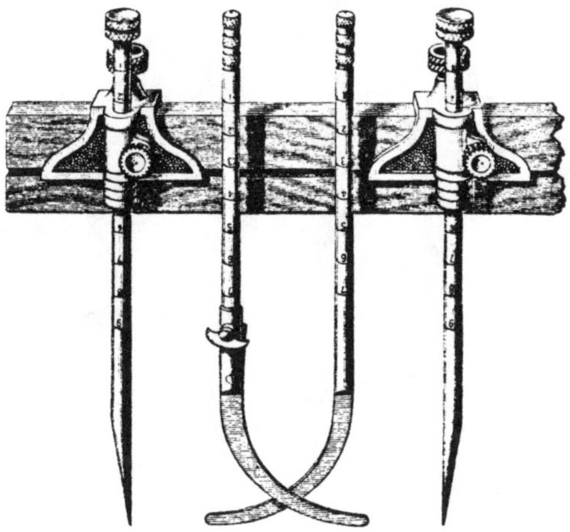

1888

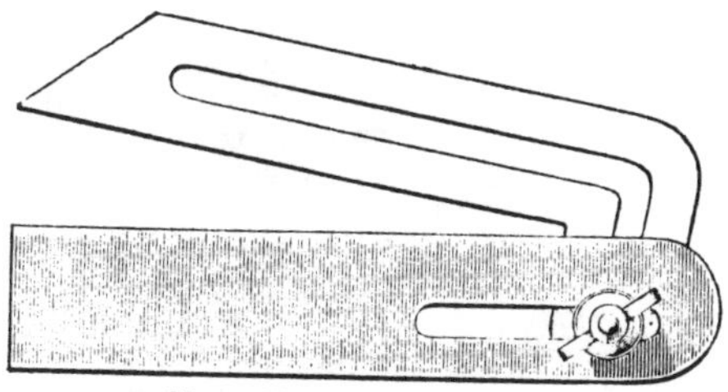

NEW UNIVERSAL BEVEL

1885

COOK'S PATENT TRAMMEL POINTS.

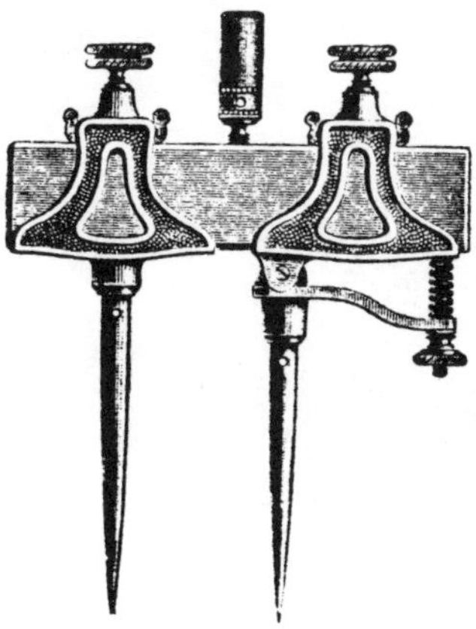

1888

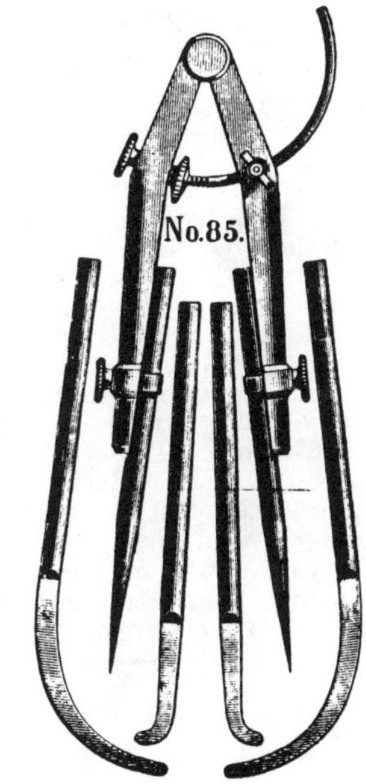

COMBINATION DIVIDERS.

1888

Combination Dividers.

L. S. Starrett, of Athol, Mass., is introducing a combination dividers for woodworkers. In the illustration given herewith, the cut shows the article complete, with the different attachments that are provided to go with it. The article may be used in the form of dividers, with either short or long extension points, and either with or without a pencil. It is also capable of being used as an outside calipers or inside calipers. and also in the form of hermaphrodite calipers. Articles of this kind have been before the public for a long time past, but this instrument, we are informed, possesses features peculiarly its own. The head and socket legs of the tool, we learn, are made from drawn bronze metal, and accordingly are hard, tough and strong. They are, in addition, finely finished. The joint is large and firm, and the quadrant is round and is fastened by an impro ved method. The steel points are thoroughly tempered. On the threaded end of the quadrant, between the adjustable legs, is a knurled nut, against the inside of which a helical spring acts. After the points are adjusted the nut may be turned back against the leg, thus locking it firmly in place. A common pencil fits either socketed leg, while an auxiliary holder is also furnished to fit the reverse end of either point. The maker claims that this is one of the most useful combinations of its kind, as well as the best made tool ever offered to the trade. In the ordinary form it is supplied with head and points 8 inches. These points, we learn, may be extended 2 inches more and in this form will describe a 24-inch circle. The longer points bring the article up to a capacity of a 34-inch circle, while extra long points, which are supplied to order, describe a 44-inch circle.

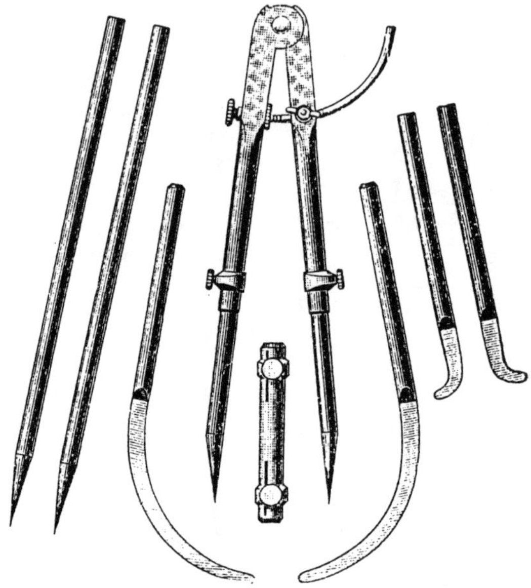

Starrett's Combination Dividers.

1888

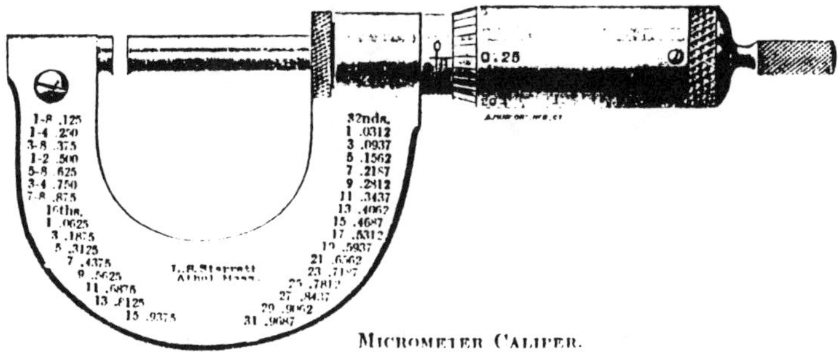

MICROMETER CALIPER.

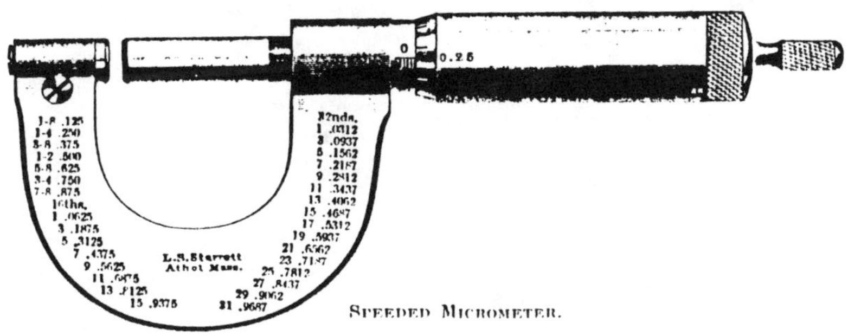

SPEEDED MICROMETER.

1889

STARRETT'S PATENT SURFACE GAUGE.

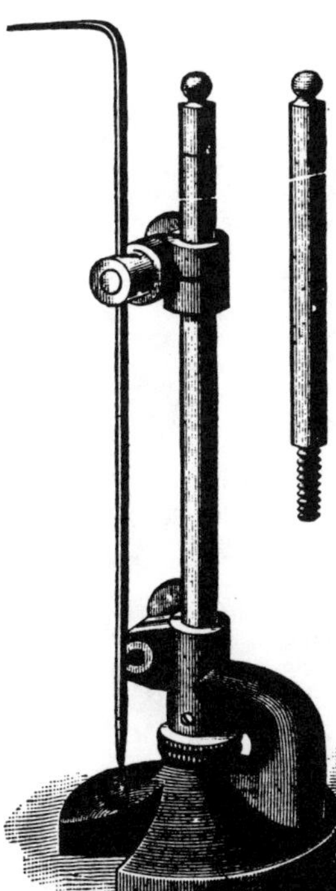

No. 52.

ADVANTAGES OVER OTHERS.

1st. Easily obtaining friction to prevent the scriber from moving, by means of a tapering eye-bolt drawn into a socket.

2nd. In having stock taken from base, allowing scriber to extend through for depth gauge, while, ***with bent end down***, work may be laid out close to base.

3rd. In the final and accurate adjustment of the scriber, the spindle passing through a nut in the base, to raise or lower the same.

—o—

This Gauge has been recently ***improved*** by ***splitting the sleeve.*** The nut on the eye-bolt draws the split together, clamping it to the spindle, making it ***impossible*** to work loose.

There is also a ***helical spring*** on end of spindle in the base, which serves to take up all ***back lash*** in the adjusting nut.

On receipt of 50 cents in stamps, we will send the improved sleeve to any wishing to substitute it for the old style.

PRICES.

No. 1,	8-inch	$2 00
No. 2,	12-inch	2 75
No. 3,	18-inch	3 00

STARRETT'S
MICROMETER SURFACE GAUGE.

This Gauge has a turned and polished base, a Micrometer adjusting nut (reading 100 and finer), and a 6 inch extension for the spindle. By means of springs and taper fitting parts of the sleeve (not shown in cut), the scriber is held by slight friction in any position while adjustments are made, and FIRMLY held by a turn of the nut. A knurled cam on the base releases and locks the spindle for adjusting.

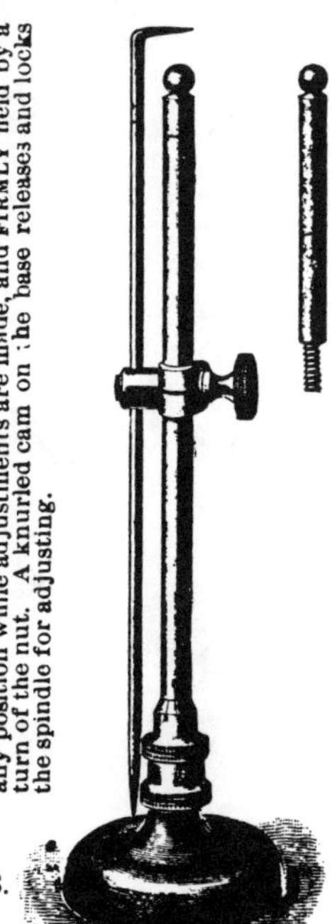

No. 53.

1891

PRICE:

8 inch without extension	$2 50
12 " " "	3 50
12 " with extension	4 00

Starrett's Patent Combination Square

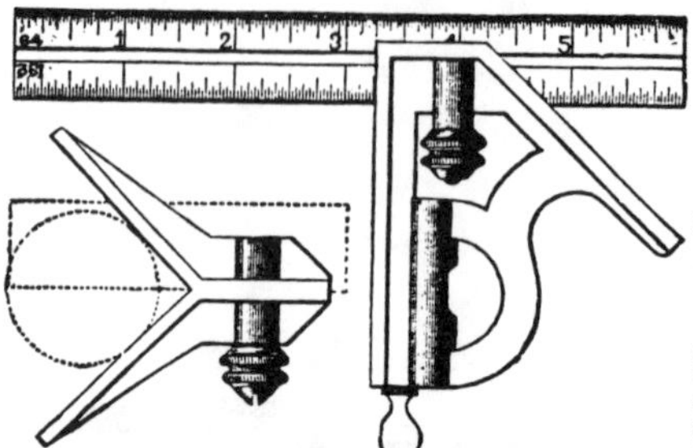

PRICE PER SET.

12 inch, Complete	$3 50
9 " "	3 00
6 " "	2 50
4 " without Center Head or Level	1 25

PRICE OF SEPARATE PARTS.

	Scale.	Stock	Center Head.
12 inch	$1 60	$1 50	$1 00
9 "	1 25	1 25	1 00
6 "	80	1 10	75
4 "	60	75	

Every Tool warranted Accurate. With the adjustable Scale, this forms one of the most convenient and useful tools ever devised for mechanics' use. One is a complete substitute for a whole set of common try squares, and is one of the best gauges ever made for transferring exact measurement or laying out work. It is also convenient for a depth gauge or to square in a mortice. For a mitre, it is perfect, while with the auxiliary centre head it forms a centering square, both inside and outside, which for convenience and accuracy has no equal.

Parties ordering extra parts, should send in by mail the piece they want stock or blade fitted to.

STARRETT'S PATENT DOUBLE SQUARE.

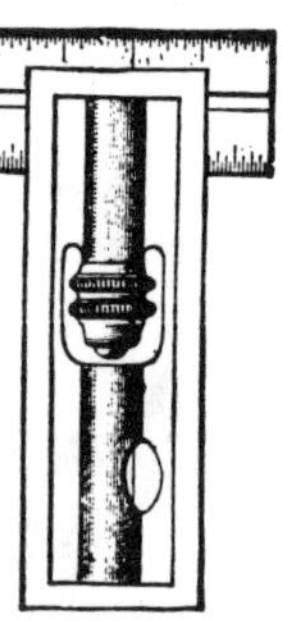

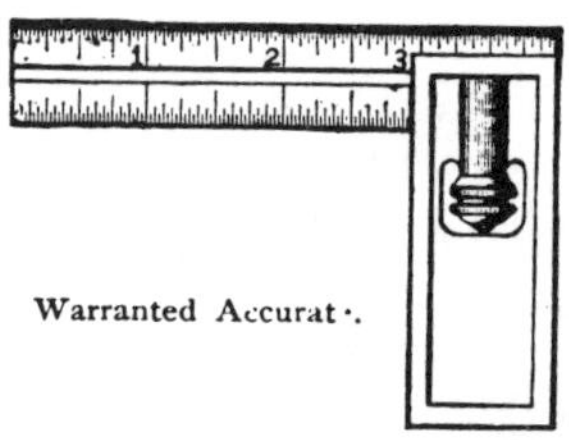

Warranted Accurate.

Price, 4 inch, without Level	$1 50
" 6 " with "	2 50
" 9 " " "	3 50
" 12 " " "	4 50

This is conceded the most practical Square for machinists' and fine tool makers' use, ever offered. The sliding scale, shortened or extended full length, makes it more valuable than a full set of the common kind, while with the *extra bevel blade represented below*, it gives both octagon and hexagon bevels. The constant accuracy of these squares is secured by the clamping bolt having a counterbored nut, within which is a helical spring, serving when the nut is loose, *to hug the blade to the seat*, which is *convex*, so that should the corners of the rule get bruised, its accuracy will not be affected thereby. It also prevents dirt or dust from working under the scale to throw it out of true.

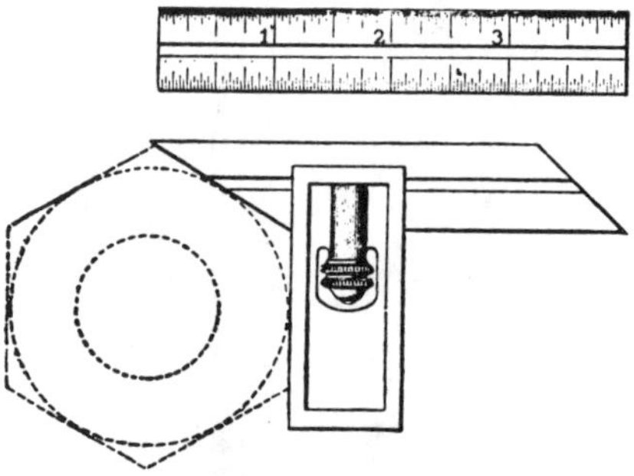

This cut shows Starrett's Patent Double Square with an extra blade; one end of which is beveled to an octagon, and the other to a hexagon, making the most simple and practical tool for obtaining these angles, yet known. These blades will be furnished to order for all sizes of double Squares.

PRICE OF BEVEL BLADES.

For 2½ inch square	$0 30
" 4 " "	40
" 6 " "	50

STARRETT'S DOUBLE SOLID STEEL SQUARE.

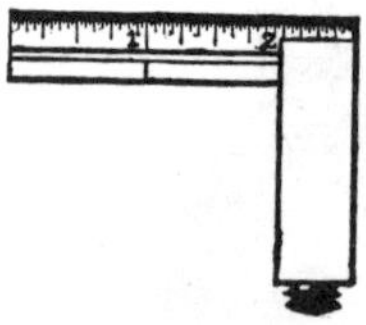

This cut represents a double, *solid steel* Square, with Patent Sliding Scale, 2½ inch, and is especially designed for fine tool makers. The rule being narrow and instantly adjusted to any length, however short, allows it to be used where it would be impossible to use any square with a fixed blade.

Price of square, as per cut	$2 00
" with either bevel or narrow blade	2 30
" " both " and " "	2 60

STARRETT'S GRADUATED SCRATCH GAUGE. 1891

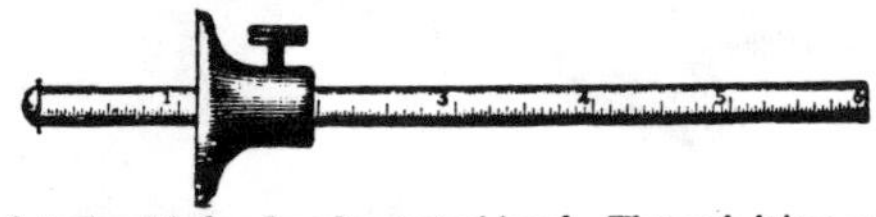

This Gauge is made of steel, with *hardened* cast steel head. Through it is a split bushing, against which the set screw acts to hold it firm. The beam is graduated in 50ths or 64ths of an inch. The marker is a thin square piece of steel, nicely tempered; this is firmly held against the end of beam, presenting four marking points, one or more of which may be ground with cutting edge for wood-work. The rear end of beam with its graduations, forms a convenient and reliable depth gauge for holes too small for a scale to enter.

Price, No. 1, 15-64 x 5 inch beam, graduated, $1.00. Not graduated, 65c.
" " 2, 5-16 x 6 " " $1.25. " 75c.

Unless otherwise ordered, we shall send those graduated in 64ths. Two extra cutters will be sent with each gauge fastened to the case. They should last for years.

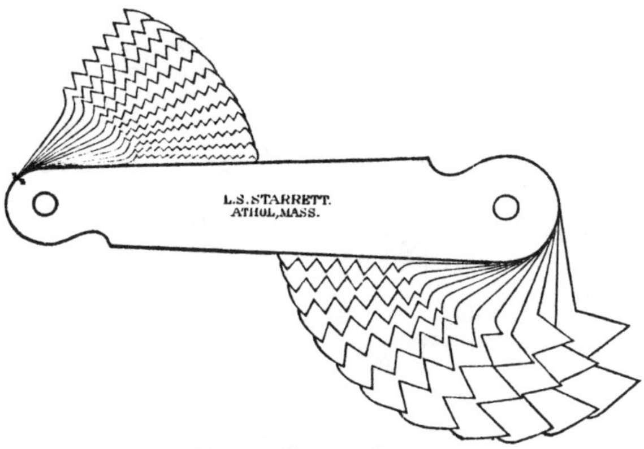

1892

SCREW PITCH GAUGE.

STARRETT'S UNIVERSAL

Surface Gauge

FOR TOOL-MAKERS.

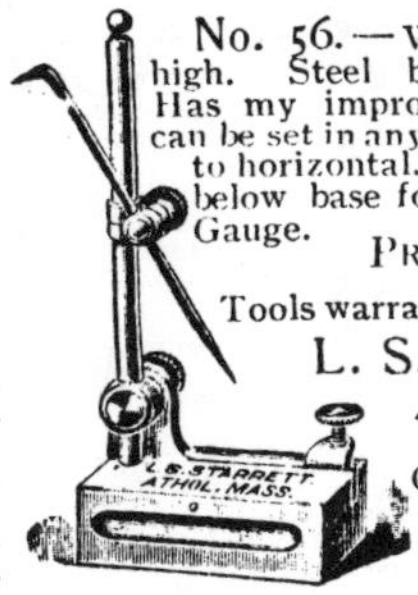

No. 56.—Weighs, 11 oz. 5 in. high. Steel base, case hardened. Has my improved Sleeve. Spindle can be set in any position from vertical to horizontal. Scriber can be used below base for Depth or Marking Gauge.

PRICE, $3.00.

Tools warranted. Catalogue free. 1893

L. S. STARRETT,

ATHOL, MASS., U.S.A.

London Agents:
Chas. Churchill & Co., Limited,
21 Cross Street,
Finsbury, E. C.

Fig. 1.—Starrett's Universal Surface Gauge.

Starrett's Gauges.

The gauges shown in the accompanying cuts are being put on the market by L. S. Starrett, Athol, Mass.

The gauge, Fig. 1, is referred to as having the following improved features, viz.· A joint at the basewhich allows the spindle and scriber to be moved back and forth and placed in any position from upright to horizontal to reach over, back of and under work, while by inclining the spindle over the work its scope for long reach is increased.

The fine adjustment is obtained by the knurled screw in the rocking bracket at the base acting against a stiff spring under the opposite end, while the joint above with the spindle may be set and rigidly held in any position desired. Two pins through the base, frictionally held, it is stated, may be pushed down by slight pressure to form a bearing to work from the edge of, or in the slots of the planer bed for lining up work, while the weight of the gauge against the bed with a little pressure is sufficient to push them back. Grooves around these pins, against which a pointed spring plunger presses, insure their being held in place either up or down. Concaved depressions milled in the sides of the base are designed to make it convenient for thumb and finger to grasp.

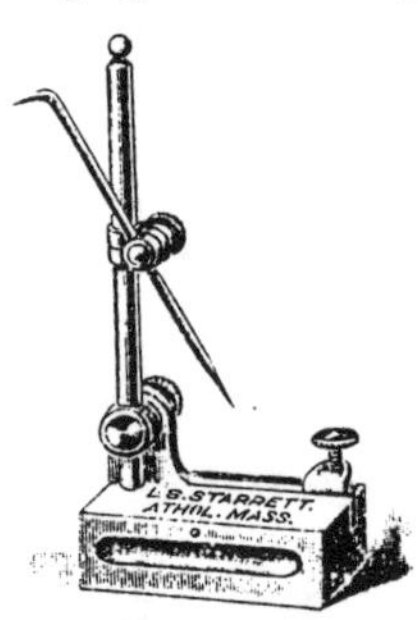

Fig. 2.—Tool-Makers' Universal Gauge.

This gauge is furnished with an improved sleeve to rigidly hold the scriber.

The gauge shown in Fig. 2 is adapted for light work and is made on the principle of the one illustrated in Fig. 1.

The base is steel, case-hardened, with depressions milled in the sides for the thumb and finger to grasp. The top side of it is slotted, and the rocking bracket is pivoted in the same. There is a stiff spring under one end of the bracket and a knurled adjusting screw in the other; the spindle jointed to this may be set and rigidly held in any position from vertical to horizontal, and the scriber placed in position to be used below its base for depth gauge, or (with bent end down) a scribing gauge. The point is made that it weighs 11 ounces and is 5 inches high, and, folding the spindle (which is 4 inches long) horizontally over the base, it may be packed in 1$\frac{3}{8}$ inches x 1$\frac{1}{8}$ inches x 4 inches space in the tool chest.

The stair gauge, Fig. 3, is to be used in connection with any carpenter's steel square, and can be adjusted to any pitch or angle desired. It is designed for cutting in rafters, braces, stairs, &c.

It is made in the shape of a steel angle $\frac{7}{8}$ x $\frac{5}{8}$, $\frac{1}{8}$ inch thick, ground straight, nickel plated, and is referred to as a most valuable addition to a carpenter's kit of tools.

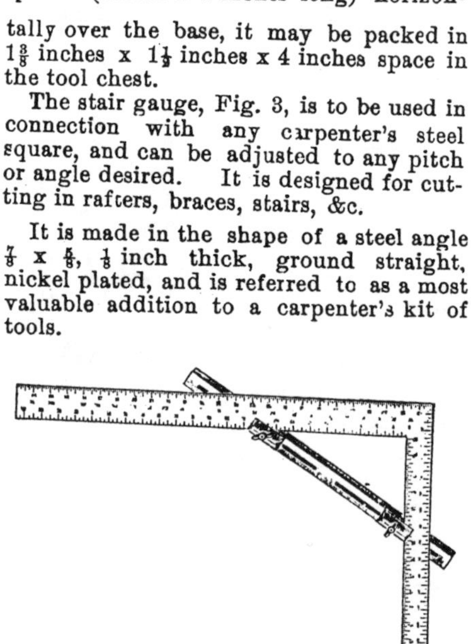

Fig. 3.—Stair Gauge.

1893

Starrett's Calipers.

L. S. Starrett, Athol, Mass., is putting on the market a line of calipers, as represented in the accompanying cuts. Figs.

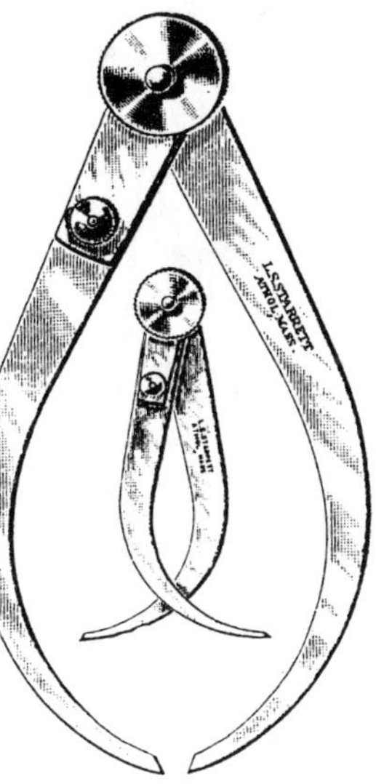

Fig. 1.—*No.* 38 *Lock-Joint Calipers.*

1 and 2 show lock joint calipers, which are made in sizes from 4 to 24 inches. In the under side of the short arm is a slot containing a stiff spring. Riveted into the middle leg and projecting through an opening in the arm is a threaded stud, on which is a knurled nut having a beveled hub; this bears against a cone in the arm and the action of the spring holds them together. Turning the nut presses them apart and adjusts the leg while the joint is locked. The spring takes up all back lash, so that the legs are firm. The calipers are referred to as being simple, light, low priced and reliable, of wide scope for both inside and outside work, and can be instantly adjusted to their full extent and as quickly locked firm in the joint and yet provided with a sensitive adjustment. They are made to supersede the old style firm joint, also the lock joint with split leg adjustment formerly made by him. The improvement consists in a socket joint made tapering and locked or released by a partial turn of the knurled disk drawing it together. A spring washer under the disk maintains an easy friction in the joint when unlocked.

The caliper shown in Fig. 3 not only has the excellent features of Nos. 38 and 39, but in addition to common use may be used inside of chambered cavities, over flanges, &c., removed and replaced without losing the size calipered. This is done by loosening the nut binding one arm to the auxiliary leaf and swinging it out or in, while the joint is locked, to clear the obstruction, then moving it back against a stop, where it will show the exact size measured. A lock-joint caliper is also made in form similar to that shown in Fig. 1. The hermaphrodite caliper, Fig. 4, has a lock-joint attachment and sensitive adjustment. It is light and stiff, with large capacity, instantly opened, closed and locked. The points are nicely tempered. This is also made with one sharp point and one curved point, as in Fig. 3.

1893

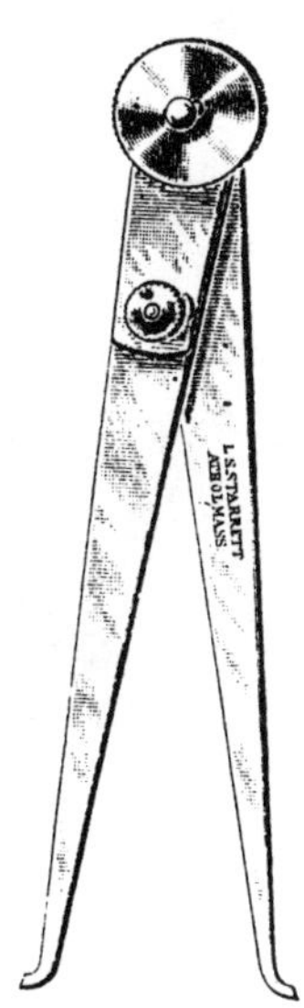

Fig. 2.--*No.* 39 *Lock-Joint Calipers.*

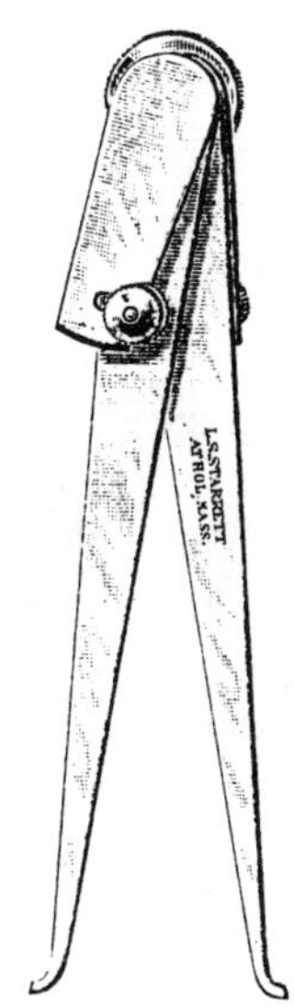

Fig. 3.—*No.* 37 *Lock-Joint Transfer Caliper.*

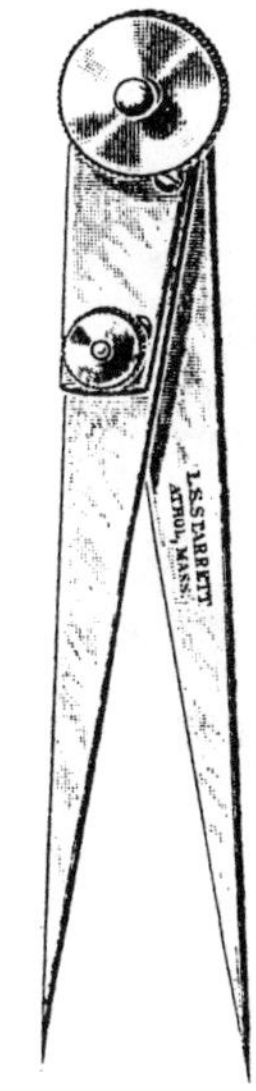

Fig. 4.—*No.* 43 *Hermaphrodite Caliper.*

Improved Firm Joint Calipers.

The accompanying cuts represent firm joint calipers, which are being manufactured by L. S. Starrett, Athol, Mass. The improvement in the calipers is described as consisting in the construction of the joint, which is so made as to be drawn together by means of a screw forcing a washer on to a squared end of the main rivet which extends through the legs, thus clamping them together between bearings so shaped as to insure a smooth and uniform friction, of more or less tension, to suit the user. The quality of these calipers is referred to as superior to that of old style riveted joint calipers.

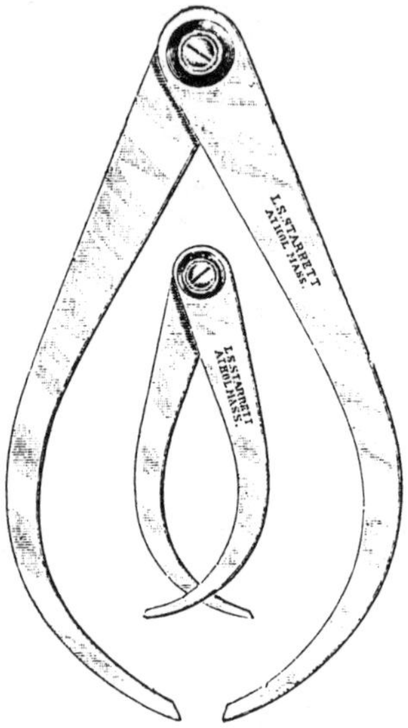

Fig. 1.—Improved Firm Joint Calipers No. 26.

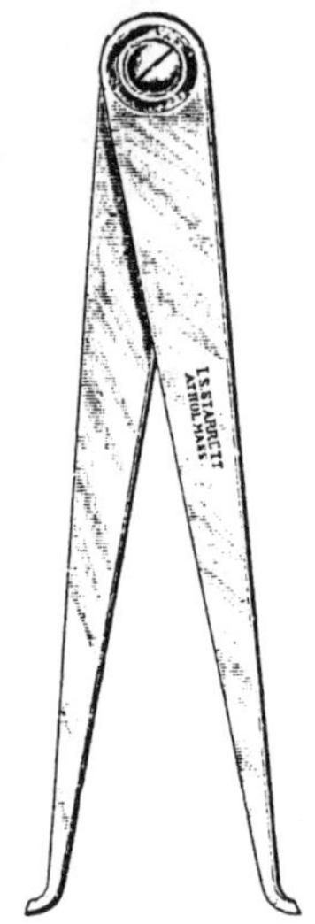

Fig. 2.—Improved Firm Joint Calipers No. 27.

Starrett's Inside Micrometer.

The accompanying cut represents an inside micrometer, put on the market every $\frac{1}{2}$ inch of its length, while the screw fills in the finer measurements. It is explained that from 4 inches to 11$\frac{1}{2}$ inches requires three extension rods. For lengths above 11$\frac{1}{2}$ inches an extension sleeve is screwed on to the micrometer head providing for a greater range to telescope longer rods and requiring less of them, four from 11$\frac{1}{2}$ inches to 30 inches, two of which are made to couple together, thus taking less room. A binding screw locks the micrometer to prevent accidental movements after the tool is set. It is stated that all wearing parts are hardened, and are also provided with an adjustment to compensate for wear.

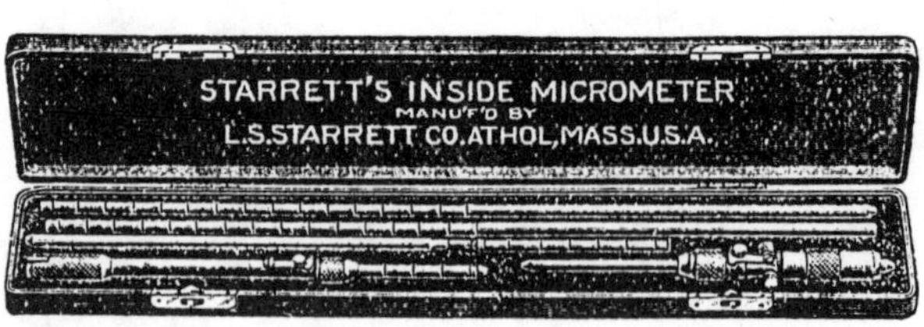

Starrett's Inside Micrometer.

1894

Starrett's Adjustable Jaw Cutnipper.

The adjustabe jaw cutnipper, shown herewith, is put on the market by L. S. Starrett, Athol, Mass. The head and

1893

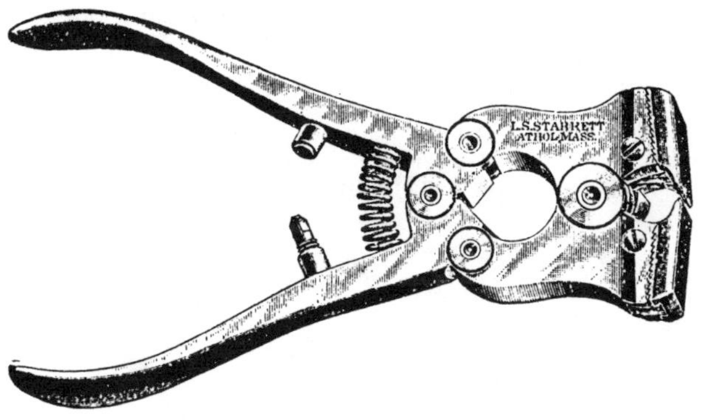

Starrett's Adjustable Jaw Cutnipper.

handles are described as being made of drop forged steel, finely finished, and all parts case hardened except the jaws, which are made from a high grade of steel, nicely tempered. The jaws are detachable, so that they can be removed, ground and adjusted when they become worn. It is pointed out that each jaw can be ground away to the extent of ¼ inoh, remaining as good as new for practical use; and when used up, if ever, new jaws can be procured. The jaws have a dovetailed slot in their under side to receive a spline which extends into a slot in the frame. A tapering screw through the frame and spline draws the jaw fimly down to a toothed seat in the frame, holding it securely. The nipper has a flat spring below the cutting edges and over the joint, forming a yielding seat for the end of the wire to press against while being cut. This spring is to obviate the breaking of the jaws, which might occur if the wire was inserted against a solid surface, thus creating a pushing out strain on the jaws when they are pressed together. Jaws warranted to stand for cutting music wire have the cutting edge ground to a steep bevel. The jaws for general use, it is stated, are not ground to so acute an edge, and are better for cutting soft wire and for ordinary purposes. The tool shown in the cut is known as No. 1, and is 6 inches long. A larger size, we are advised, will be made later.

1894

Universal Surface Gauge.

L. S. Starrett Company, Athol, Mass., are making the tool makers' Universal surface gauge No. 56, illustrated in Fig. 1. It is well adapted for light work. The base is steel, well finished and case hardened, with depressions milled in the sides for the thumb and finger to grasp. The top side of it is slotted, and the rocking bracket is pivoted in the same. Under one end of the bracket is a stiff spring, and a knurled adjusting screw at the other.

Fig. 1.—Tool Makers' Universal Surface Gauge No. 56.

The spindle jointed to this may be set and rigidly held in any position from vertical to horizontal, and the scriber placed in position to be used below its base for depth gauge, or (with bent end down) as a scribing gauge. A V-shaped groove in the end and bottom adapts it for use on cylindrical work. There is a small hole in the clamp next to the base, in which the scriber may be used for

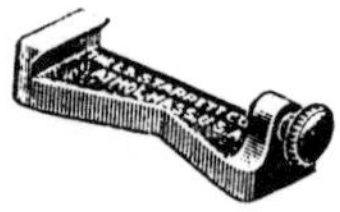

Fig. 2.—Auxiliary Guide Piece for Clamping to Gauge Base

light work, the spindle being removed. Fig. 2 shows an auxiliary guide piece furnished, which can be clamped to the base above described. The gauge weighs 10 ounces and is 5 inches high. Folding the spindle (which is 4 inches long) horizontally over the base, it may be packed in 1⅜ x 1½ x 4 inches space in the tool chest.

1898

Six-Inch Micrometer.

Among the new goods recently offered to the public by the L. S. Starrett Company, Athol, Mass., is a patented 6-inch micromoter, which is here illustrated. It will measure round work up to 4¼ inches and flat work to 6

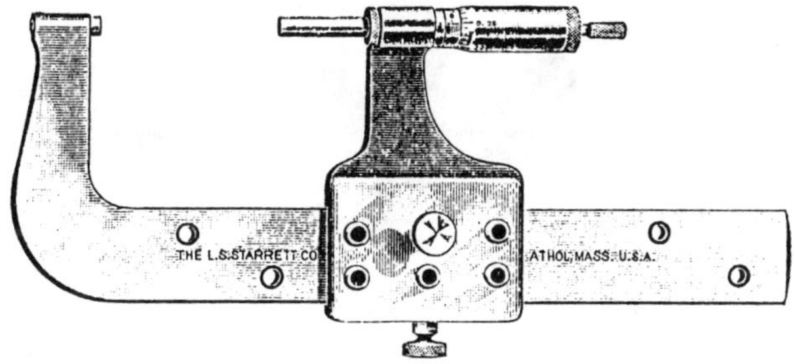

Six-Inch Micrometer No. 128.

inches, inclusive. It weighs 21 ounces and is rigid and accurate. It can be instantly set to exact position, from 1 to 6 inches, by inserting the tapering plug. A marked feature of the implement is a set of six independent holes through both the movable part and the beam. Each hole is bushed with hardened steel bushings, ground and lapped to fit the tapering plug, which locates to exactness the various inch settings.

Engineers' and Plumbers' Level.

The L. S. Starrett Company, Athol, Mass., are introducing the level herewith shown. The instrument, which is known as No. 133, includes an adjustable incline level, a fixed level and a plumb. The hinged tube inside the working face of the frame, carrying a level glass, is adjustable to the graduated scale, and shows any incline by thirty-secondths of an inch, or less, to 2 inches to the foot

Engineers' and Plumbers' Level.

without interfering in the least with the plumb or level. A longitudinal groove in the seat of the frame, not shown in the cut, adapts it to rest on a cylindrical shaft or pipe, as well as on flat surfaces, making it convenient to determine the pitch of drain pipes, &c. The instrument is supplied with either ground or plain glasses.

1898

Adjustable Bench Levels.

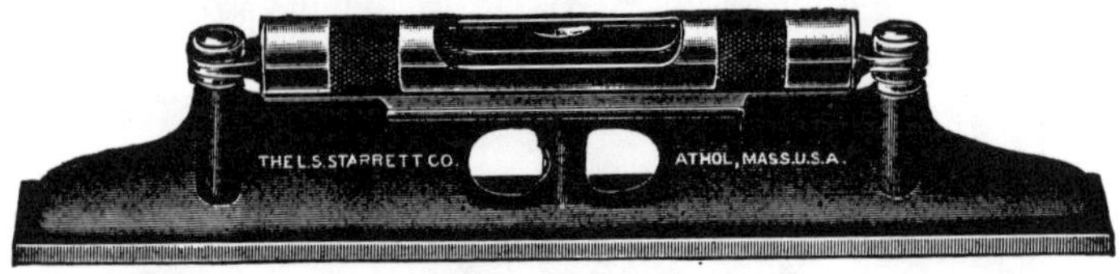

With ground and graduated vials,—accurate and very sensitive.

These Levels are so constructed that they can be accurately adjusted, and when so adjusted are not liable to get out of truth, the vials being set in tubes having solid ends which are firmly clamped to the base. The tubes are nickel plated, the bases japanned or nickel plated. The outer tube may be turned so as to protect the glass when not in use. For sizes and prices see our catalog.

CATALOG OF FINE MECHANICAL TOOLS, MILLING, GEAR AND OTHER CUTTERS SENT ON REQUEST.

THE L.S. STARRETT Co., Box 59, ATHOL, MASS., U. S. A.

NEW YORK OFFICE: 126 LIBERTY STREET.

FOREIGN AGENTS:

LONDON—Tools, Chas Churchill & Co., Ltd., 15 Leonard Street, Finsbury, E. C.
Cutters, Chas. Neat & Co., 112 Queen Victoria Street, E. C.
BRUSSELS—Schuchardt & Schütte, 17 Rue de Fossé aux Loups.
BERLIN—Schuchardt & Schütte 59 Spandauer Strasse.
SYDNEY—Edge & Edge, 250 Pitt Street.
YOKOHAMA—F. W. Horne, Box 174.
VIENNA—Schuchardt & Schütte, VII, Bez. Breitegasse, 17.
BUDAPEST—Székely & Káldor, Erzsebet-Körut 9-11.
LA CHAUX DE FONDS—W. Hummel Fils, 30 Rue Peopold Robert.

1899

Spring Screw Driver.

The L. S. Starrett Company, Athol, Mass., and 126 Liberty street, New York, have just put on the market a series of screw drivers, or, more specifically, screw placers, as here illustrated, in three sizes, and known to the trade as No. 152. This finely made tool is of high grade steel, knurled, polished and nickeled. The series are lettered A, B and C, the former being the large one, B intermediate and C small. They are respectively 7, 5 and 5 inches long, actual size, and 3-16, 5-32 and ⅛ inch thick, the width of the blades being approximately the same. The main purpose of this tool is to compress the spring blades and by introducing them into the slot of a screw place it in any required position, when, after giving two or three turns where the hole is made by awl or other instrument, the screw can be driven to its place in the ordinary way with hand or bit screw driver. This device is mainly for work where the screws are so small they cannot be manipulated expeditiously by hand, or in machine parts, clocks, electrical instruments and other small work where a reach has to be made not possible with the fingers.

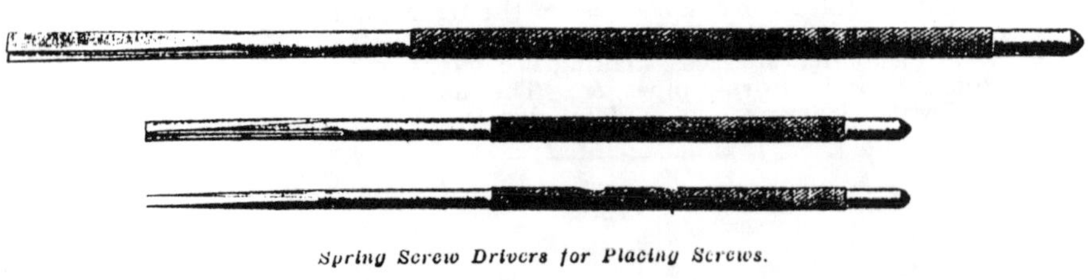

Spring Screw Drivers for Placing Screws.

1901

Measuring Bar and Rule Clamps.

The L. S. Starrett Company, Athol, Mass., and 126 Liberty street, New York, have just put on the market

Fig. 2.—Steel Rule Clamps No. 299.

two forms of clamps for joining wood and steel rules together, as here illustrated. Fig. 1 is known as a measuring bar clamp, is numbered 69 and nickel plated. The clamps are 1 inch square inside and are to be used with two wooden bars about 1 x ½ inch, of any desired length. This combination is designed as a convenience in the way of adjustable measuring rods for carpenters' use, as well as extension beams for their No. 59 trammels.

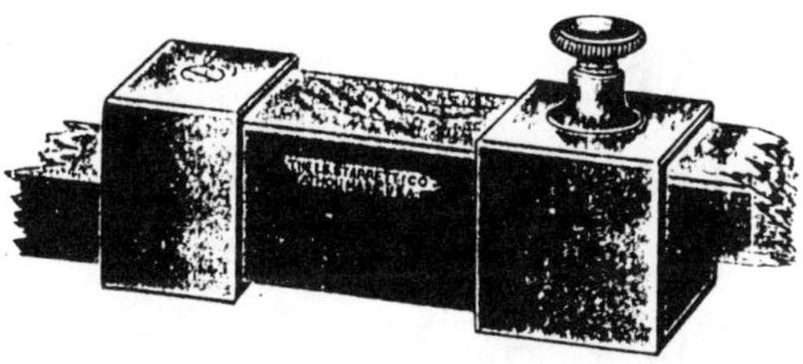

Fig. 1.—Measuring Bar Clamps No 69.

Figs. 2 and 3 illustrate their No. 299 rule clamps. The device is designed to clamp two steel rules together, end to end, making one long rule. The rules may be of the same or different widths up to 1¼ inches. It is especially valuable to mechanics whose tool chests will usually not hold rules longer than 12 inches.

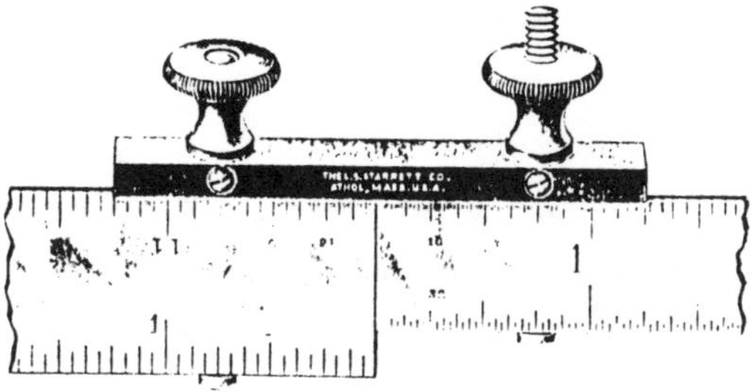

Fig. 3.—Method of Using Clamps

1901

Stair Gauge Fixtures.

L. S. Starrett Company, Athol, Mass., and 126 Liberty street, New York, have put on the market Starrett's stair gauge fixtures No. 111, as here illustrated. These fixtures, which are nickel plated, can be readily clamped to a carpenter's steel square to form a gauge for various purposes. In size they are 1⅞ inches long over all. The outline sketch in the engraving show some of the various uses to which the gauge can be put. Sketch A shows the

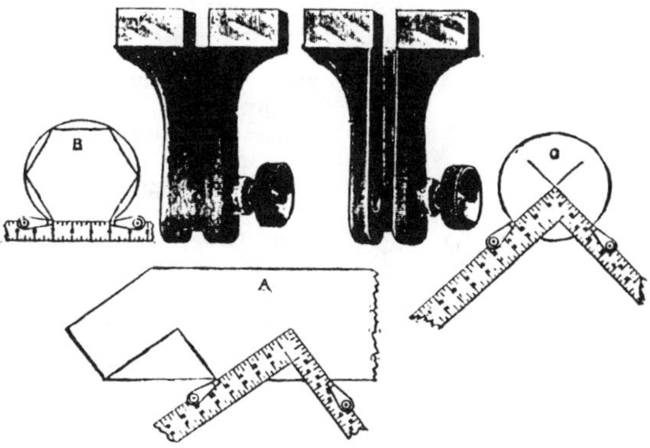

Starrett's Stair Gauge Fixtures.

application of the gauge for laying out a stair stringer; sketch B laying off hexagon angles; sketch C used as a center gauge or in quartering a circle. Especial attention is called to the moderate price, compact form and other advantages not possessed by ordinary stair gauges.

1901

Universal Dividers.

The L. S. Starrett Company, Athol, Mass., and 126 Liberty street, New York, have just put on the market the Universal Dividers, No. 89, here shown. The adjustable scriber holder is reversible and carries either a fine tempered steel point or a pencil lead held in a split socket by a knurled nut. With the holder turned outward work can be laid close to corners; turned inward, the points may be brought close together to scribe the smallest circle. With the 4-inch beam regularly furnished, 7½ inches and under may be scribed. An

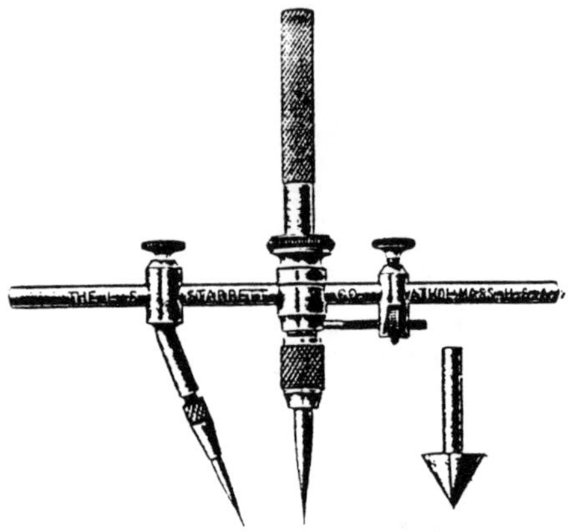

Universal Dividers No. 89.

auxiliary beam, 13 inches long, is supplied when specially ordered, with which a 25-inch circle may be drawn, while for exceptional occasions an ordinary steel rod of any length and suitable diameter may be used for almost any radius. The V center point may be substituted for the regular point seen in the tool for getting quickly the exact center in scribing around a drilled hole. A fine adjustment is obtained by tightening the small milled top nut, loosening the center nut and adjusting by means of the lower vertical nut. The tool, with a 4-inch beam and V center point, weighs 1½ ounces, the illustration being about five-eighths size.

1901

Mercury Plumb Bobs.

L. S. Starrett Company, Athol, Mass., and 126 Liberty street, New York, have recently begun the manufacture of an improved form of mercury plumb bobs, as here illustrated. The marked characteristics of this tool are great weight in proportion to size, low center of gravity, small diameter, thus offering slight resistance to wind, hardened and ground point, reel at top to hold line when not in use, and a device at the extreme upper part for securing the end of cord. Parts of the body are knurled, and the remainder polished, the entire surface being nickel plated. The mercury contained in the barrel or cylinder body is held in place by an inside slotted cap, which works in a thread on the inner side of the cylinder. The bob is made in 4, 5, 5½ and 6 inch lengths, the diameters in the order named being respectively ½, ⅝, ⅞ and 1 inch, weighing 3½, 6, 12 and 16 ounces each. In this connection the company issue illustrated descriptive slips, envelope size, with the imprint of the dealer on, to those who request them.

Mercury Plumb Bob No. 87.

1902

Micrometer Gauges.

The L. S. Starrett Company, Athol, Mass., and 123 Liberty street, New York, have recently put on the market some improved forms of gauges, of which the two here illustrated are examples. Fig. 1 represents paper gauge micrometer No. 223, with ring, which is used for measuring the thickness of paper, cardboard, sheet rubber, &c. The disks are placed on the anvil and spindle so that measurements can be taken without compressing the articles measured. It will measure all sizes from zero to 11-32 inch by thousandths of an inch. The illustration shows the gauge with knurled end ratchet, the purpose of which is to enable the user to know instantly when the disks meet the article to be measured and gives the operator the advantage of a more delicate touch, as when the disk meets with resistance the ratchet turns without forcing the disks further along. The same gauge is made without the ring attachment at a less price and is known as No. 225. Another of the same series is the No. 223M, graduated to read in hundredths of a millimeter, while still another is the same as the No. 225, only graduated to read to hundredths of a millimeter. All of the above styles are made with or without ratchet, those without ratchet costing 50 cents each less on the list. Fig. 2 illustrates United States Government micrometer gauge No. 127. These gauges are designed to meet the requirements of the Government in making big guns and other work in the Ordnance Department of the Government shops, where they are now used. The frames are cut from steel plates, nicely finished. The sides are covered with hard rubber, put on with brass screws, preventing inaccuracy through expansion caused by change in temperature when held in warm hands. The micrometer screw adjusts 1 inch, reading 1-1000 inch, and is provided with patent lock nut. The different length tail spindles, forming anvils, are interchangeable and have positive stops to set against their socketed seats. The adjusting collars on these anvils have notches to facilitate the removal of dirt which would prevent them from setting accurately against the seat. The contact ends of the spindles are slightly convex, to prevent catching on cylindrical work. The gauges can be furnished with ratchet stop or speeded screw thumb piece, as desired. A similar gauge is made, No. 127M, only graduated in metric measurement for recording by hundredths of a millimeter.

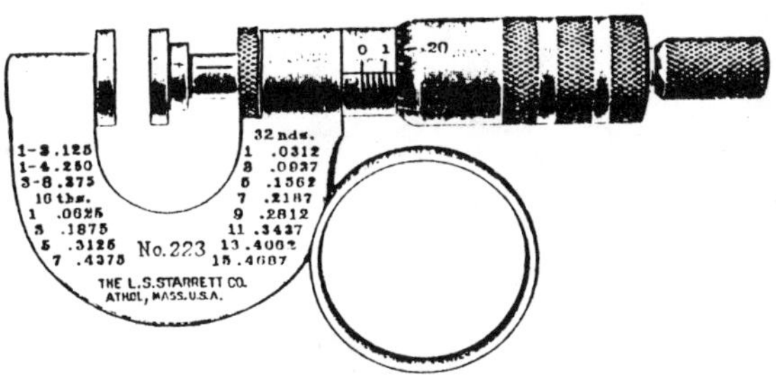

Fig. 1.—Paper Gauge Micrometer.

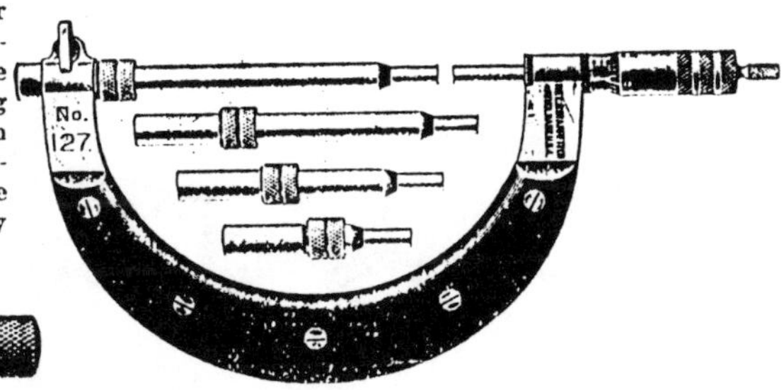

Fig. 2.—United States Government Micrometer Gauge.

Universal Test Indicator.

The L. S. Starrett Company, Athol, Mass., and 123 Liberty street, New York, are now introducing a recent addition to their large line of mechanical tools in the form of a Universal test indicator, No. 64, as here shown. Its purpose is to test and show the imperfections or degrees of exactness of inside, outside or surface work, errors often occurring through vibration of machinery

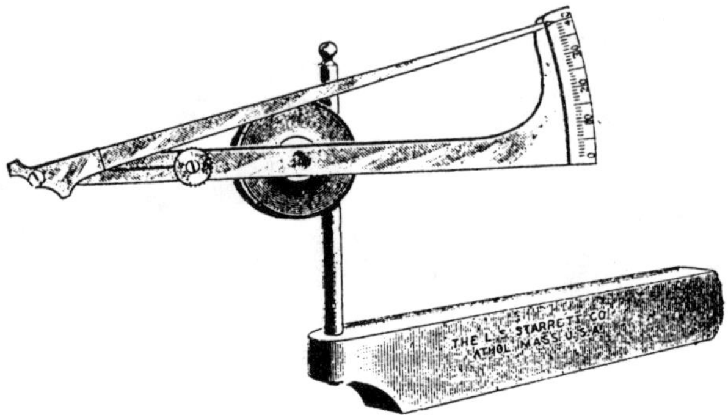

Fig. 1.—Universal Test Indicator No. 64.

or fault of lathe, &c. It can be instantly attached to the spindle or to the needle of any surface gauge and used in connection with it to show the slightest variation in thousandths. It may be clamped to a flat or round support varying in size from a surface gauge needle up to ⅜ inch diameter, flat or round. A special holder, as here shown, is designed to go in the tool post of a lathe, adapting it for use to show the accuracy of all kinds of lathe work turning, chucking or locating and centering work on face plate. The head of the needle has three working points, equidistant from its fulcrum, so the 1902

Fig. 2.—Same Indicator Applied to Surface Gauge.

telltale needle will vibrate, reading in thousandths, when the work contacts with either point—in front, above or below it. When in front the spring operating the telltale needle needs to be reversed to throw the point of the needle up instead of down, as when used above or below the work. This is accomplished instantly by a slight turn of the milled disk to which the vibratory spring is attached. The working parts of the head are hardened and as a surface or test gauge for accurate work it is especially recommended by the manufacturers. Fig. 2 shows the indicator as attached to one of their newer surface gauges.

Starrett Screw Drivers Nos. 550 and 551.

The accompanying cuts represent patent screw drivers put on the market by the L. S. Starrett Company, Athol, Mass. Screw driver No. 550 has a knurled hard wood handle, 1¾ inches in diameter, large enough to fill the hand and give leverage. Its steel shank has a socketed end to which is fitted a set of three screw driver tips of different sizes, adapted to screw heads, from very small up to ⅜-inch. The tips are shaped and tempered to give

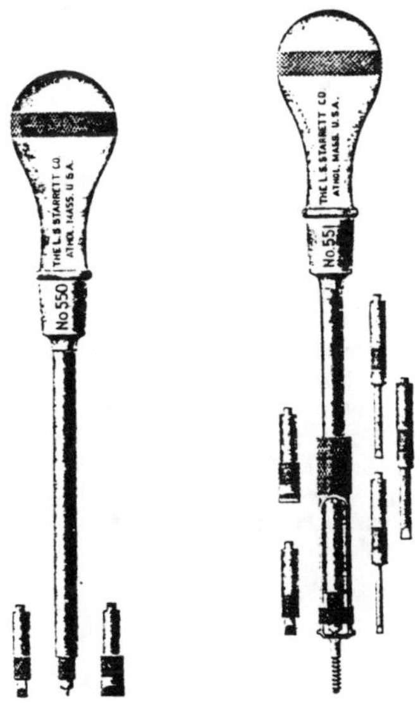

No. 550. No. 551.

Starrett Screw Drivers.

the greatest strength. Either size may be instantly withdrawn and another inserted, thus supplying a full set of screw drivers at a comparatively small cost. The screw driver is 10 inches long. Tool No. 551 is the same as the one described, with the addition of a sleeve with spring fingers, which slide upon the shank, and a set of brad awls, which may be used interchangeably with the screw driver bits. The ends of the fingers grasp the head of the screw, draw it back and hold it in firm contact with the screw driver, so that the screw can be driven home straight and true without slipping from the head, and in places where it would be difficult to start screws with a common driver. The fingers also hold the brad awls from pulling out. The changing of one tool for another may be done almost instantly. Slipping the finger sleeve up against a stop and sliding the knurled ring closes them on the screw head and holds it tightly. The brad awls and screw driver tips are put up in a neat case and can be carried in the pocket.

Hub Gauge No. 430.

The gauge shown herewith is for use in measuring the length of pulley hubs, wagon wheel hubs, thickness of

Hub Gauge No. 430.

iron plate through holes, &c. The gauge will measure all lengths to 7½ inches, and can be inserted through a ⅝-inch hole. The gauge is put on the market by the L. S. Starrett Company, Athol, Mass.; New York store, 123 Liberty street.

Starrett Draftsmen's Scale.

The scale herewith illustrated has tilting studs, so placed that each of its four corners, with different graduations, will come in contact with the paper by its own gravity when resting on the studs, with the back edge raised at an angle of about 30 degrees. The scale is made in two numbers, graduated on each of the four corners in parts of inches, as follows: No. 405, 10ths,

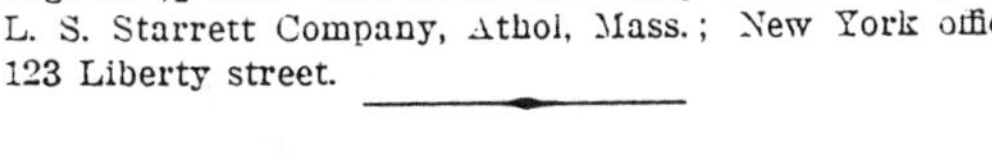

Starrett Draftsmen's Scale.

40ths, 50ths, 100ths; No. 405A, 8ths, 16ths, 32ds, 64ths. Another scale, No. 405M, is graduated in the metric system, one edge of each side in millimeters, the other edge in ½ mm. The scales are being introduced by the L. S. Starrett Company, Athol, Mass.; New York office, 123 Liberty street.

1903

The Universal Dial Test Indicator No. 196.

The L. S. Starrett Company, Athol, Mass., are putting on the market the dial test indicator shown in Fig. 1. It is referred to as being simple, easily read and very sensitive, and is described as follows: The slightest pressure upon the contact point produces a movement of the hand on the dial. The circumference of the dial is divided into 125 equal spaces, each one representing a movement of the contact point of one-half thousandth of an inch. One revolution of the hand therefore indicates 1-16 inch, and two revolutions 1/8 inch, which is the capacity of the instrument. The dials are figured in two different ways: Style A is marked 0, 5, 10, 15, &c, to 62½, the figures denoting thousandths, and is most useful in the greater forward movement, measuring, indexing, spacing, &c.; style B is marked from 0 to 31¼ to right and left, and is best for general use. By bringing the contact point against the work just enough pressure to give the hand one full turn, then setting it at 0, an opportunity is given for one full revolution of the hand to both right and left of 0, showing a rise or drop in the work and the amount of the variation. A valuable feature is the adjustable dial. By turning the knurled rim the dial may be instantly moved to bring the 0 mark to any point desired in relation to the hand. Each indicator is fitted with friction joint and removable 3-inch rod, adapting it for use in any position, at the top, bottom or side of the work; also with three hardened and ground contact points adapted for different classes of work. There is also a safety collar to prevent damage from an upward jam or crowding, and a small, hardened screw connecting the collar and contact point, with a V groove in the center. This little screw is strong enough to stand all the pressure necessary, but is intended to break under too much strain or side jam, and so prevent greater injury to the instrument. The broken ends are easily removed, and a new screw inserted. The special tool post and sleeve, as shown in Fig. 2, are useful in lathe work. For general work the indicator is adapted for use with the company's 9-inch or 12-inch surface gauges. On lathe, planer, milling machine and in setting up machinery, it is remarked that this tool will be found indispensable when once used.

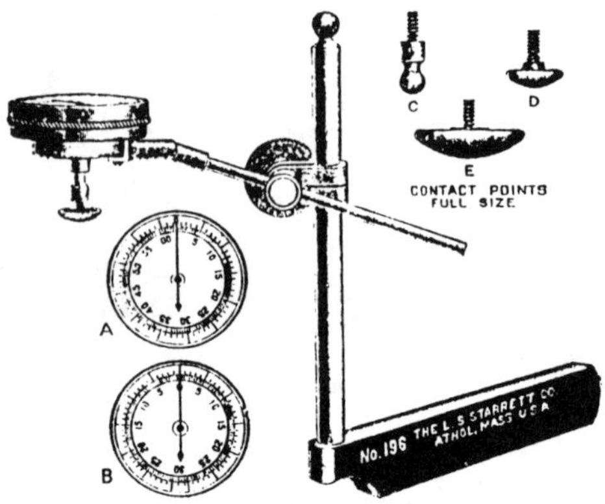

Fig. 1.—Universal Dial Test Indicator No. 196.

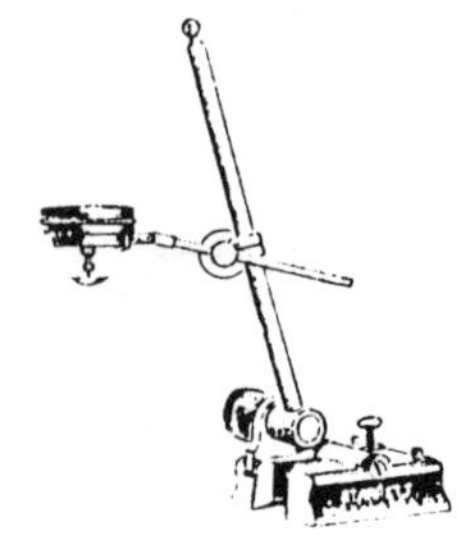

Fig. 2.—Special Tool Post.

1903

Protractor No. 19.

The L. S. Starrett Company, Athol, Mass., is placing on the market the protractor shown herewith. It is graduated in degrees from 0 to 90, both ways. The blade is 6 inches long, and by means of the company's patent lock joint is set firmly by a slight turn of the nut. The back of the tool is flat. The tool is alluded to as accurate and as convenient for setting bevels, for transferring angles, for use as a small T-square, and for a large number of other uses by machinists, draftsmen or mechanics.

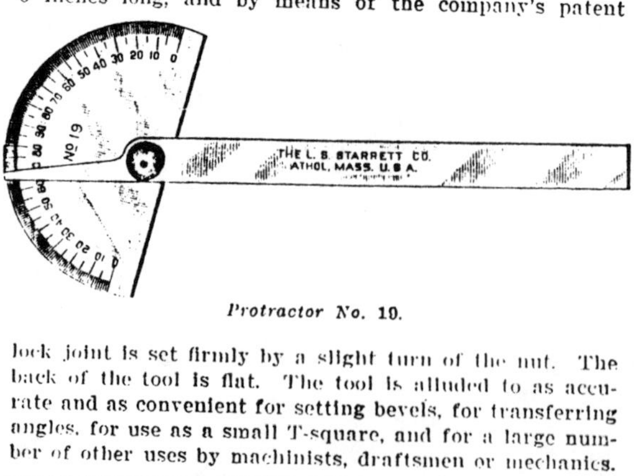

Protractor No. 19.

1903

Protractor No. 16.

The L. S. Starrett Company, Athol, Mass., is placing on the market the protractor shown herewith. The blade closes in the stock either way against a stop, making a

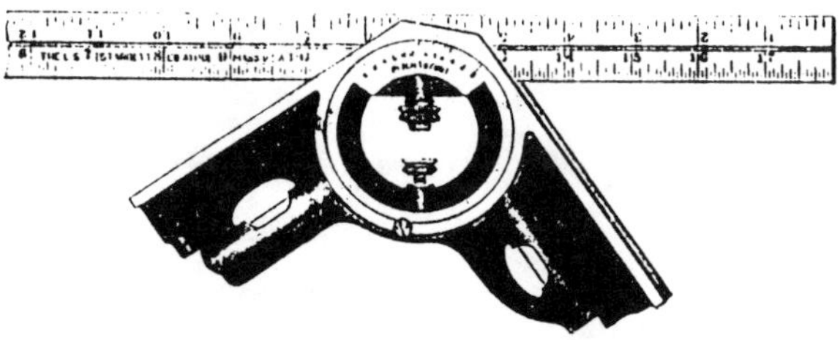

Protractor No. 16.

perfect square, plumb and level. A 12, 18 or 24 inch blade may be used in connection with the stock, and with a 24-inch blade it weighs 1¾ pounds. The turret is graduated on both sides, one in degrees, the other to show pitch to the foot, so that the blade may be set by the graduation for laying off angles to any degree or any pitch, and the opposite branch of the stock will be right to lay out the complementary angle without mental calculation or error for valley roofs, bridge work, stair gauges, &c. The levels are so arranged that work can be leveled up to any degree or pitch underneath or on top of a roof, rafter, stair stringer, &c. As a square or protractor with the sliding blade it can be used in places where a fixed blade cannot, and is a substitute for a whole kit of squares, it is explained, from the shortest to the full length of blade, making a depth gauge for squaring in mortises and transferring measurements. It may be used in place of the carpenter's old time square, with the advantage of being packed in a chest without taking up so much room. Without the blade the stock may be used in contracted places as a 6-inch level and plumb, while with an 18 or 24 inch blade a level and plumb of corresponding length is obtained. It is remarked that, taken altogether, the tool makes a kit that will be appreciated by the progressive mechanic. The stock is sold with or without the blade.

1905

Double Joint Blacksmith's Calipers.

The L. S. Starrett Company, Athol, Mass., and 132 Liberty street, New York, is manufacturing the firm double joint calipers for blacksmiths, No. 173, here illustrated. It is 22 in. long, over all, one caliper having a capacity of 6 in. and the other 12 in. The difference in length of arms serves to prevent confusion in use where there are but slight variations in diameters and a marked advantage is the fact that with this tool both rough and finished sizes can be determined comfortably and with

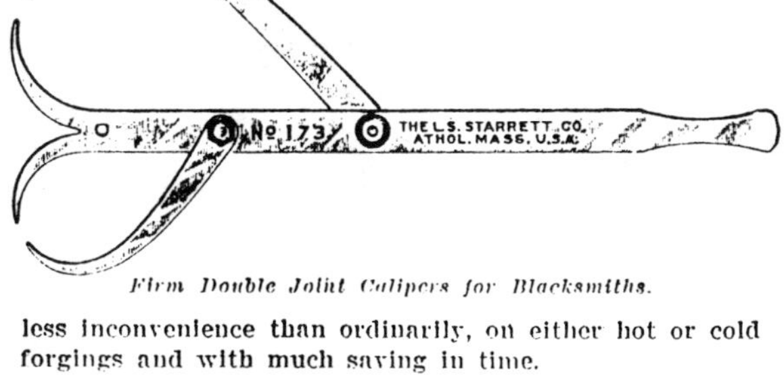

Firm Double Joint Calipers for Blacksmiths.

less inconvenience than ordinarily, on either hot or cold forgings and with much saving in time.

1908

Micrometer Depth Gauge.

The gauge shown in the accompanying cut is designed for measuring the depth of grooves, holes or irregular parts. It has ¼-inch movement of screw, reading in thousandths; and with two ½-inch and one 1-inch standard collars to slip off or on the spindle, 2½ inches, reading in thousandths can be obtained. The split nut is covered and protected, the company state, by their patent graduated sleeve, which not only protects the nut from dirt, but provides a quick and accurate way of taking up wear and adjusting the micrometer to insure correct reading. The sleeve, being held by a stiff friction, may be rotated by a spanner wrench, which accompanies each gauge, so that the zero lines will coincide for correct reading. The head is about 3-16 inch thick and 2⅝

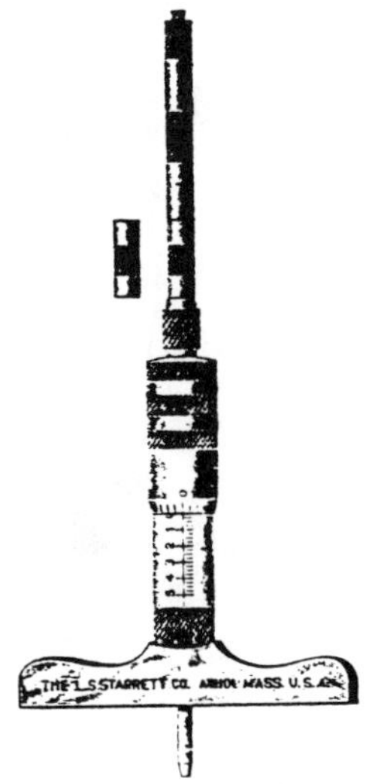

Micrometer Depth Gauge.

inches long; this and the point of the measuring rod are hardened. The head carries with it a knurled set screw for locking the spindle to prevent changing after being set. The gauge is offered by the L. S. Starrett Company, Athol, Mass.; New York store, 123 Liberty street.

1903

Magazine Screwdriver.

The L. S. Starrett Company, Athol, Mass., and 132 Liberty street, New York, is placing on the market Starrett's patent magazine or telescoping screwdriver, here

Fig. 1.—Magazine Screwdriver.

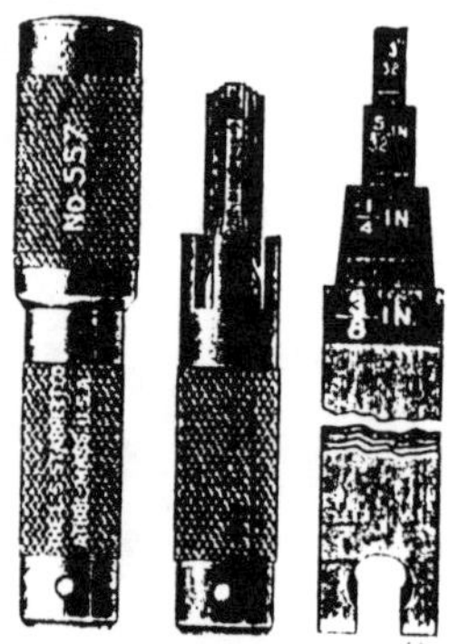

Fig. 2.—Screwdriver Closed and Its Component Parts.

shown, both ready for use and in detail. It offers for pocket use a range of four blades, strong enough for effective service and yet compact in dimensions. The screwdriver as represented in Fig. 1, is 6⅜ in. long, ¾ and ⅝ in., respectively, diameter of barrels, and 4 in. long over all, with blades inside. All blades are of the same dimensions at the heel, but have widths at other ends of 3-32, 5-32, 8-32 and 12-32 in. The point is also made that the smaller blades may frequently be used to make holes in wood for screws as well as to drive them home. The four blades are carried in the small grooved barrel, in which there is a flat, curved spring, extending lengthwise of the cylinder, which not only keeps the pieces from falling out, but from rattling in carrying about. Each blade is held in the socket by pressing a spring button, so that the base of the blade may enter, when releasing the pressure causes the larger diameter of pin to move forward and entirely fill the circular space. The larger part of the handle is rigidly kept from turning by means of the four rectangular openings made to receive a pin riveted to the inner side of the larger cylinder, and frictionally from coming off by the accurate taper fit without the use of screws to bind or hinder.

1908

Draftsmen's T-Square and Blacksmiths' Rule.

The L. S. Starrett Company, Athol, Mass., and 123 Liberty street, in charge of L. H. Briggs, has just added to its large stock of mechanics' fine tools the draftsmen's T-square and blacksmiths' hook and handle rule here shown. The T-square, No. 171, Fig. 1, is very light, made of steel, polished and nickeled, and is well suited to use in manual training schools, colleges, &c., as well as by the professional user. The head is struck up from sheet steel, the face of which is ground straight, as is the reverse side, for use either way. The blade, 0.045 inch thick and ⅞ inch wide, is spring tempered and riveted on top of the head, so that the blade forms a bearing to

Fig. 1.—Draftsmen's Nickeled T-Square.

guide a draftsman's triangle, &c., back past the face of the head, not striking against it, as would be the case if the blade was sunk flush with the head. The instrument is both light and accurate, the 14-inch size weighing only 5 ounces, and 5½ and 6 ounces for the 16 and 18 inch sizes, respectively, the three sizes all having 8-inch heads. Fig. 2, Blacksmiths' hook and handle rule, No. 465, is made from hard rolled sheet brass 1-10 inch thick and 1 1-16 inches wide, with heavy graduations and figures, graduated from the end in sixteenths of an inch on one

Fig. 2.—Blacksmiths' Hook and Handle Rule.

side and from the inside of the hook in sixteenths of an inch on the other, thus adapting the rule for taking correct measurements from either the outside edge of a hot piece of iron or from the inside when held against a corner. The graduations extend 12 inches, the handle is flat and the over all length is 16¾ inches.

1906

Quick Adjusting Micrometer.

The L. S. Starrett Company, Athol, Mass., and 132 Liberty street, New York, has put on the market the Quick Adjusting micrometer, here illustrated. The notable feature of the tool is the rapidity with which it can be opened or closed to any point within its capacity of 1 in., without, it is asserted, impairing its accuracy or sensitiveness. This advantage is apparent when it is realized that 40 complete revolutions of the screw

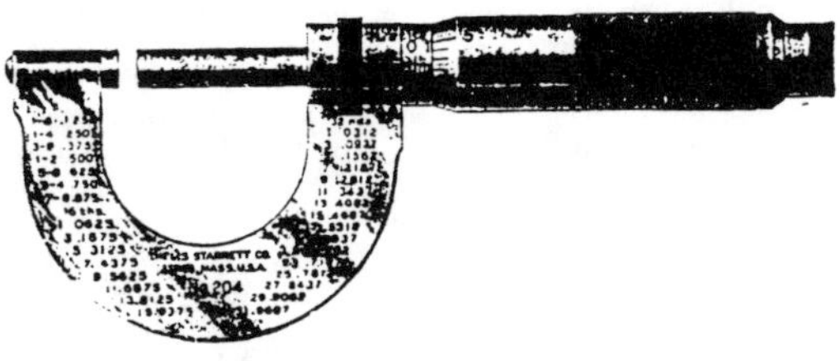

Quick Adjusting Micrometer.

must be made to open or close an ordinary 1-in. micrometer its full length, consuming in time approximately 20 sec. In manipulating this micrometer a finger pressure against the end of the plunger immediately releases the nut, disengaging it from the screw, when any adjustment within an inch may be instantly made. Discontinuing the pressure causes instant engagement of nut with screw, when fine adjustments may be made in the customary way. This micrometer likewise has the Starrett patented sleeve and new lock, as well as the ratchet stop, the latter, it will be recalled, serving to prevent injury to the tool from turning too strongly on the article being measured, as the knurled thumb piece begins to ratchet as soon as contact occurs. Two styles of the Quick Adjusting micrometer are made, both of 1 in. capacity, No. 204 recording by thousandths and No. 205 by ten thousandths of an inch. The micrometers are sent singly in leatherette cases at a slight additional charge for case.

1908

Tap Wrench and Adjustable Parallels.

The L. S. Starrett Company, Athol, Mass., and 132 Liberty street, New York, has added tap wrench No. 91 and adjustable parallels, No. 154, to its lines of mechan-

Fig. 1.—Tap Wrench, No. 91.

ical tools. The tap wrench is of new design and has other novel features. It is designed to hold securely square or round shank taps, and inside the knurled adjusting screw a spring, connected with the plunger, holds it back and causes instant movement with the screw. Fig. 2 illustrates a set of adjustable parallels to be used in connection with milling, planer and shaper vises, tak-

Fig. 2.—Adjustable Parallels, No. 154.

ing the place, it is said, of the large number usually required. They are also designed for leveling up work on planer, drill press, &c. Their capacity is from ⅜ in. to 2¼ in., and they are made in six sizes, the lengths of which run from 1¾ in. to 5 1-16 in., all 9-32 in. thick.

1908

Pocket Screw Driver.

The L. S. Starrett Company, Athol, Mass., and 132 Liberty street, New York, has just brought out the Pocket screw driver, No. 560, for electricians and others, here shown. It is similar to the No. 557, except that the handle is covered with hard rubber for insulation from electrical currents, and is ribbed lengthwise to insure a firm grip in use. The four blades, each 3-32, 5-32, ¼ and ⅜ in. wide are carried in the handle, and prevented by spring pressure from rattling or loss when the telescope handle is detached from the barrel. The handle is ⅞ in. in diameter, and the blade holding section ¾ in. in

Insulated Pocket Screw Driver for Electricians.

1909

diameter, one fitting the other on a taper, friction tight, but quickly separable by a twist.

Starrett Tools.

The L. S. Starrett Company, Athol, Mass., and 132 Liberty street, New York, is marketing several fine mechanics' tools, which possess some improvements, al-

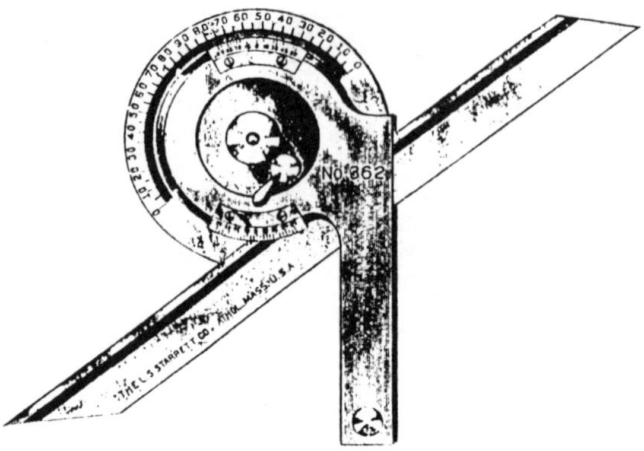

Fig. 1.—Universal Bevel Protractor with Vernier, No. 362.

though similar in a general way to kindred goods previously made. The Universal Bevel Protractor, with vernier, No. 362, Fig. 1, is made with verniers read-

Fig. 2.—Level Sight Attachments, with Sight Holes, One with Cross Wire.

ing five minutes, or one-twelfth of a degree, which are so placed with relation to the graduated half-circle as to make the Protractor readable by vernier in any po-

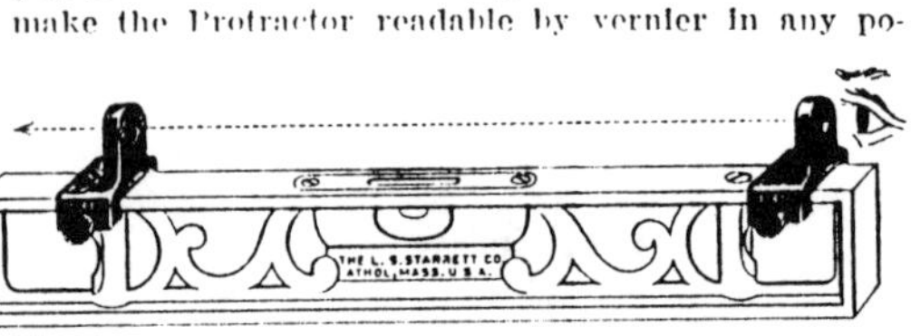

Fig. 3.—Level Sight Attachments in Position on a Level, No. 131.

sition. The protractor stock is 4 in. long, and has either a 7 or a 12 in. blade, ½ in. wide. With the 7-in. blade the weight is but 6 ounces. The disk is graduated in degrees from 0 to 90 each way, and rotates the entire circle on a central stud inside the case. The blade, clamped by an eccentric stud against the edge of the disk, may be slipped back and forth its full length, or turned at any angle around the circle and rigidly clamped at any point. Other features are the placing of the figures on vernier close to the lines for easy reading and the central locking nut. By a slight turn of the nut the Protractor is firmly held in position. There is also an attachment, not shown, for grinding short or long tapers to any degree or pitch. Figs. 2 and 3 show level sight attachments as sold and method of using on a level. They are so made as to be readily slipped on and off the top side of the Starrett iron levels. Both have sight holes, one with a cross wire to line accurately from top of and parallel with the level. Sighting through the holes enables one to use the common level for leveling a plot of ground from a fixed point at long range. The attachments are made to fit 6, 9, 12, 18 and 24 in., No. 132 levels. Figs. 4, 5 and 6 relate to lines of tool makers' callipers and dividers, made from round stock with legs drawn down, making them hard and stiff. The fulcrum stud is hardened, bows are extra strong, screw and nut accurately fitted, all finely finished and made with solid nut only. The No. 275 outside calliper, Fig. 4, is made in five sizes, 2 to 6 in., inclusive. The No. 274, inside calliper, Fig. 5, is offered in the same sizes and takes the same list. The dividers, No. 277, are also made in 2, 3, 4, 5 and 6 in. sizes, and listed uniformly with the Nos. 275 and 274.

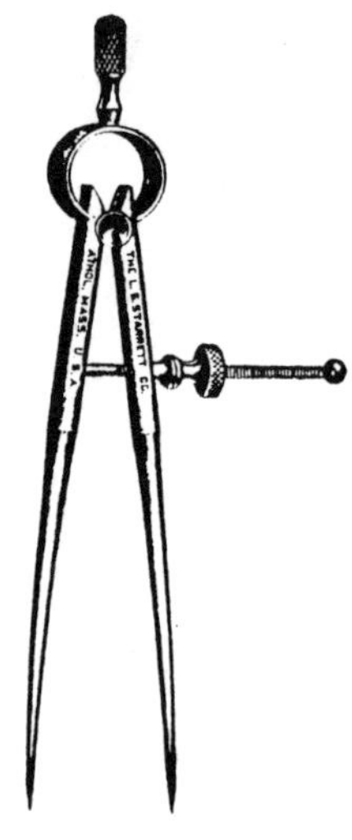

Fig. 6.—Dividers.

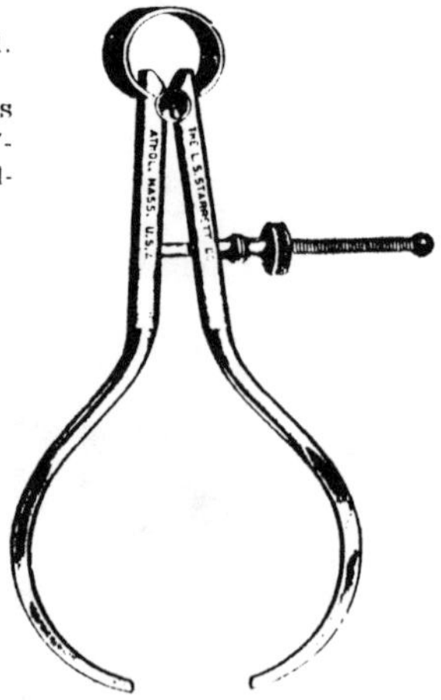

Fig. 4.—Outside Calipers.

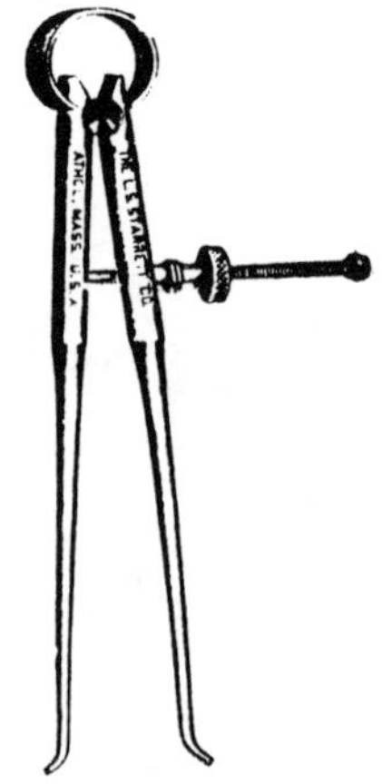

Fig. 5.—Inside Calipers.

1909

Fine Mechanics' Tools.

The L. S. Starrett Company, Athol, Mass., and 132 Liberty street, New York, which is constantly adding to an already large line of mechanics' fine tools, is offering those shown herewith. Fig. 1 illustrates a group of toolmakers' buttons, with screws and washers for jig work, No. 494. The buttons are hardened and ground to standard size, .400 x ½ in., and are used to locate holes to be chucked and bushed for jigs where positive accuracy is required. In use the jig piece is laid out, prick punched,

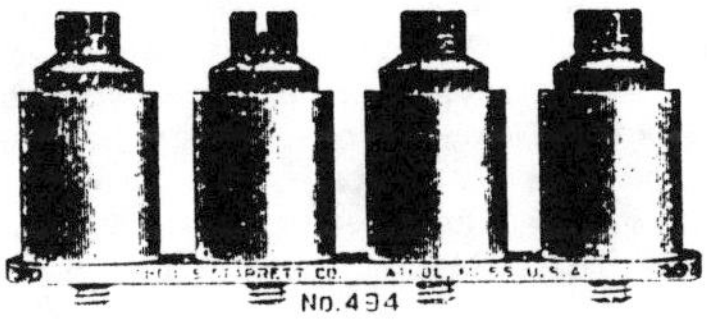

Fig. 1.—Toolmakers' Buttons with Screws and Washers for Jig Work.

drilled and tapped for button screws, smoothing off the burr raised by tapping. The buttons are then fastened on, the pieces strapped to an angle iron and placed on a surface plate, where, by the use of a surface gauge, hight gauge or other instruments, the buttons (with holes larger than the screws), are brought to the desired location. The angle iron with button pieces is then strapped to the lathe face plate, bringing one of the buttons to run true with the center, aided by a test gauge. This accomplished, the button is removed and the hole is chucked and reamed. The same procedure is repeated with the remaining buttons, until all the holes are chucked. The buttons are furnished screwed to a small plate, which serves as a convenient holder for them when

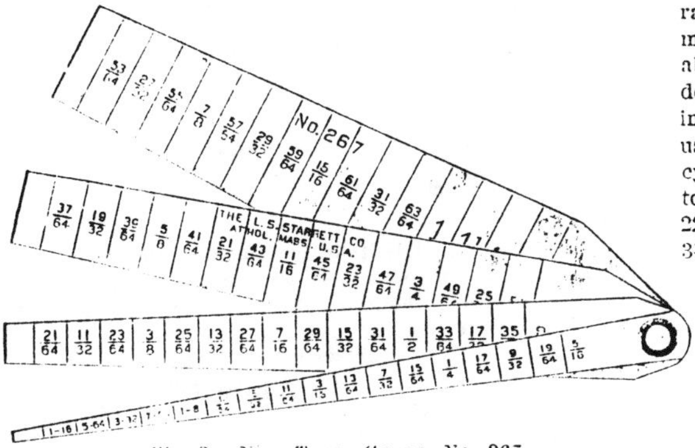

Fig. 2.—New Taper Gauge, No. 267.

not in use. Fig. 2 reproduces the new taper gauge, No. 267. The thin leaves are tapering, the width varying by 1-64 in. to every ¼ in. of length. They are graduated in ¼ in. and figured to read in fractions of an inch from 1-16 in. up to 1 1-16 in. The gauge is designed for brass and steel tube manufacturers and other users for inside measurements. It is likewise convenient for mechanics with which to measure the width of slots and size of holes in nuts drilled for tapping. Another service is

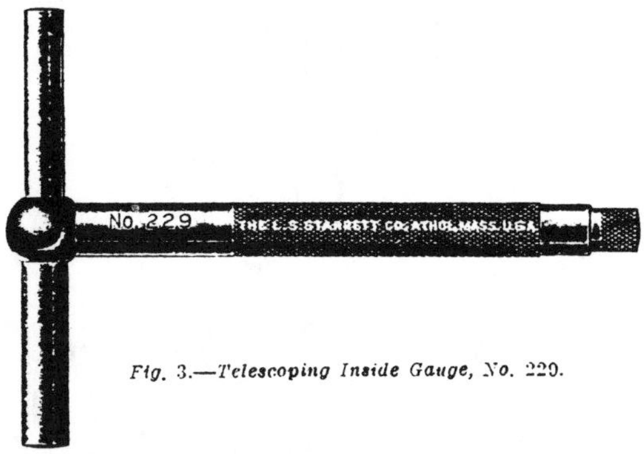

Fig. 3.—Telescoping Inside Gauge, No. 229.

for setting calipers to sizes within its capacity. The gauge is furnished in both plain and nickel finishes. Fig. 3 is a view of the telescoping inside gauge, No. 229. These are instruments from which the exact size of holes or slots may be taken by an outside caliper or micrometer, so that shrink and close or loose fits varying in thousandths or less may be made and measured. The tool is used by compressing the telescoping head and locking the plunger by a slight turn of the knurled screw in the end of the handle, then inserting the head inside the hole, when by releasing the lock the plunger will expand across the hole to a fit. The next operation is to lock the plunger by a slight turn of the screw, withdraw and caliper over the ends of the head with a micrometer, which will give the exact size of the hole. The ends of each telescope head are hardened and are made on a radius of the smallest hole it will enter. These instruments are recommended by the company as more reliable than ordinary leg calipers on account of the tendency of caliper legs to spring and of the points to catch in blowholes or other depressions. They also serve a useful purpose in fitting cylinders to holes or holes to cylinders. The gauges are made to enter holes from ½ to 6 in., the capacities being 229 A from ½ to ¾ in.; 229 B, ¾ to 1¼ in.; 229 C, 1¼ to 2⅛ in.; 229 D, 2⅛ to 3½ in., and 229 E, 3½ to 6 in.

1909

Engineers' Gauge.

The L. S. Starrett Company, Athol, Mass., and 132 Liberty street, New York, has put on the market the engineers' taper, wire and thickness gauge, No. 245, here illustrated. It is particularly serviceable to marine engineers, machinists and others requiring a set of gauges compactly arranged. The taper gauge shows the thickness by sixty-fourths to 3-16 in. on one side, and on the reverse side is graduated as a rule 3 in. of its length, reading in eighths and sixteenths of an inch. The wire gauge, English standard, shows on one side sizes numbered from 19 to 36, with two extra slots, one 1-16, the other ⅛ in., and on the reverse side contains the decimal equivalents expressed in thousandths. This gauge also has nine thickness or feeler gauge leaves, approximately 4 in. long, each having a thickness of 0.002, 0.003, 0.004, 0.006, 0.008, 0.010, 0.012, 0.015 and 1-16 in., all folded within the case, which is 4¾ in. long.

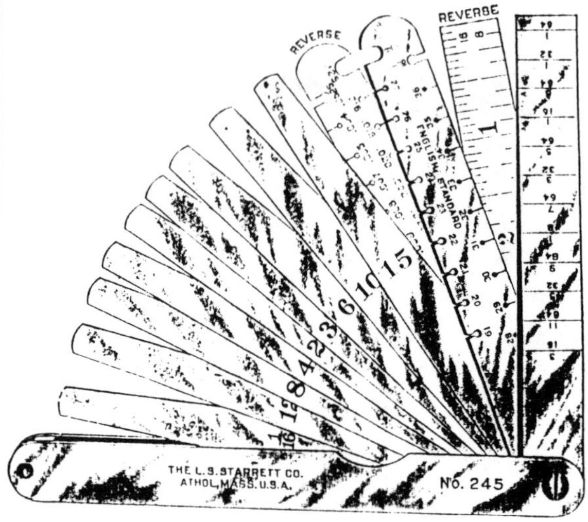

Engineers' Taper, Wire and Thickness Gauge.

Draftsmen's Protractor.

The L. S. Starrett Company, Athol, Mass., and 132 Liberty street, New York, has devised a high grade draftsmen's protractor No. 361, as here shown. It is made of sheet steel, nickel plated, graduated in degrees, figured to read from either right or left, and has vernier to read in 5 min. The three straight edges are graduated in inches and sixteenths, the long member measuring up to 6 in. The tool will lie flat on a surface and has a knurled lock nut which is serviceable for lifting. To reverse angles without resetting the user places the opposite straight part of the stock against the T square or straight edge of a drawing board, which will give the same angle right or left. By loosening the lock nut the parts turn freely, thus permitting of quick, easy adjustment to degrees when the sections may be relocked. The outer dimensions of the tool are 7¼ x 4 in., as set for packing singly in double pasteboard box or leather case.

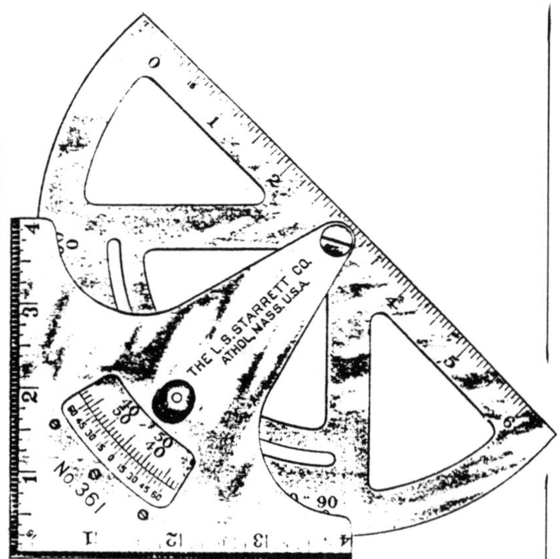

Draftsmen's Nickeled Stee. Protractor.

1909

Attachment for Combination Squares.

The L. S. Starrett Company, Athol, Mass., and 132 Liberty street, New York, has recently brought out an attachment for combination squares, No. 289. It is made to fit the 12, 18 and 24 in. blades of the company's Nos.

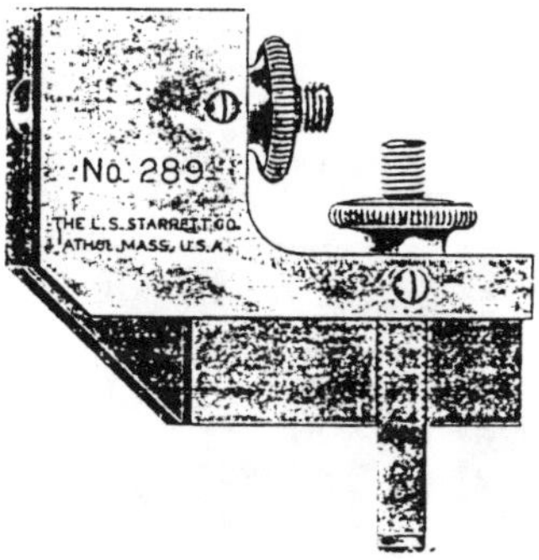

Fig. 1.—Attachment for Starrett Combination Square, No. 289.

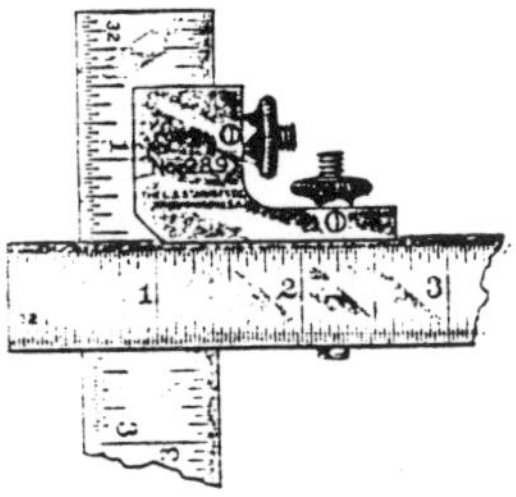

Fig. 2. — One Method of Using the Attachment.

11, 23 and 33 squares, and can be used with any of its regular rules as wide as 1 in. or with the flat steel square No. 21, for laying out keyseats, &c. Fig 1 is a reproduction of the attachment proper, Fig. 2 representing one of numerous uses, a variety of which are illustrated in the company's literature describing the attachment.

1909

Planer and Shaper Gauge.

The L. S. Starrett Company, Athol, Mass., and 132 Liberty street, New York, is making the planer and shaper gauge, No. 246. It is made of steel, drop forged, and is designed for obtaining different hights on a planer. The gauge will measure from ½ to 5½ in. It is likewise serviceable on other tools; for example, on machines where slots are being milled. The gauge can be inserted in the slot and by sliding the block on the wedge a perfect fit is obtained. Then the tool may be withdrawn and the exact measurement ascertained by using a micrometer. When an accurate dimension is desired the tool may be first set by the micrometer and used as a

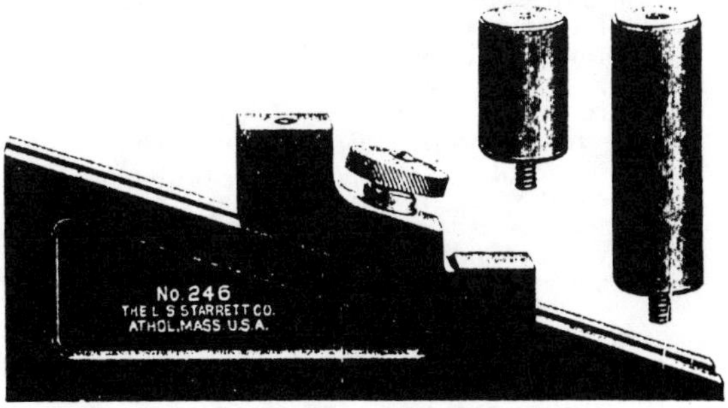

Planer and Shaper Gauge.

1909

standard gauge. The instrument is case hardened, finely finished and useful for planer and milling machine work.

STARRETT'S PATENT RATCHET WRENCHES

EFFECTIVE JANUARY 31, 1920

No. 443

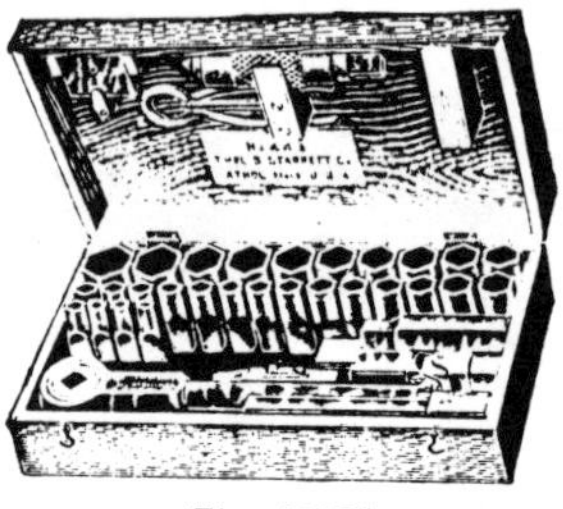

Fig. 4886A

For engineers, machinists, automobilists, the garage, etc.

In places difficult of access or in cramped quarters where a swing through a long arc is impossible, the ordinary monkey wrench or S wrench is out of the question.

The twenty-seven small engravings around the outside of the illustration at the right, represents the hexagon steel sockets furnished in set, varying in sizes by 32nds. From $\frac{1}{16}$-inch to 1-inch, also $1\frac{1}{32}$, $1\frac{3}{32}$, $1\frac{5}{32}$ and $1\frac{9}{32}$-inches. The set has also two square steel sockets, $\frac{13}{32}$ and $\frac{21}{32}$-inch.

The wrench consists of a ratchet with reversible pawl and a long wrench handle. With the wrench is furnished an extension to reach into otherwise inaccessible places. Also a universal joint for turning nuts or bolts when it is impossible to get the wrench on at right angles to the ends of the bolt; a spark plug socket for use on automobiles; a drilling attachment which takes standard square shank drills from $\frac{1}{8}$-inch to $\frac{1}{2}$ inch diameter; a screwdriver with reversible end; together with several adjustments to go with the drilling attachment.

This ratchet wrench is of particular value to engineers and chauffeurs who have to work about machinery crowded into small space or around hot engines. The sockets for the wrench will turn nearly any standard hexagon nut or bolt. Will not mar finished surfaces.

COMPLETE SET

Price, No. 443A............each **18.00**

WITHOUT DRILL FIXTURE

Price, No. 443B............each **14.40**

Sent complete, unless otherwise ordered.

EXTRA PARTS

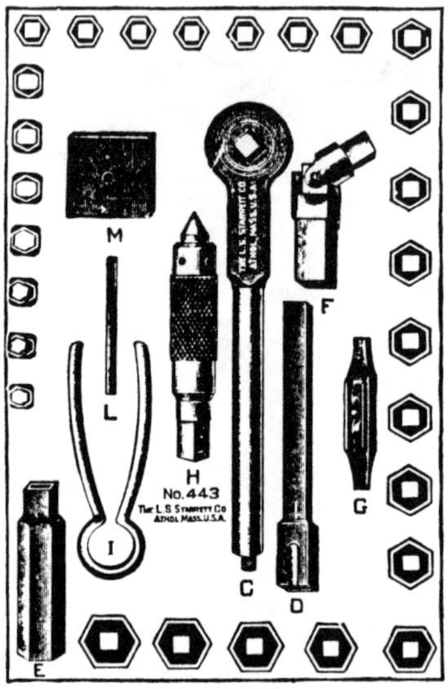

Fig. 4886B

No. 443C

Ratchet wrench, with reversible pawl.
Price........................each **4.20**

No. 443D

Extension to fit part A.
Price........................each **.90**

No. 443E

Spark plug socket.
Price........................each **.40**

No. 443F

Universal joint.
Price........................each **1.80**

No. 443G

Screwdriver.
Price........................each **.50**

No. 443H

Drilling attachment.
Price........................each **3.60**

No. 440I

Holder or friction wrench.
Price........................each **.30**

No. 440L

Drift pin.
Price........................each **.15**

No. 440M

Thrust plate.
Price........................each **.20**
Sockets, except Spark Plug.... " **.25**

618 **MACHINISTS SUPPLY CO.**

STARRETT PATENT EXTENSION PLIERS

No. 240

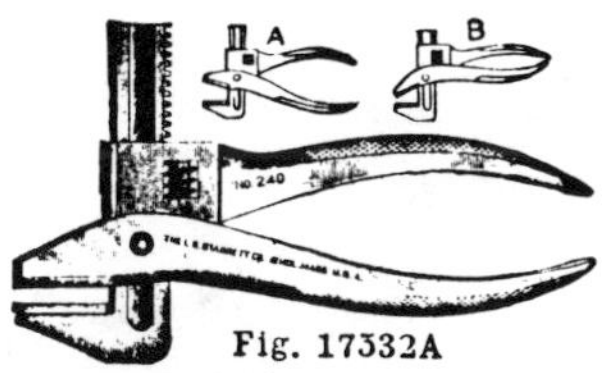

Fig. 17332A

Common pliers having a jaw and handle in one continuous piece, and the two pieces jointed, do not give satisfactory results on large work.

It is evident that when the jaws of ordinary pliers are wide open they spread the handles so far apart it is difficult to get a good grip on them.

And unless the surfaces of the piece to be held in the jaws happen to come parallel with them, it will be impossible to get a good grip at the jaws.

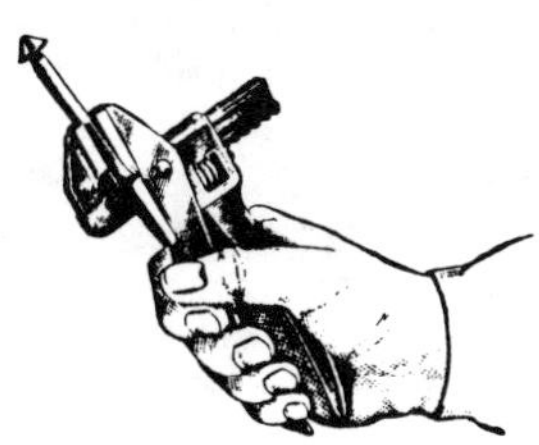

To overcome these difficulties, we offer the Starrett Extension Plier.

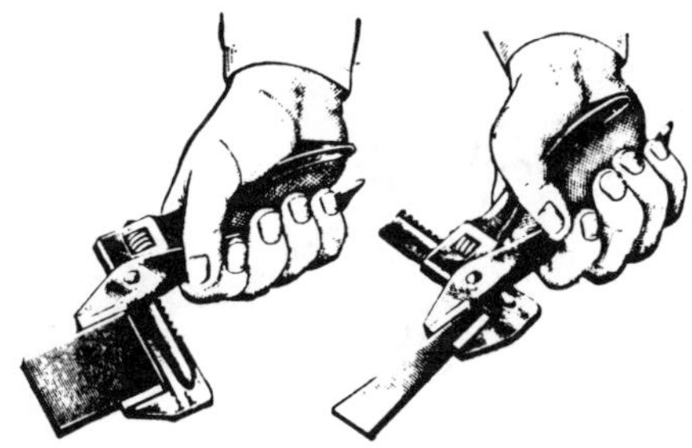

The distinctive features of these pliers is that one of the jaws may be adjusted to suit any piece, straight or tapering, from zero up to $1\frac{1}{4}$ inches.

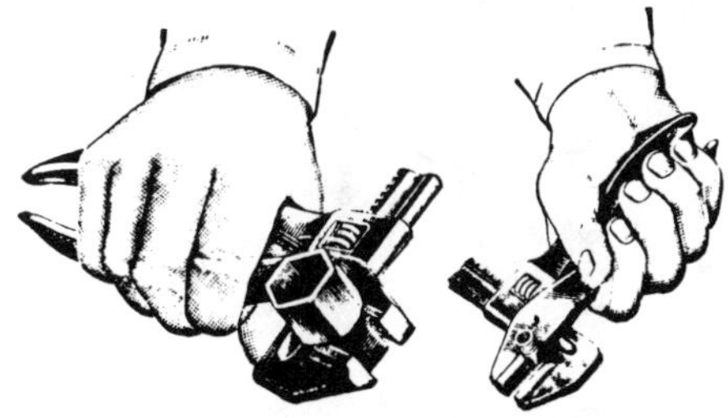

These pliers will turn any nut, pipe or bolt within their limits, and will grip a headless nail, bolt or wire.

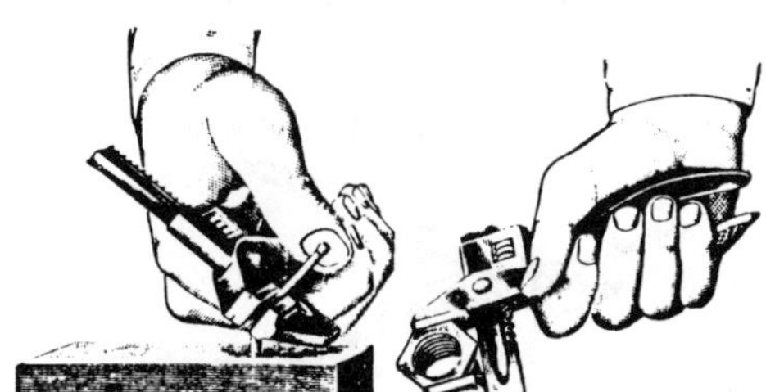

For working about an automobile, or in fact any machinery having small parts, this tool will be found one of the handiest in the whole kit. It is also useful about the home.

Style of Finish........	Plain	Nickeled
Price.............each	2.40	2.70

STARRETT'S TOOLS

EFFECTIVE JANUARY 31, 1920

No. 437 CUT NIPPERS

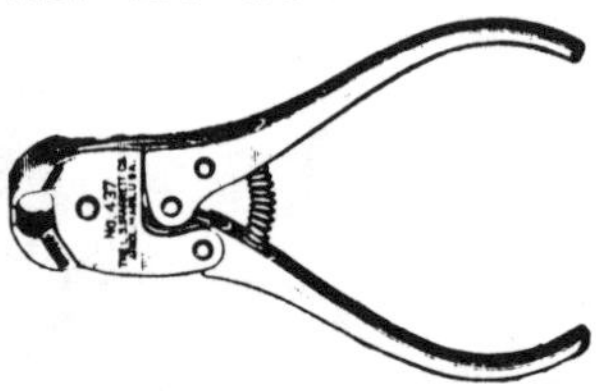

Fig. 18965A

Similar to No. 199, except that it is a heavier and stronger tool. Wire can be cut at extreme end of the jaws. Nippers open $\frac{3}{32}$ inch. Length of nippers over all, 5½ inches.

Price, Cut Nippers..........each **3.00**
" Jaws..............per pair **1.80**
" " ....each **.90**

No. 235 ADJUSTABLE JAW TILE CUT NIPPERS

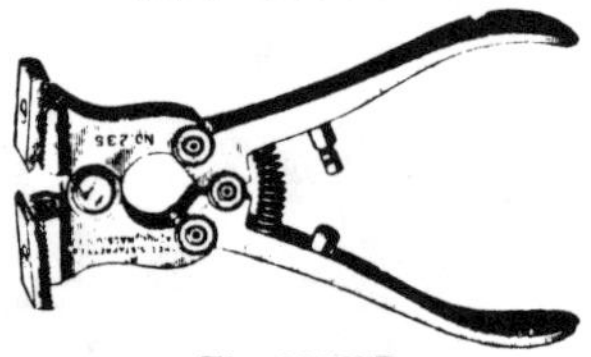

Fig. 18965B

These nippers are the same as No. **1**, except the frames are cut out to allow the jaws to be adjusted for wide opening as shown in illustration, thus fitting them to be used in cutting tile, for which purpose they are recommended. Made in two sizes.

Price, 5½-inch..............each **3.30**
" 7 " .... " **3.90**

No. 276 CUT PLIERS

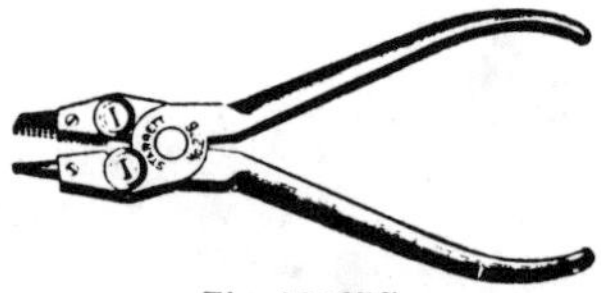

Fig. 18965C

Combination round and flat nose pliers. Circular rotating wire cutters with sharp edges.

Price......................each **2.10**

EXTRA PARTS

Price, Round Points.........each **.20**
" Flat " .... " **.25**
" Cutters...........per pair **.30**

No. 199 CUT NIPPERS

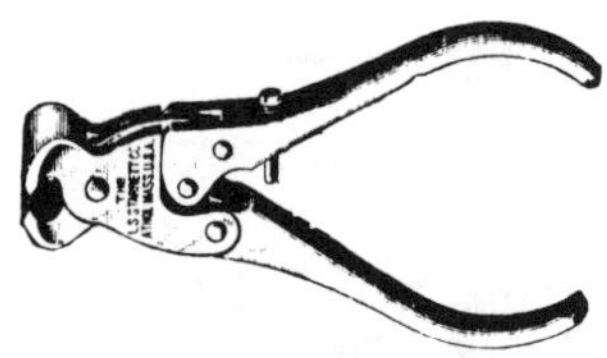

Fig. 18965D

This nipper is used for cutting bicycle spokes, etc. The cutting jaws conform to the inside of a bicycle rim, and will cut off the spokes just as close as required. All parts are interchangeable, therefore, if a jaw should break a new one can be obtained.

Price, 5-inch................each **2.70**
" Jaws..............per pair **1.20**
" " ....each **.60**

No. 1 ADJUSTABLE JAW CUT NIPPERS

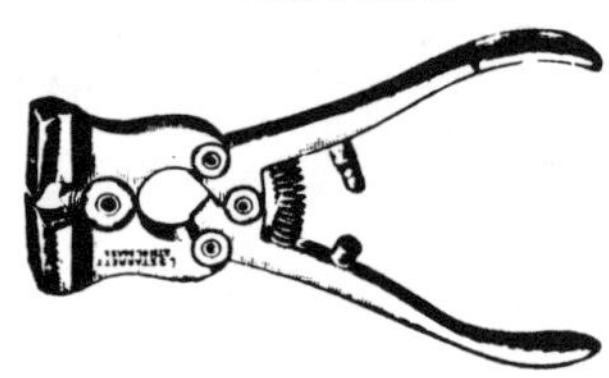

Fig. 18965E

These nippers are made from a high grade of steel, nicely tempered. Those warranted to cut music wire have their cutting edges ground to a short, steep bevel, while those of common use have their cutting edges ground more acute, work easier, and are preferable for cutting softer wire or for general use.

Price, 5½-inch *M.........each **3.30**
" 5½ " **C.......... " **3.30**
" 5½ " †B.......... " **3.30**
" 7 " M, C or B.. " **3.90**

*M (for music wire). **C (for common use). †B (for bicycle use).

JAWS

Fig. 18965F

Price, Extra Jaws, either M, C or B, Which Should be Designated as Above................per pair **.60**
Price, Screws for Jaws....per dozen **.30**
" Splines " " .... " **.30**

J. STEVENS & CO.

J. STEVENS & CO. Chicopee Falls, MA

Founded by Joshua Stevens (1814-1907), William B. Fay (?- 1896) and James Taylor (1829-1904). The partnership was formed in 1864 to make a small pocket pistol which had been patented by Stevens on September 6, 1864 (No. 44123). In 1865 a line of spring calipers was added. By 1867, the company had 20 employees making pistols, dividers and calipers.

The census of 1870 shows Stevens making 7,000 pistols with a value of $20,000, 5,000 calipers and dividers with a value of $2,500 and 4,000 levels with a value of $8,000 in that year. Total employment in 1870 was 58, with an annual payroll of $20,000.

In 1880, Stevens was offering forged spring calipers, helical coil spring calipers (based on the T.C. Page and G.W. Hadley patent of February 8, 1870, No. 99698), Hathaway's patent combination gauge (R. Hathaway patent of December 4, 1866, No. 60186) and "Reliable" right and left hand screw calipers (C.A. Fairfield patent of July 21, 1863, No. 39281). Business was good enough in 1880 to carry out expansion plans. The old grist mill which had served as the first plant was torn down and a new building, with twice the floor space, was put up on the site.

The Stoddard caliper and divider, patented by Oscar Stoddard on August 27, 1872 (No. 130879), a double-lip countersink, patented May 12, 1868 (No. 77819), by C. Krebs, and cutting nippers, patented October 30, 1883 (No. 287589), by Joshua Stevens, were offered by 1883.

In 1885, Stevens advertised outside wing, firm joint calipers, the new "Exact" surface gauge (patented by Charles P. Fay September 23, 1884, No. 305515 and assigned to Stevens), forged spring calipers and dividers with patent washer (Oliver Warfield patent of November 20, 1883, No. 288961), and a patent combination gauge (formerly called Hathaway's combination gauge).

Edward R. Markham, in a 1902 article published in *American Machinist* magazine (see Appendix IV), notes that Stevens began using the "Richardson" quick adjusting nut for calipers about 1885. This was the Frederick Thomas patent of March 22, 1881 (No. 239072), which had been assigned to C.F. Richardson of Athol, MA. Markham, who had worked at C.P. FAY & CO. until Fay sold out in 1887, joined Stevens that year to run its fine tool department.

In 1886, the partnership was dissolved and the company was incorporated as:

J. STEVENS ARMS & TOOL CO. Chicopee Falls, MA

Capital for the new corporation was set at $40,000. The capital amount was set by valuing the assets of J. STEVENS & CO. at $40,000 and selling them to the new J. STEVENS ARMS & TOOL CO. J. Stevens and William B. Fay had 120 shares each, George S. Taylor had 10 shares, James Taylor 110 shares and I.H. Page 40 shares.

In 1888, Stevens introduced a new line of detachable bow spring calipers and dividers made under Joshua Stevens's patents of March 9, 1886 (No. 337531), October 26, 1886 (No. 351474) and February 14, 1888 (No. 377867 & 377868). The square leg version was named the "Ideal" and the flat leg version "Leader". Stevens continued to offer the "Reliable" calipers, firm joint calipers with or without a wing, the "Exact" surface gauge and the combination gauge.

The following announcement was made in the 1888 catalog.

> "Having been the pioneer manufacturers in this country of spring calipers and dividers, we take pleasure in announcing, that we have placed upon the market a New Style with Detachable Bow and Quick Adjusting Nut, and all practical mechanics who have examined same, speak in highest

> terms of it. We shall continue the manufacture of our large line of forged calipers, dividers, and other fine machinist tools. [The forged calipers were made as late as 1895]
>
> We shall have our Cheap Grade, called the "Leader" Dividers and Calipers ready in a short time."

In 1890, the "Stoddard" caliper and divider was offered in a new version which incorporated changes in the fine adjustment mechanism patented July 22, 1890 (No. 432578), by Joshua Stevens. This was to be Stevens' last tool development. Other new offerings in 1890 included transfer calipers based on the Thomas C. Hendey patent of August 10, 1886 (No. 346915), parallel dividers based on the George M. Pratt patent of February 23, 1886 (No. 336868), an adustable scratch gauge with fine adjustment, and a line of steel rules.

It seems strange that the steel rules, such a commonly found tool from other manufacturers, are so rare. None have been located to date. It may be that Stevens was selling some other company's rules and, despite the catalog illlustration, they were not marked with the Stevens name.

In January, 1896, I.H. Page bought the shares held by Stevens and Taylor and assumed control of the company, acting as president and treasurer. Charles P. Fay, son of the recently deceased William B. Fay, left L.S. STARRETT to became vice president and superintendent. G.E. Page was elected assistant treasurer and secretary. Fay began his career at J. STEVENS & CO., leaving in 1883 to join L.S. STARETT for a year, then operating his own company, C.P. FAY & CO., from 1884 until rejoining L.S. STARRETT in 1887.

The following announcement appeared in the J. Stevens Arms & Tool Co. catalog of 1898:

> Chas. P. Fay, so well known as an inventor and designer of calipers, dividers, and fine tools has acquired the interest of his late father in the J. Stevens Arms & Tool Co., where he learned his trade and brought out his first calipers and dividers. Our tool department is under his supervision.

Tools listed in the 1898 catalog were: lock joint calipers and dividers, Fay's patent firm joint calipers, bench surface gauge (same as Hathaway's patent of December 4, 1866, No. 60186), IDEAL surface gauge (the EXACT surface gauge with a new name), micrometer surface gauge (Edward Markham's patent of September 30, 1890, No. 437308), Fay's patent tool maker's surface gauge, extension dividers, combination dividers, centering disk, Fay's patent (October 5, 1897, No. 591348) IDEAL spring calipers and dividers, Fay's Patent LEADER calipers and dividers, scribers, depth gauges, bicycle spoke nippers, parallel dividers, trammel points, universal bevels, adjustable scratch gauges, tool-maker's patent center squares (no patent has yet been found for this tool), Fay's patent (January 10, 1899, No. 617501) center indicators, tool-maker's clamps, micrometer protractors, Fay's patent (January 1, 1901, No. 665316) angle indicators, shot plumb-bobs, center punches, and bull-dog wrenches.

It is notable that the Fay's patent tool maker's surface gauge appears almost identical to the L.S. STARRETT tool maker's surface gauge introduced in 1893 and covered by a patent issued jointly to Charles P. Fay and L.S. Starrett on March 17, 1896 (No. 556703). It appears that Fay had an arrangement with Starrett to use some of the patents Fay had received while working at L.S. STARRETT.

Catalog D, issued by the SAWYER TOOL CO., Fitchburg, MA, in 1900, included the complete line of Stevens calipers and dividers. Sawyer, at that time, did not make calipers so the arrangement was a good one for both companies.

The departure of the J. STEVENS ARMS & TOOL CO. from the tool business was noted in the following announcement from *Iron Age* magazine of November 19, 1903.

The L.S. STARRETT CO., Athol, Mass., have bought the entire fine Machinist's Tool business of the J. STEVENS ARMS & TOOL CO., Chicopee, Falls, Mass., including manufactured stock, stock in process, special tools for manufacturing and good will. The Firearms part of the Stevens business having grown so rapidly that the company found it necessary either to erect more buildings, curtail their product, or dispose of some department, and the L.S. STARRETT CO. having made a satisfactory offer for the Tools, the sale of this portion of the company's business was closed on the 13th inst., the transfer to take place on the 20th. The J. STEVENS ARMS & TOOL CO., will immediately occupy the space this vacated in increasing their Firearms output, making them one of the largest manufacturers in the line in the country. They will also enlarge their production of Automobiles, among the makers of which they are taking a leading place.

A 1904 supplement to the number 17 Starrett catalog lists the #174 tap wrench, #195 double-lip countersink and the #199 cut-nipper which are obviously tools previously made by Stevens.

For a detailed look at all the Stevens tools offered in 1890, a reprint of the 1890 J. STEVENS ARMS & TOOL CO. catalog is available from: Martin J. Donnelly Antique Tools, P.O. Box 281, Bath, NY, 14810.

First plant of J. Stevens & Co.

Joshua Stevens
1814-1907

Both photos are from Stevens Pistols and Pocket Rifles, by Kenneth L. Cope

Stevens' New "Exact" Surface Gauge.

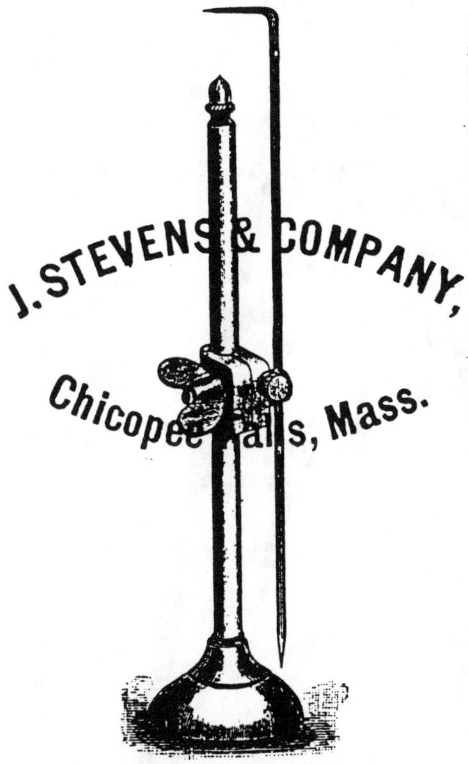

No. 59.

PATENT APPLIED FOR.

Price per dozen, small size, - - -	$24 00
Price per dozen, large size, - - -	30 00

1882

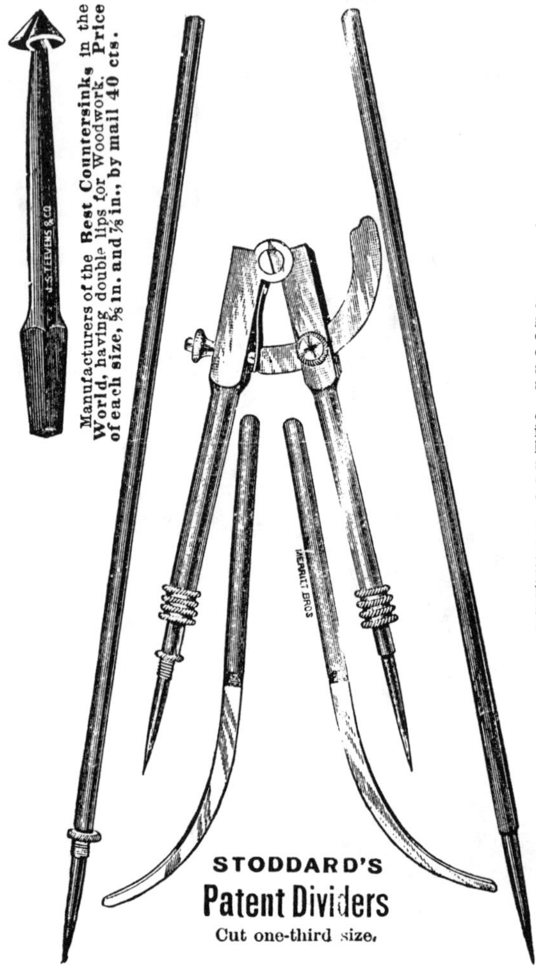

THE STODDARD

CALIPER AND DIVIDER.

The Best and Most Handy Tool for a Carpenter Made,

and no one who has once used them will go without them. As the cut shows, it consists of a pair of heavy hinged divider legs swinging on a slide with spring attached, which by the two screws can be regulated to any width.

To these legs are fitted two sets of divider points, one pair 5½ and one 14 inches long. One of each pair of divider legs has a reversible point which unscrews, and by being reversed a pencil can be placed in same and used for drawing circles.

Also, we have a pair of caliper legs which fit the same stock, and as they will caliper 16 inches, are invaluable for turning pulleys, and, in fact, all large caliper work in a shop. With the long divider legs a 5-foot circle can be struck. These tools are made of brass and finely finished.

Made in two Styles, Plain or Nickeled.

Price of plain, postage paid, $3.50.
Price of nickel, " " 4.00.

For Illustrated Catalogue of Fine Tools, send to

J STEVENS & CO.,

P. O. Box 2300,

CHICOPEE FALLS, MASS.

1885

No. C 392

PATENT HELICAL COIL SPRING CALIPERS.

2 inch	Each,	$0 75
2½ "	"	75
3 "	"	75
4 "	"	1 00
5 "	"	1 00

No. C 393

PATENT HELICAL COIL SPRING DIVIDERS.

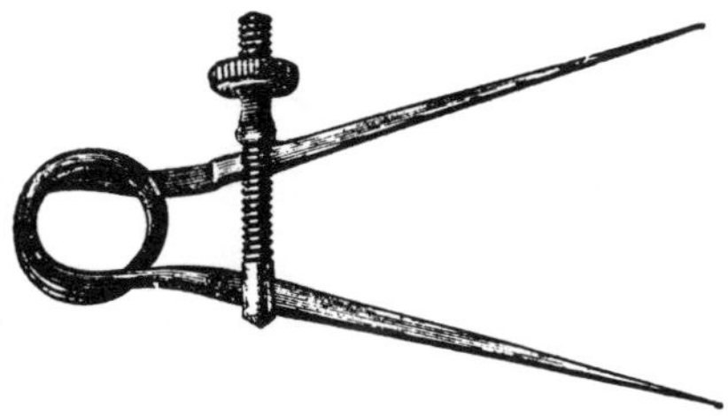

PRICE EACH.

3½ inch	$0 75
4 "	1 00

No. C 394

FIRM JOINT CALIPERS.—Outside and Inside.

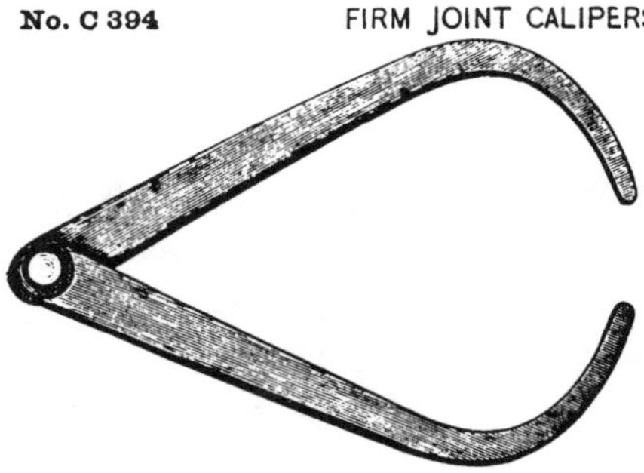

Outside.

No. C 395

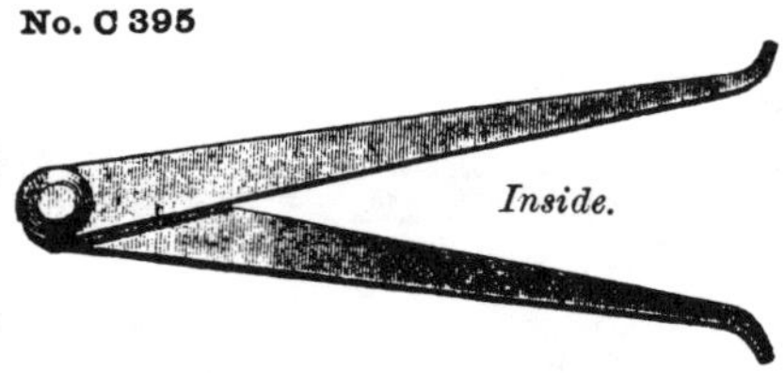

Inside.

Prices for Nos. C 394 and C 395.

Inches.	3	4	6	8	10	12
Price..	$0 40	$0 50	$0 75	$0 85	$1 00	$1 25

1885

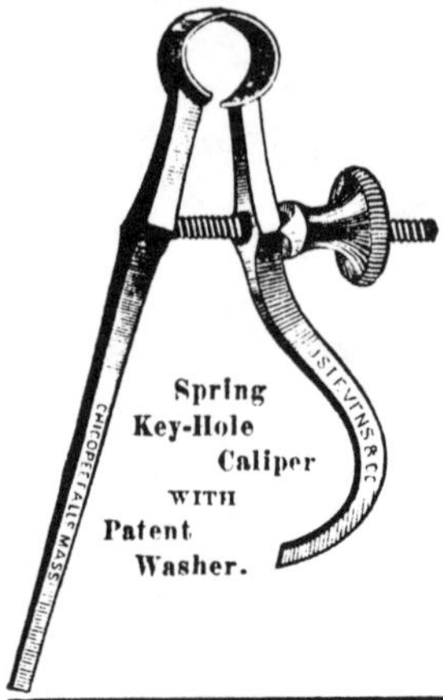

J. STEVENS & CO.,

P. O. BOX 1200,

Chicopee Falls, - Mass.,

MANUFACTURERS OF

Spring Caliper and Dividers,

FINE MACHINIST TOOLS
AND FIRE ARMS.

Our SHOOTING GALLERY RIFLE

IS THE FAVORITE EVERYWHERE.

SEND FOR ILLUSTRATED CATALOGUE.

Patent Double Lip Countersink.

The FASTEST and BEST CUTTING TOOL MADE.

J. STEVENS & Co.,

P. O. BOX 1200,

Chicopee Falls, Mass.,

—MANUFACTURERS OF—

SPRING CALIPER AND DIVIDERS,

Fine Machinist Tools and Fire Arms.

Our SHOOTING GALLERY RIFLE

IS THE FAVORITE EVERYWHERE.

SEND FOR ILLUSTRATED CATALOGUE.

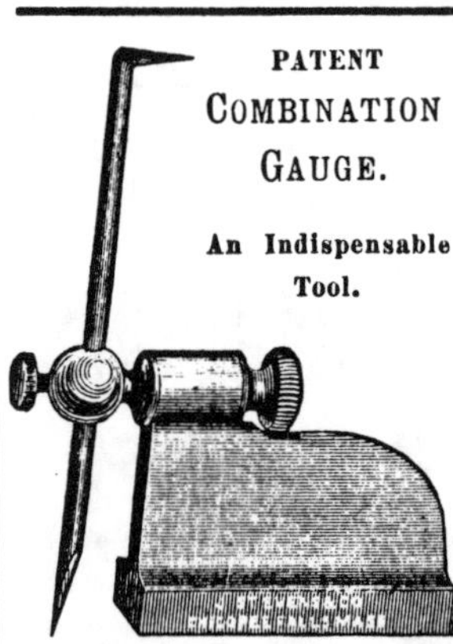

Cut shows only one of the positions in which it may be placed.

J. STEVENS & CO.,

P. O. BOX 1200,

Chicopee Falls, - Mass.,

MANUFACTURERS OF

Spring Caliper and Dividers,

FINE MACHINIST TOOLS
AND FIRE ARMS.

Our SHOOTING GALLERY RIFLE

IS THE FAVORITE EVERYWHERE.

SEND FOR ILLUSTRATED CATALOGUE.

All 1885

SPRING CALIPERS AND DIVIDERS

FINE MACHINISTS' TOOLS & FIREARMS

J. STEVENS & CO.,
CHICOPEE FALLS, MASS.

P.O. BOX 1200. Manufacturers of

Our Shooting Gallery Rifle is the Favorite Everywhere.

STEVENS' NEW "EXACT" SURFACE GAUGE.
The Best Gauge Made.

SEND FOR ILLUSTRATED CATALOGUE

1885

OUTSIDE CALIPERS.

Stevens' Fine Adjusting and Transfer.

No. 100.

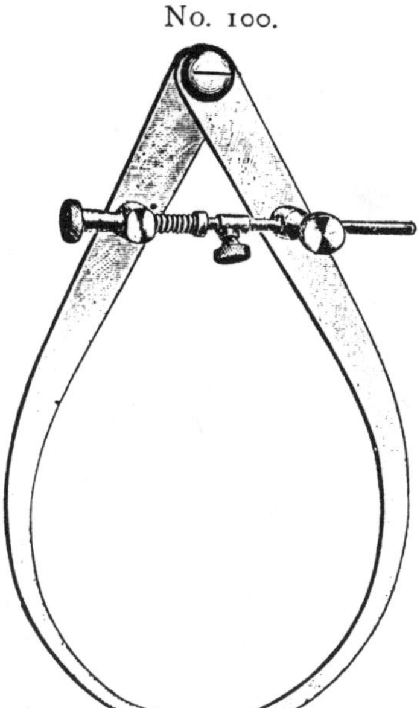

Inches..	6	8	10	12
	$1.50	1.75	2.00	2.25

These tools combine in themselves a Fine Adjusting Firm Joint and Spring Caliper which can be instantly opened or closed and held very rigid in any position.

By means of the Nut it can be adjusted to very fine measurement.

The adjusting power of the screw is far greater than any mechanic will want to use on any ordinary work. The Transfer Stop allows the tool to be set at any required point, and then opening or closing as the case may be, the tool can be removed from work and then set to the Stop. This feature is especially valuable on the Inside Caliper as a recess on the inside of a piece of work may be calipered and the Transfer Stop set against Stud, calipers closed and removed from work and then set to same size again. To show the usefulness of this tool, see how thick the glass is in your window, as this was what first called our attention to make a Transfer Caliper.

INSIDE CALIPERS.

Stevens' Fine Adjusting and Transfer.

No. 105.

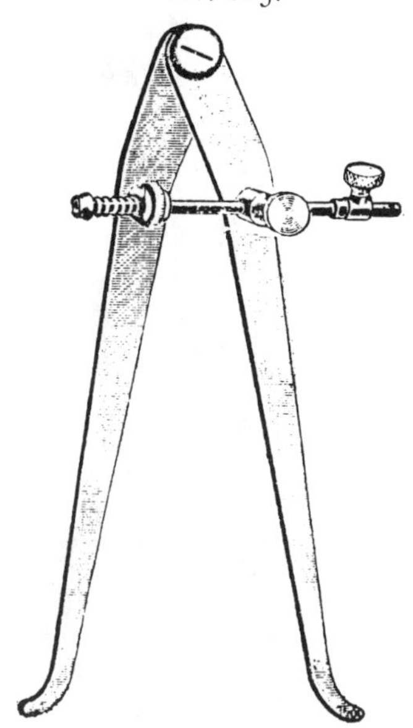

Inches,	6	8	10	12
	$1.50	1.75	2.00	2.25

1890

PATENT SPRING DIVIDERS.

Stevens' Ideal, No. 66.

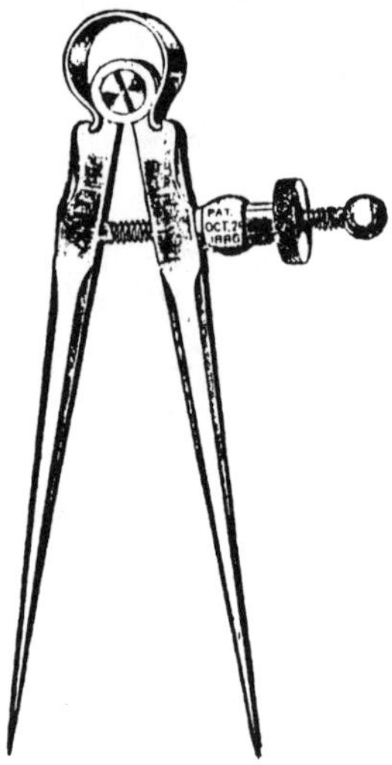

3 in. $1.00
4 " 1.25
5 " 1.25
6 " 1.50

PATENT SPRING KEY-HOLE CALIPERS

Stevens' Ideal, No. 68.

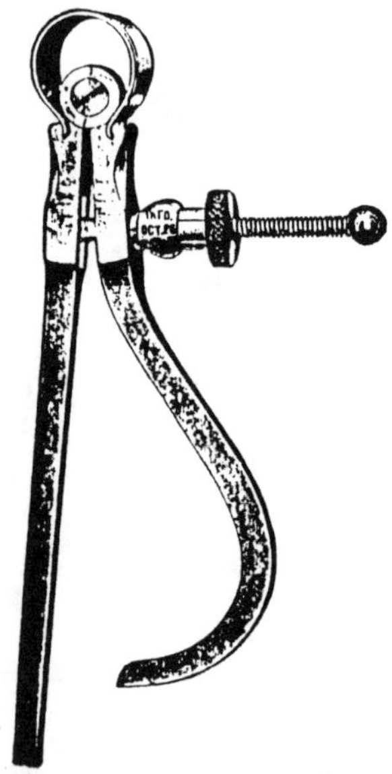

3 in. $1.15
4 " 1.50

Diagram of the Stevens' Ideal

PATENT NUT AND WASHER COMBINED.

Showing Sectional Parts.

PATENT SPRING DIVIDERS.

Stevens' Ideal, No. 67.
With Thumb Attachment.

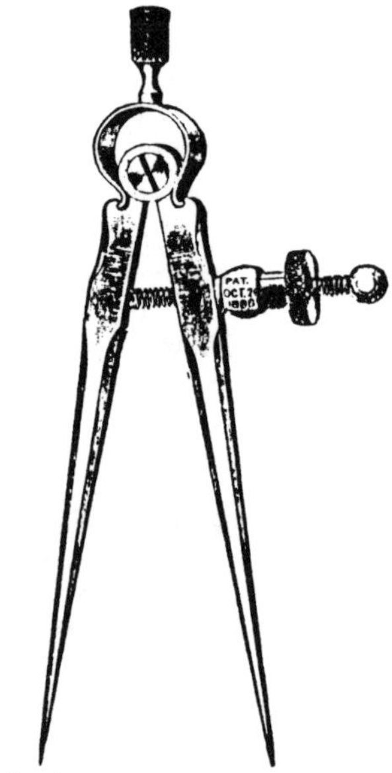

Inches..	3	4	5	6
	$1.15	1.40	1.50	1.75

DUPLICATE PARTS.

Leg (either one) . . . . 35c.
Spring . . . . 25c.
Joint Stud . . . . 25c.
Screw . . . . 10c.
Tip for Screw . . . . 10c.
Ideal Nut . . . . 25c.
Hood for Ideal Nut . . . . . . 10c.
Washer " " . . . . . . 05c.
Thumb Attachment . . . . . . 15c.
Solid Nut . . . . 15c.
Joint Screw (on first model) 10c.
All parts are interchangeable.

PATENT SPRING INSIDE CALIPERS.

Stevens' Leader, No. 72.

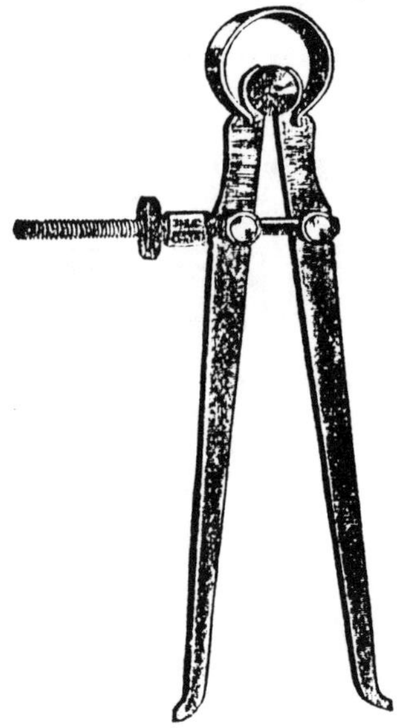

Inches....	4	5	6
	75c.	80c.	85c.

PENCIL DIVIDERS.

Stevens' Ideal, No. 69.

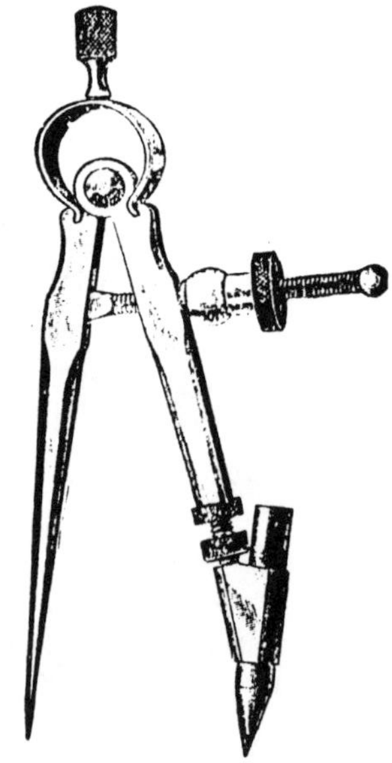

3 inch only $1.75

1890

PATENT SPRING CALIPERS.

Stevens' Leader, No. 70.

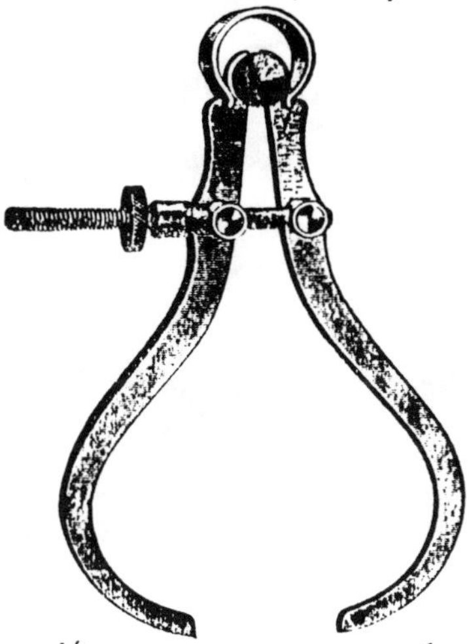

Inches..	2½	3	4	5	6	8
	$0.65	.70	.75	.80	.85	1.00

PATENT SPRING SCREW THREAD CALIPERS.

Stevens' Leader, No. 71.

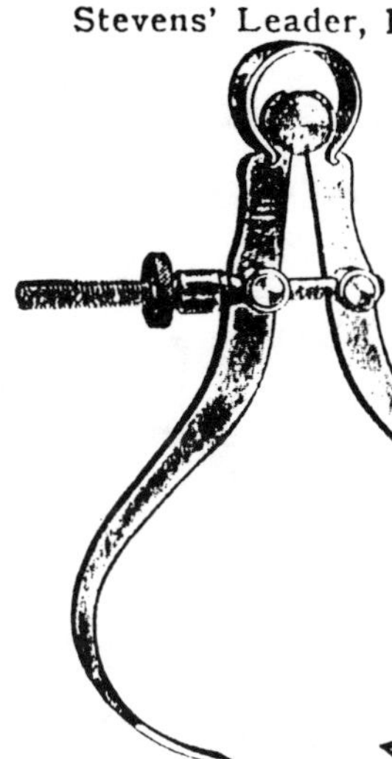

3 in...75c. | 4 in..90c. | 5 in..$1.00

PATENT SPRING KEY-HOLE CALIPERS

Stevens' Leader, No. 73.

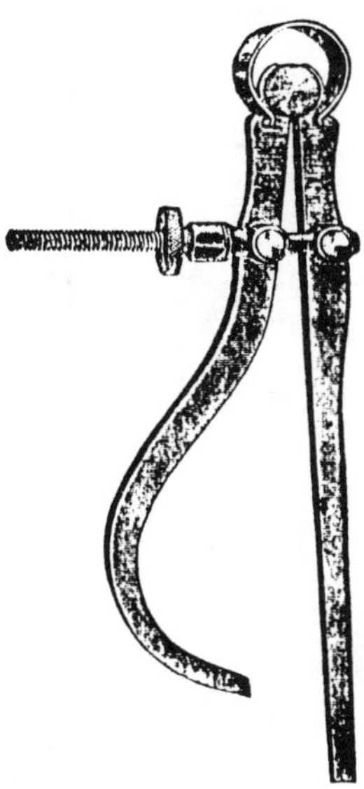

Inches......	3	4
..	70c.	75c.

PATENT SPRING DIVIDERS.

Stevens' Leader, No. 74.

Inches..	2½	3	4
	$0.65	.70	.75
Inches..	5	6	8
	$0.80	.85	1.10

PATENT SPRING DIVIDERS.

Stevens' Leader, No. 75.

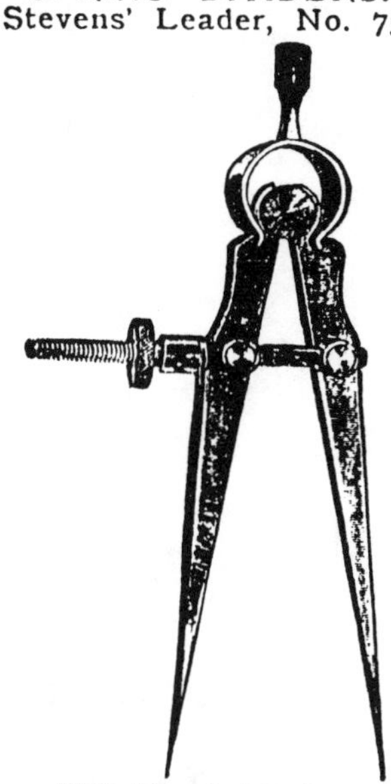

With Thumb Attachment.

Inches..	2½	3	4
	$0.80	.85	.90
Inches..	5	6	8
	$0.95	1.00	1.25

Leader Duplicate Parts.

Leg (either one)	25c.	Ideal Patent Nut	25c.
Spring	25c.	Hood for Ideal Nut	10c.
Joint Stud	25c.	Washer " "	05c.
Screw	10c.	Thumb Attachment	15c.
Solid Nut	15c.	Screw Tip for Ideal Nut	10c.
Washer	10c.	Screw Stud	10c.

All parts are interchangeable.

1890

144 THE MACHINISTS' SUPPLY CO., CHICAGO.

Stevens' Firm Joint Calipers.

Made of Very Finest Quality of Steel, and the Best Goods in the Market. All Highly Polished.

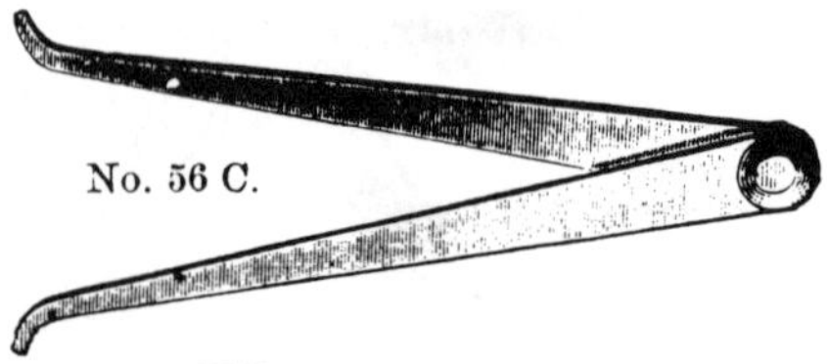

No. 56 C.

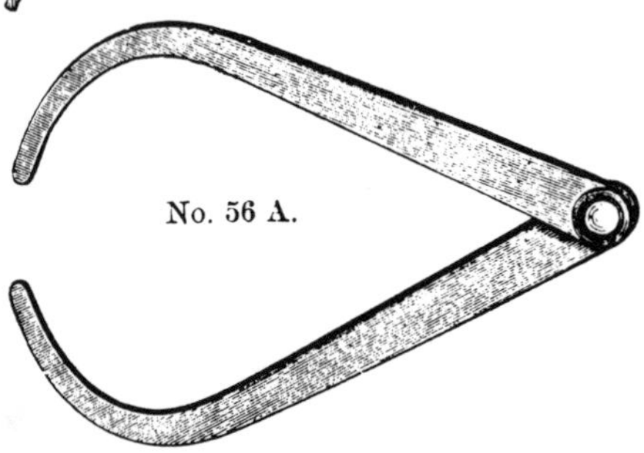

No. 56 A.

Prices, for Either Inside or Outside.

In.	Each.
3	$0 35
4	45
5	55
6	65
8	75
10	85
12	1 00

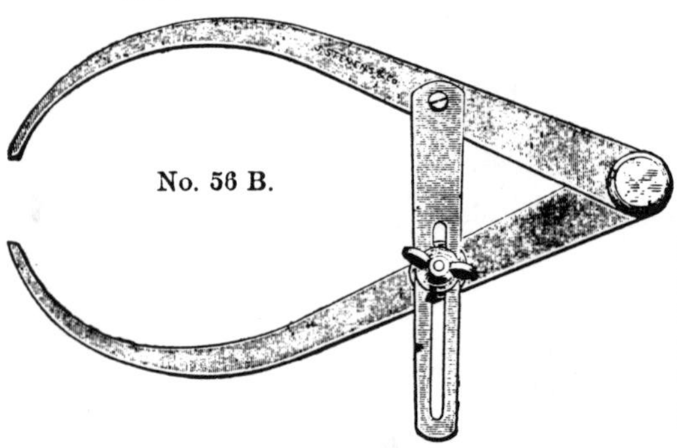

No. 56 B.

With Wing and Set Screw.

Prices, for Either Inside or Outside.

In.	Each.
8	$1 00
10	$1 10
12	1 25
14	1 50
16	1 75
18	2 10
20	2 50

STEVENS' NIPPERS.

NO. 90. PRICE LIST.

5 inch, each	$1 50	Extra Jaws, each	$0 50
6 " "	1 75	" " "	50

Both 1891

CALIPERS.

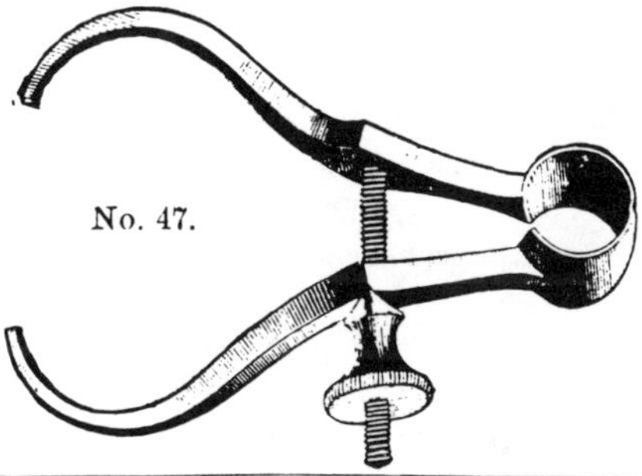

No. 47.

Steven's Spring Calipers.

EXTRA FINE STEEL AND TEMPER.

2 inch.............Price, each $1 00
2½ " .... " " 1 00
3 " .... " " 1 00
4 " .... " " 1 25
5 " .... " " 1 25
6 " .... " " 1 50

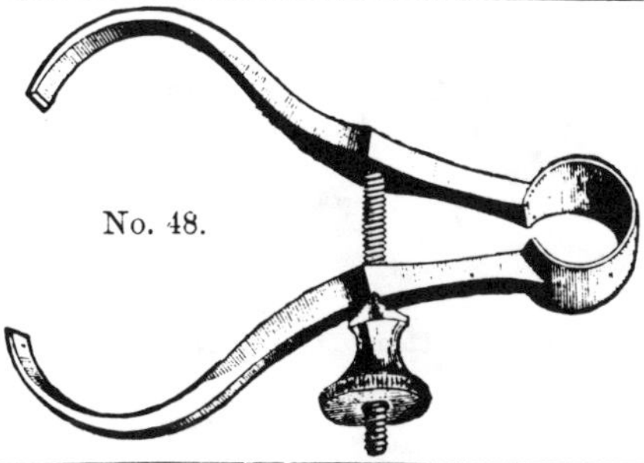

No. 48.

Steven's Spring Screw Calipers.

A new article for calipering screw threads and taps.

EXTRA FINE STEEL AND TEMPER.

3 inch.............Price, each $1 00
5 " .... " " 1 25

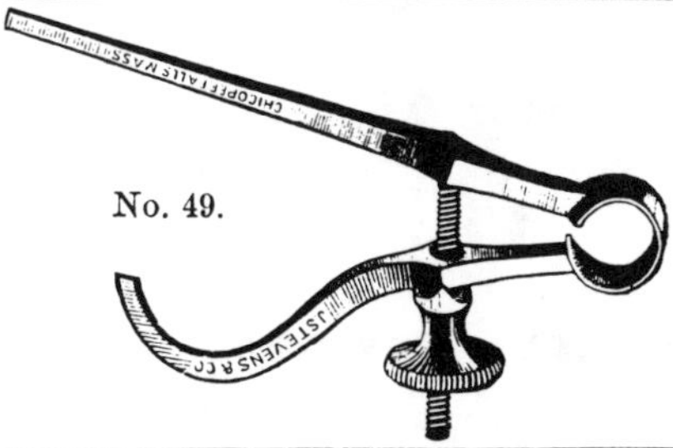

No. 49.

Steven's Spring Key-Hole Calipers.

A NEW ARTICLE.

EXTRA FINE STEEL AND TEMPER.

3 inch...........Price, each $1 15
4 " .... " " 1 50

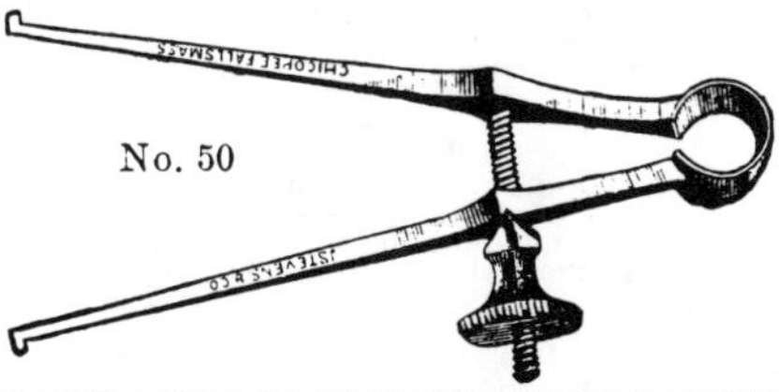

No. 50

Steven's Inside Spring Calipers.

A NEW ARTICLE.

EXTRA FINE STEEL AND TEMPER.

5 inch.............Price, each $1 25

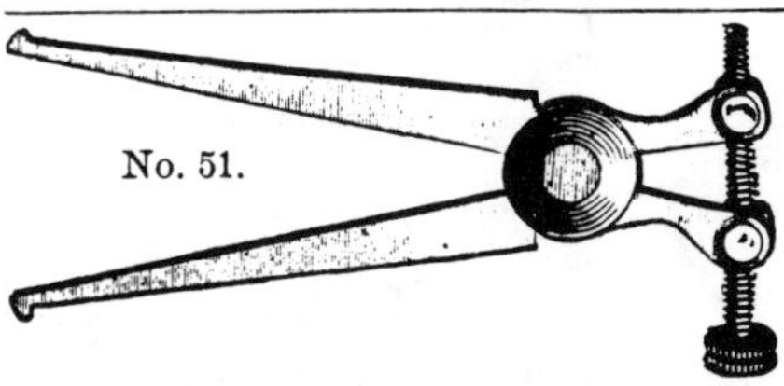

No. 51.

Steven's Inside Screw Calipers.

A NEW ARTICLE.

EXTRA FINE STEEL AND TEMPER.

5 inch.............Price, each $1 25

STEVENS' SPRING DIVIDERS.

No. 52.

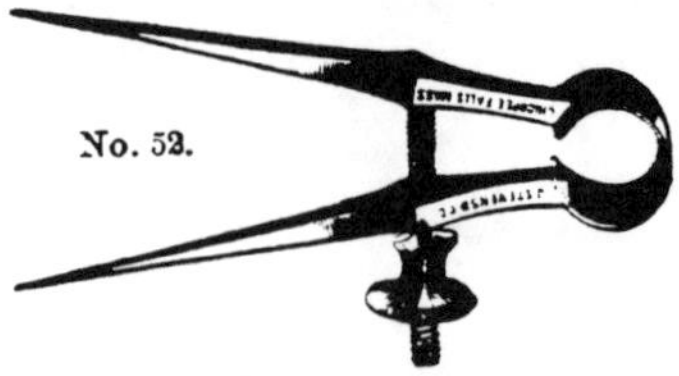

EXTRA FINE STEEL AND TEMPER.

2	inch, Price each	$1 00
3½	" " "	1 00
4	" " "	1 25
5	" " "	1 25
6	" " "	1 50

STEVENS' SPRING DIVIDERS (With Thumb Attachment).

No. 53.

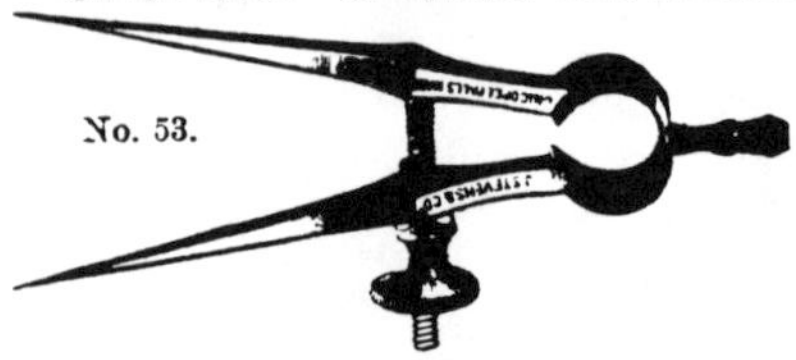

EXTRA FINE STEEL AND TEMPER.

2	inch, Price each	$1 15
3½	" " "	1 15
4	" " "	1 40
5	" " "	1 50
6	" " "	1 75

STEVENS' RELIABLE DIVIDERS.

No. 60.

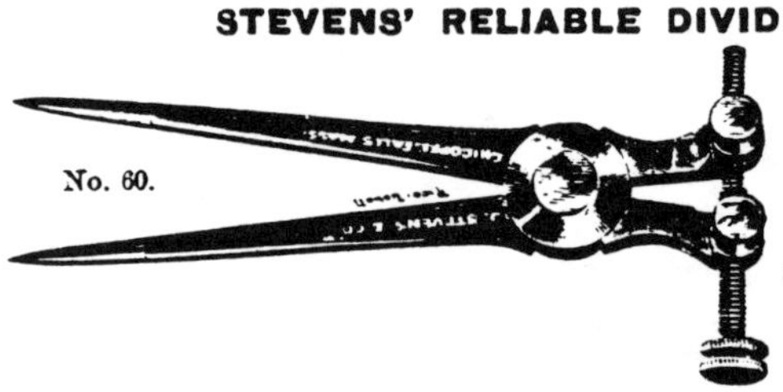

EXTRA FINE STEEL AND TEMPER.

5 in., Price each..$1 50

STEVENS' RELIABLE DIVIDERS (With Thumb Attachment).

No. 61.

EXTRA FINE STEEL AND TEMPER.

5 INCH.

Price each, $1 75

STEVENS' PARALLEL DIVIDER.

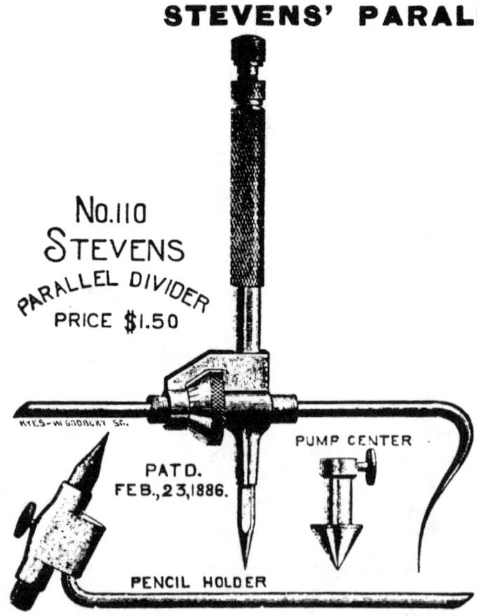

We do not claim this tool as a substitute for the ordinary divider so long in use by mechanics. It has a field of its own. In laying out work on a surface as it ordinarily comes from the planer it will be found invaluable, the point of pressure being directly over the stationary point of tool. With ordinary care there will be no trouble with the tool jumping from center punch mark.

Everything being so rigid and compact about it there is no tendency for the marking point to follow any irregularities there may be on the surface being laid out.

It is very easy to set, the needle being graduated to fractions of an inch, and by means of micrometer device the tool may be moved to a thousandth of an inch.

It can be used with pencil and pen if desired.

A Pump center can be furnished with tool by means of which a circle can be scribed on work laid out around a drilled hole, or BUSHING in a drill jig or other similar work.

The marking parts are made of the best of tool steel and the points carefully hardened. Regular size will draw 7 inch circle.

Price, $1.50. Long bar, 24 inch circle, 50c. Pump center, 25c. Lead pencil holder, 50c.

Both 1891

HATHAWAY'S COMBINATION GAUGE.

1.

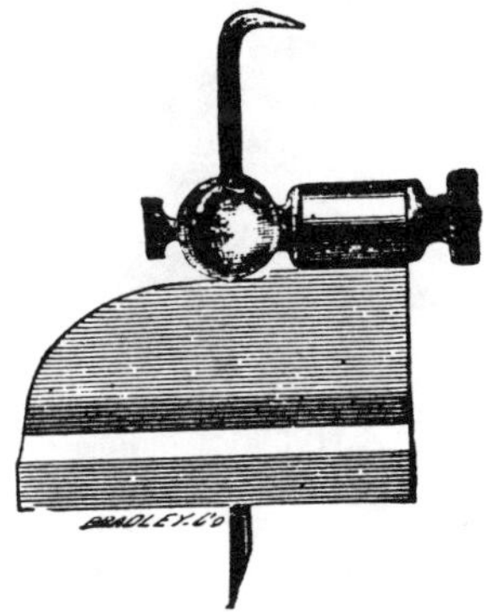

2.

NO. 58.

The various uses to which this little instrument can be applied, place it, in the estimation of all who have used it, as one of the most useful tools known to mechanics.

FIG. 1 represents a Surface Gauge, capable of being set with the greatest accuracy, and, being strong, is not liable to spring or change from its position. In this form it can also be used to find the center of pieces for Lathe Work. Place the Gauge upon a surface plate, bring the piece to be centered in contact with the scratch, and the center will be accurately determined, thus saving the trouble of revolving the piece upon centers. FIG. 2 represents the same as a Depth Gauge: this is done by unscrewing the binding nut and reversing the barrel, thus making a really valuable tool for all purposes of planing where perfect accuracy is indispensable, such as Dovetailing, Fitting Slide Rests, etc. Should the depth to be gauged exceed the length of the pointer, it can be replaced by a piece of Stubs' Wire

PRICE EACH $2.00

STEVENS' SCRIBER.

No 80. 7 inches long over all................Price, each 40 cents.

1891

Bench Surface Gauge,

—FOR—

MACHINISTS, TOOL MAKERS, MODEL MAKERS, DIE SINKERS AND OTHERS.

New No 58
Stevens Improved
Bench Surface Gauge with
Fine Adjustment
Price $2.00

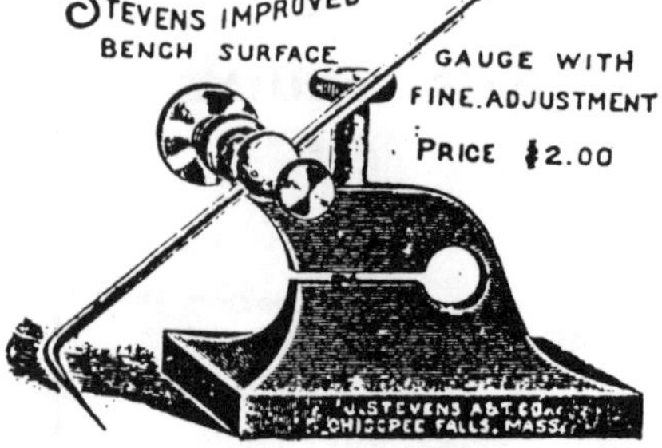

The various uses to which this little instrument can be applied place it, in the estimation of all who have used it, as one of the most useful tools known to mechanics. We have lately remodeled it making it a much more practical tool than it was before. The base has been made much broader, the clamping devise for holding the needle has been made to hold much stronger, and a fine adjustment has been added, making it by far the most desirable tool of its kind on the market. Some of the uses to which it can be put: It can be used on the planer or milling machine to set two or more parts of work at the same height. It may be used on a surface plate in laying off the centre of a piece of work, or in laying out jigs, gauges, dies, templets, models, etc. It can also be used as a depth gauge, and as a scratch gauge.

Stevens
New
No. 58
Gauge
Price $2.00

This cut shows our No. 58 tool in another position.

1891

Reliable Tools.

This is only one of many.

Send for 1894 catalogue.

PROPER PRICES.

J. STEVENS ARMS & TOOL CO.,
Box 232, CHICOPEE FALLS, MASS.

1891

Tool Makers' Surface Gauge.

(FAY'S PATENT.)

No. 60.

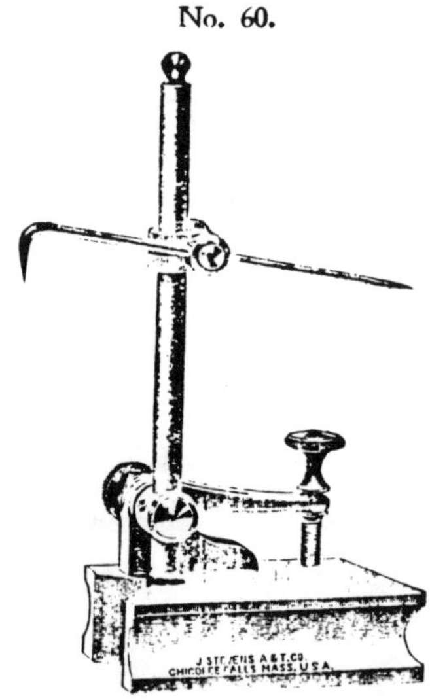

This Gauge is especially designed for fine tool work. It is an "All Steel Tool" standing about 5 inches high. Its base is hardened and ground flat on the bottom. It has a square cut in its front end, adapting it for round work.

A cutter is screwed on the end of the spindle, making a good scratch gauge. The spindle can be set and used in any position within 90 degrees of a complete circle. It occupies a space 4 inches long by 1¾ inches high, when the spindle is parallel with the base. Made in one size only.

Price $2.50.

Improved Soldering or Tinning Acid.

Muratic Acid 1 lb.; put into it all the zinc it will dissolve and 1 oz. of Sal Ammoniac, then it is ready for use.

1898

Stevens' Tools.

Centre Punches.

3 Sizes.

We use nothing but the finest drill rods and every punch is fully guaranteed. 1-4, 5-16 or 3-8 in. diameter.

Sure Grip Wrench,

No. 163.

A handy steel tool that will hold taps, reamers, drills or any tool that can be put in it. Will take either round, square or oval tools.

Send for our catalogue showing full line of fine tools.

J. Stevens Arms & Tool Co.

P. O. Box 5709,
CHICOPEE FALLS, MASS., U. S. A.

1900

STEVENS TOOLS

STANDARD

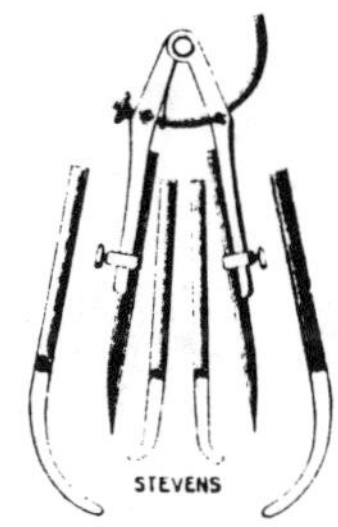

STANDARD

EXTENSION DIVIDERS.

There is a large demand for this tool. Have you them in stock?

The most popular and quickest selling line of machinist's tools on the market.

SEND FOR CATALOG.

J. Stevens Arms & Tool Co.

P. O. Box 5709.
CHICOPEE FALLS, MASS.

1901

FOR LATHE WORK.

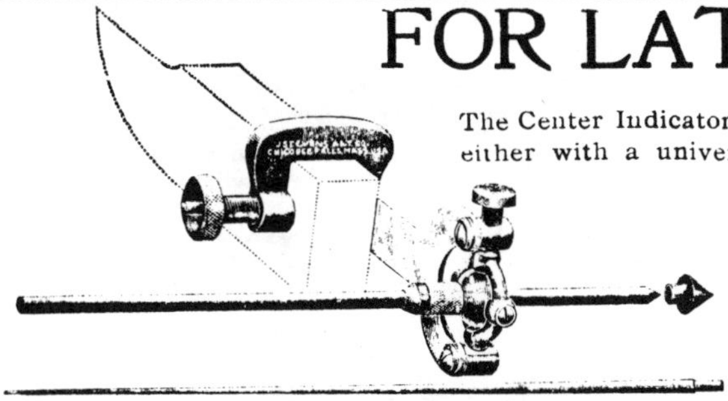

Center Indicator (Fay's Patent.)

The Center Indicator, shown herewith, can be operated either with a universal or oscillating movement, and will meet the requirements of finest work. Pointer tip of lancewood in place of steel. Steel clamp for fastening the indicator to shank end of any lathe tool.

Price $2.50

SEND FOR COMPLETE CATALOGUE OF FINE TOOLS.

J. Stevens Arms & Tool Co., P. O. BOX 5709, CHICOPEE FALLS, MASS, U. S. A.

1900

STEVENS
TOOLS
ARE
STANDARD

CALIPERS.
DIVIDERS.
GAUGES.
SCRIBERS.
SPOKE NIPPERS.
BEVELS.
INDICATORS.
CLAMPS.
REAMERS.
PUNCHES.
PLUMB-BOBS, Etc.

SEND FOR OUR CATALOG AND PRICES.

J. Stevens Arms & Tool Co.

P. O. Box No. 5709,
CHICOPEE FALLS, MASS., U. S. A.

STEVENS
ALL STEEL TOOLS
ARE
STANDARD FOR QUALITY.

Bench Surface Gauge No 58. Price, $2.00.

Send for our Tool Catalog.

J. STEVENS ARMS AND TOOL CO.,

P. O. Box 5709,
CHICOPEE FALLS, MASS.

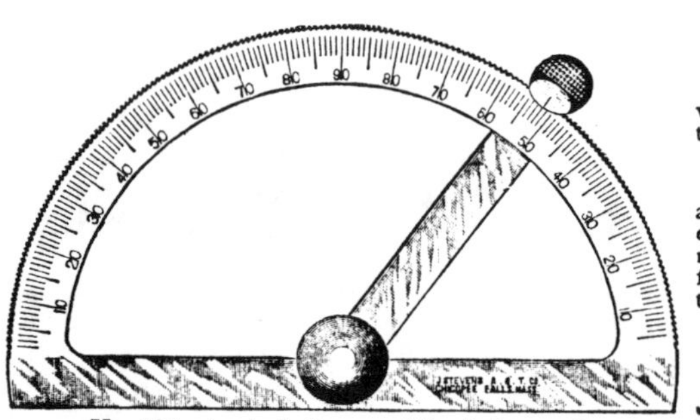

No. 151—Angle Indicator. (Fay's Patent.)

A NEW TOOL.

This tool is invaluable for use in connection with a universal bevel for transferring or taking any angle.

It will produce a given degree accurately.

It is nicely graduated, 0 to 90, from both ends and well finished. One should be owned by every mechanic, tool-maker, machinist, cabinet-maker, carpenter, draftsman and tinsmith. In fact any one who makes use of angles will find this tool just what he wants.

PRICE, POSTPAID, ONLY $1.00
For Sale By Leading Dealers.

Write for our complete Catalogue of Fine Tools.

J. STEVENS ARMS & TOOL CO.
P. O. Box 5709. CHICOPEE FALLS, MASS., U.S.A.

All 1900

SMALLER MACHINIST'S TOOL COMPANIES

ADELL BROTHERS MFG. CO. Orange, MA

Operated by John Adell who was granted a patent for a thread gauge, August 29, 1911 (No. 1002019). The Adell Brothers Mfg. Co. made the thread gauge and a new design of machinist's clamps formed from sheet metal. Both products were later made by the L.S. STARRETT CO., Athol, MA, which bought the company about 1913. John Adell, who had probably worked at L.S. STARRETT CO. when he was granted four joint patents with Laroy S. Starrett from 1906 to 1908, rejoined Starrett.

ALMOND, T.R. Brooklyn, NY

Founded by Thomas R. Almond (1855-1906) in 1879 to make drill chucks. In 1904 Almond sold out to Carleton A. Hubbell, who incorporated the business as:

ALMOND MFG. CO., T.R. Brooklyn, NY, later, Ashburnham, MA

In 1907, Carleton A. Hubbell, president of the company, moved it to Ashburnham, MA. Hubbell was also president of the SAWYER TOOL MFG. CO. by 1910, and moved Sawyer to Ashburnham in 1912. In 1917, the company began manufacturing micrometers based on the Henry S. Hubbell micrometer patent issued on May 21, 1918 (No. 1267075). Hubbell was also granted a micrometer patent on December 26, 1922 (No. 1440080). Both his patents were assigned to the T.R. Almond Mfg. Co. Steel rules marked with the company name have been found, but these may have been for advertising only. By 1930 the company was no longer making micrometers. A last remnant of the company existed as late as 1970 as the D'NOMLART CO. (T.R. ALMOND spelled backwards), still operated by the Hubbell family.

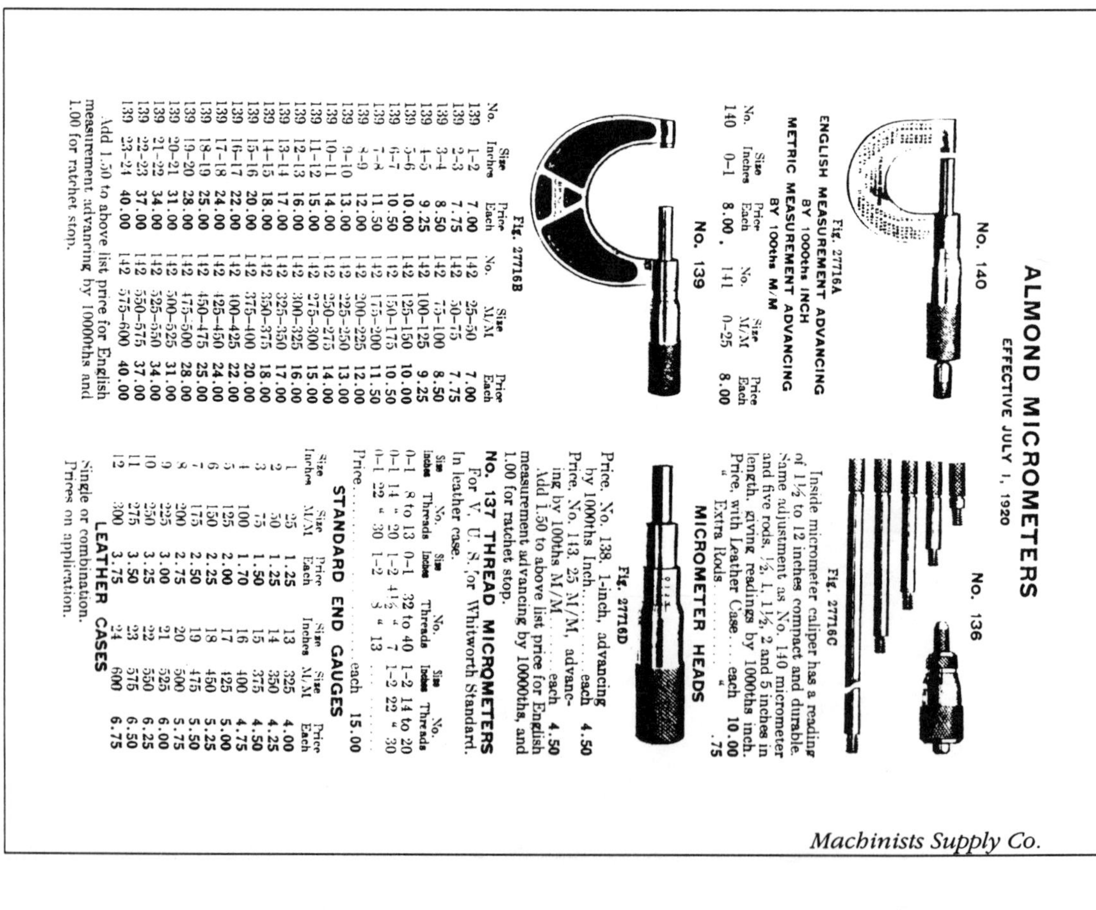

ALMOND MICROMETERS

EFFECTIVE JULY 1, 1920

No. 140

Fig. 27716A

ENGLISH MEASUREMENT ADVANCING BY 1000ths INCH

METRIC MEASUREMENT ADVANCING BY 100ths M/M

No.	Size Inches	Price Each	No.	Size M/M	Price Each
140	0–1	8.00	141	0–25	8.00

No. 139

Fig. 27716B

No.	Size Inches	Price Each	No.	Size M/M	Price Each
139	1–2	7.00	142	25–50	7.00
139	2–3	7.75	142	50–75	7.75
139	3–4	8.50	142	75–100	8.50
139	4–5	9.25	142	100–125	9.25
139	5–6	10.00	142	125–150	10.00
139	6–7	10.50	142	150–175	10.50
139	7–8	11.50	142	175–200	11.50
139	8–9	12.00	142	200–225	12.00
139	9–10	13.00	142	225–250	13.00
139	10–11	14.00	142	250–275	14.00
139	11–12	15.00	142	275–300	15.00
139	12–13	16.00	142	300–325	16.00
139	13–14	17.00	142	325–350	17.00
139	14–15	18.00	142	350–375	18.00
139	15–16	20.00	142	375–400	20.00
139	16–17	22.00	142	400–425	22.00
139	17–18	24.00	142	425–450	24.00
139	18–19	25.00	142	450–475	25.00
139	19–20	28.00	142	475–500	28.00
139	20–21	31.00	142	500–525	31.00
139	21–22	34.00	142	525–550	34.00
139	22–23	37.00	142	550–575	37.00
139	23–24	40.00	142	575–600	40.00

Add 1.50 to above list price for English measurement advancing by 10000ths and 1.00 for ratchet stop.

No. 136

Fig. 27716C

Inside micrometer caliper has a reading of 1½ to 12 inches compact and durable. Same adjustment as No. 140 micrometer and five rods, ½, 1, 1½, 2 and 5 inches in length, giving readings by 1000ths inch.
Price, with Leather Case. . . . each 10.00
" Extra Rods. . . . . " .75

MICROMETER HEADS

Fig. 27716D

Price, No. 138, 1-inch, advancing by 1000ths Inch. . . . . each 4.50
Price, No. 143, 25 M/M, advancing by 100ths M/M. . . . . . each 4.50
Add 1.50 to above list price for English measurement advancing by 10000ths, and 1.00 for ratchet stop.

No. 137 THREAD MICROMETERS

For V, U. S. or Whitworth Standard. In leather case.

Size Inches	No. Threads	Size Inches	No. Threads	Size Inches	No. Threads
0–1	8 to 13	0–1	32 to 40	1–2	14 to 20
0–1	14 " 20	1–2	4½ " 7	1–2	22 " 30
0–1	22 " 30	1–2	8 " 13	. . .	. . .

Price. each 15.00

STANDARD END GAUGES

Size Inches	Size M/M	Price Each	Size Inches	Size M/M	Price Each
1	25	1.25	13	325	4.00
2	50	1.25	14	350	4.25
3	75	1.50	15	375	4.50
4	100	1.70	16	400	4.75
5	125	2.00	17	425	5.00
6	150	2.25	18	450	5.25
7	175	2.50	19	475	5.50
8	200	2.75	20	500	5.75
9	225	3.00	21	525	6.00
10	250	3.25	22	550	6.25
11	275	3.50	23	575	6.50
12	300	3.75	24	600	6.75

LEATHER CASES

Single or combination.
Prices on application.

Machinists Supply Co.

ALVAN MFG. CO. Newark, NJ

Maker of Universal test indicators in 1912. In 1913, the firm offered a height gauge attachment for a combination square.

ALVAN HEIGHT GAGE

The illustration shows an attachment for a combination square which adapts this type of tool for use as a height gage. This attachment is a product of the Alvan Mfg. Co., 306 14th Ave., Newark, N. J. It consists of a hardened and ground jaw or blade which is secured to the scale of the combination square by an eccentrically located taper locking stud which is carried by a semicircular saddle piece. In adjusting the blade of the height gage attachment to any required position, the stud is partially loosened. When the saddle piece is secured in this way, a steady movement can be secured which

Alvan Height Gage Attachment for a Combination Square

enables the attachment to be brought to exactly the required position without delay. Ample compensation is provided for any variation in the width of the blade of the square on which the attachment is used.

When this attachment is used as a height gage for testing and gaging surfaces or scribing lines, the stock of the combination square constitutes the base of the gage. Used in this way, the attachment adapts an ordinary combination square for such work as laying out, scribing lines, transferring dimensions, testing and gaging surfaces, etc., on pieces set on a surface plate, planer, shaper or milling machine. A pair of these attachments used in conjunction with the square stock still further increases the scope and usefulness of the tool.

Where two of these attachments are mounted on the blade of a combination square, they form a beam caliper and can be used for making either inside or outside measurements. In laying out from a given edge or shoulder on surface work, this attachment is of particular value.

Machinery, May 1913

UNIVERSAL TEST INDICATOR

The universal test indicator illustrated below is made by the Alvan Mfg. Co., Newark, N. J. This indicator has hardened cone-pivot bearings which are kept in constant uniform adjustment by spring thrusts that prevent lost motion and consequently errors in reading. These cone bearings also

Universal Test Indicator

make the indicator very sensitive, and it is claimed that no adjustment is ever required.

The indicator proper is attached to the toolpost holder by a universal joint, which is frictionally held in any desired position. By means of this joint the indicator can be set at any angle with relation to the shank. Either side of the contact point can be used, and the spring tension is reversed by half a turn of a button on the end of the barrel. This indicator multiplies error in the ratio of 50 to 1, and has a capacity up to 0.020 inch.

Machinery, July 1912

AMERICAN GAUGE CO Newark, NJ

Maker of limit snap gauges and bench centers in 1919.

AMERICAN STANDARD GAUGE & TOOL WORKS Philadelphia, PA, later, Wilmington, DE

Founded by John Richards who received the following patents; Standard Gauge, October 15, 1867 (No. 69953); Caliper Gauge, October 15, 1867 (No. 69954); Bench Micrometer, September 2, 1879 (No. 219307) and May 28, 1889 (No. 404305). The firm made a variety of fixed and standard gauges based on the Richards patents. In 1879, the company was taken over by the BETTS MACHINE CO. and moved to Wilmington, DE, where production continued. In 1895, the BETTS MACHINE CO., which was primarily a machine tool builder, sold the gauge business.The new owner then reorganized as the TAYLOR-RICE ENGINEERING CO.

American Standard Gauge & Tool Works,
WILMINGTON, DEL.
Makers of Implements for Standard Measurements.

Crescent Gauge. Flat Bar Gauge.

Send for New Pamphlet. JAS. A. TAYLOR & CO.

American Machinist, 1894

(see next page)

American Machinist, 1886

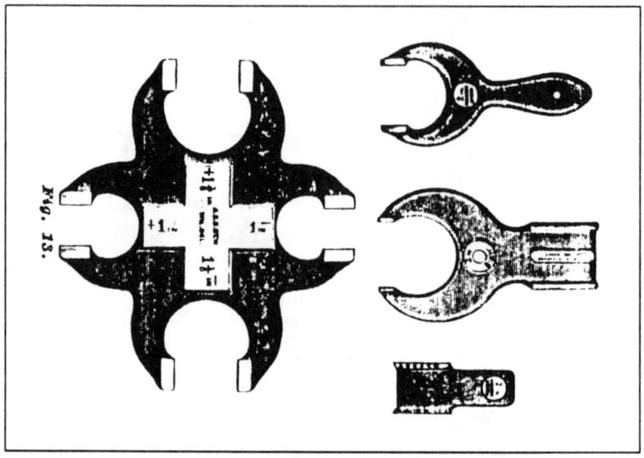

AMERICAN WATCH TOOL CO. Waltham, MA

Formed by Ambrose Webster (?-1894), John Whitcomb and Henry Fisher in 1876.The company made bench lathes and wire gauges in 1880. By 1912, they were also making universal test indicators. Webster was one of the early authorities on the use of precision measuring tools and was the author of more than a dozen articles in *American Machinist* and *Machinery* magazines, one of which will be found in Appendix I. The company merged with the Walter N. Wade Co. in 1918 to become part of the Wade-American Co.

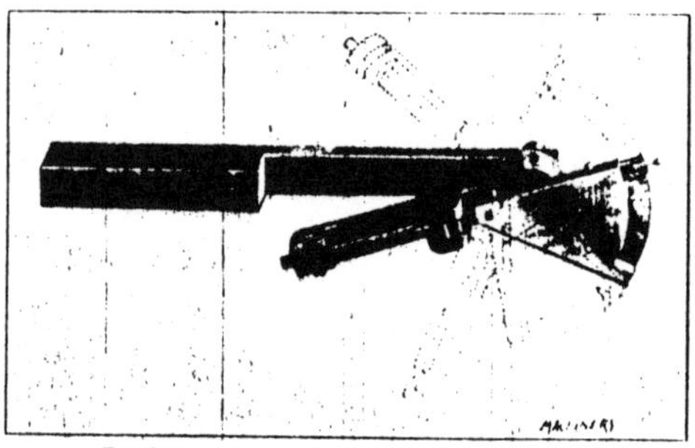

Test Indicator made by American Watch Tool Co.

Machinery, April 1912

Fig. 89. Dial Depth Gage
Courtesy of American Watch Tool Company, Waltham, Massachusetts

Bench Micrometer. The gage shown in Fig. 89 is called a bench micrometer, and there are several good makes on the market. This gage can be used in place of snap, ring, and depth gages. When used as a thickness gage, the work is laid on the table which is adjustable up or down, and in operation the gaging point actuated by the hand lever is brought to bear on the work, and the thickness of the work is indicated by the pointer on the dial in thousandths of an inch.

AMES & CO., B.C. Waltham, MA

Founded in 1898 as ALCORN & AMES by John Alcorn and Bliss C. Ames (1867-1948). In 1901, Ames bought out Alcorn. The first products were a tool maker's bench table and small bench lathes, both developed by Ames. By 1903 Ames began making a wide variety of measuring gauges and dial indicators, continuing until the company closed c1960. Warren Ames, son of B.C. Ames, was elected president when the firm was incorporated in 1922 and, with his brother Ira, bought out his father in 1923. From 1918 to 1938, Warren Ames was granted 16 patents for dial indicators and other gauges.

ANDERSON, ABRAHAM Moline, IL

Inventor and maker of a machinist's and patternmakers' protractor patented October 24, 1905 (No. 802674). This tool is a very complex device combining the features of a bevel protractor, center square and scriber.

ASHCROFT, HUGH E. Arlington, NJ

Maker, in 1887, of the "Acme" center square patented November 14, 1882 (No. 267481) by Charles L. Bellamy.

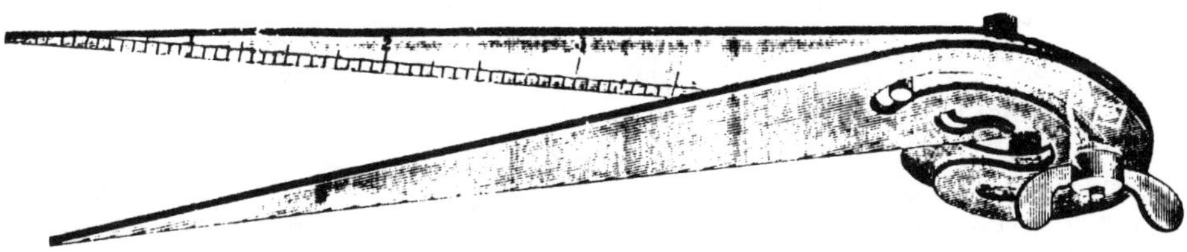

THE ACME CENTER SQUARE.

American Machinist, 1887

ASHLAND TOOL WORKS Ashland, PA

Makers, in 1899, of an angle indicator offered in german silver with transparent celluloid blades for draftsmen and in all nickel plated steel with a level attached for machinists.

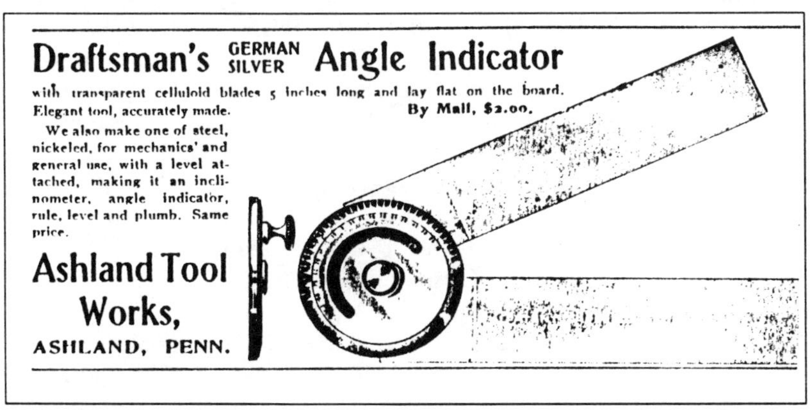

American Machinist 1899

ATCHESON TOOL & DIE CO. Columbus, OH

Introduced the "Keller" dial test indicator in 1920. The indicator was made under a patent granted to Clarence J. Keller, April 8, 1919 (No. 1299838).

ATHOL MACHINE CO. Athol, MA

Organized June 3, 1868, with $25,000 capitalization, increased to $50,000 in 1869. Officers were: George T. Johnson, president, and J. Sumner Parmenter, treasurer. D.W. Houghton, L.S. Starrett, L.D. Rich and Abner Smith were directors. The company was formed for "the special purpose of making Starrett's food chopper and other articles of Mr. Starrett's invention". The first product was a food chopper which had been patented May 23, 1865 (No. 47875) by L.S. Starrett. They later added a washing machine patented by Starrett on June 27, 1865 (No.48458), and a food press patented April 15, 1873 (No. 137803).

Starrett began developing tools about 1876 and received his first tool patents May 8, 1877 (No. 190636), and February 25, 1879 (No. 212760). The patents were for a quick adjusting wrench which was never made. In 1878, Starrett, who had just developed his first machinist's tool, a combination square patented May 6, 1879 (No. 215024), was forced out of the company. Litigation regarding the invention of the square continued for some years after the break. Starrett eventually won. (See Appendix III for details)

About 1882 the owners of the ATHOL MACHINE CO. either founded or bought the STANDARD TOOL CO. of Athol. An 1889 reference states that the STANDARD TOOL CO. was "run in connection with Athol Machine Co." The two companies shared the same officers. The relationship between the STANDARD TOOL CO. and the ATHOL MACHINE CO. is illustrated by the following excerpt from an article published in the *Athol Transcript* for February 6, 1883. The article dealt with a lawsuit filed by L. S. Starrett against the Athol Machine Co. for infringement of Starrett's combination square patent.

> "—the defense was a general denial, and that the Athol Company was not liable under a contract by which the company's power, machinery, etc., were let out to another firm [STANDARD TOOL CO.] for the manufacture of tools, that firm agreeing to pay as rent 10% of their net sales. The plaintiff [L. S. Starrett] insists that the whole contract is a device to shield the Athol Company as infringers. There is evidence tending to show that the company has made tools for the express purpose of manufacturing these try-squares and that a few were made before the contract went into effect."

ATHOL MACHINE CO.'s products in 1891 were: food choppers, food presses, flower stands, household knife shapeners, Wright's animal tethers, and vises. At that time they were also listed as "selling agents" for the STANDARD TOOL CO. of Athol. Officers in 1891 were: George T. Johnson, president; A. Bangs, secretary; and Stephen H. Bellows, superintendent. D. Appleton Newton, who had been general manager and treasurer, died in 1889. Bellows was the inventor of the group and received the following patents: Machinist's Gauge, September 16, 1873 (No. 142761) (Bellows was living in Middletown, CT, at the time); Monkey-wrench, June 8, 1880 (No. 228437), (examples may be marked either STANDARD TOOL CO. or ATHOL MACHINE CO.); Combination Square, November 22, 1881 (No. 249821) (advertised by STANDARD TOOL CO. in 1889); Combination Bevel and T Square, March 11, 1884 (No. 295100); Caliper Rule, April 15, 1884 (No. 297056) (assigned to STANDARD TOOL CO.); Calipers, June 3, 1884 (No. 299722); Micrometer Gauge, May 21, 1889 (No. 403726); Beam Calipers, May 13, 1890 (No. 427956); Micrometer Gauge, July 28, 1891 (No. 456875); Spirit Level, May 11, 1897 (No. 582517); Micrometer Gauge, October 18, 1898 (No. 512601); Surface Gauge, May 23, 1901 (No. 672634) (assigned to Athol Machine Co., examples, however, are marked STANDARD TOOL CO.). George T. Johnson, president, died in 1893 and was succeeded by his son William G. Johnson who served until his death in 1899.

The 1896 Athol Machine Co. catalog listed "vises, try-squares, depth gauges, bevels, and a variety of measuring and gaging tools." Except for the vises most or all of these tools would have been sold under the STANDARD TOOL CO. name.

In 1902, controlling interest was sold to three men from Orange, MA, Frank S. Ewing, Dresser T. Bates and Stephen E. French. French became president, Bates, vice president, and Ewing, secretary and treasurer.

In 1905, Laroy S. Starrett quietly bought control from French and Bates, eventually acquiring all the stock. Starrett announced that the company would not be merged with the L. S. STARRETT CO. but would be operated separately and would continue to make vises, meat choppers and mechanic's tools. New officers were: F. A. Ball, president; L. S. Starrett, vice president and treasurer; and F. E. Wing, clerk. All were officers of the L. S. STARRETT CO.; F. A. Ball was Starrett's son-in-law.

After Starrett gained control the STANDARD TOOL CO. part ceased operation and the machinist's tools made under the ATHOL MACHINE CO. name were simply a lower priced version of tools made by the L. S. STARRETT CO. Their new level and "PREMIER" caliper lines, introduced in 1905, were almost identical to the Starrett equivalents.

Catalog number 30, issued in 1906, lists vises, machinist's tools, and grindstone frames. The 1916 catalog lists both Athol and Starrett vises, "Premier" spring calipers and dividers, firm joint calipers, steel rules, squares, bevels, vises, etc.

The Laroy S. Starrett vise patent of September 12, 1922 (No. 1428996) was assigned to the ATHOL MACHINE CO. This was to be Starrett's last patent and was issued after he died on April 23, 1922. About 1925 the company was reorganized as:

ATHOL MACHINE & FOUNDRY CO. Athol, MA

In 1930 the company was making vises, grinders, and semi-steel castings. Capital was $500,000 with employment at 150. A. H. Starrett was secretary and treasurer. The company ceased operation in 1963 and the building was demolished in 1978.

Machinists' Swivel Clamp and Iron Level.

The Athol Machine Company, Athol, Mass., has recently supplemented its line of mechanics' fine tools by the addition of the machinists' swivel clamp and double plumb iron level here illustrated. The machinists' swivel

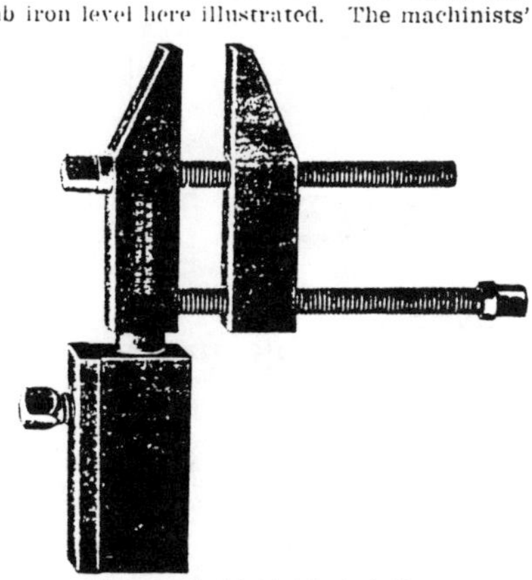

Fig. 1.—Machinists' Swivel Clamp.

clamp, Fig. 1. is a convenience especially adapted to vise work, where it is often desirable or necessary to change position of work to make it more accessible to file or other tool. The holder can be inserted in the vise jaws and so remain, a swivel motion being obtained by loosening the set screw and relocking clamp portion in an altered posi-

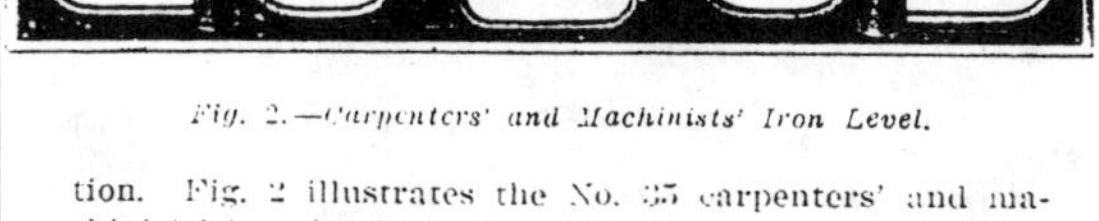

Fig. 2.—Carpenters' and Machinists' Iron Level.

tion. Fig. 2 illustrates the No. 35 carpenters' and machinists' iron levels, with double plumb. Sizes 6, 9 and 12 inches long have square ends, the 18 and 24 inch sizes differing slightly on the ends in appearance only. The base is grooved for cylindrical work, both groove and base being accurate.

Iron Age, March 9, 1905

Spring Calipers and Adjustable Dividers.

The Athol Machine Company, Athol, Mass., manufacturer of lines of mechanics' fine tools, has just introduced the Premier outside spring calipers and extension dividers here illustrated. Fig. 1 shows characteristics of a line embodying the same principles, applied also to inside spring calipers, spring dividers, two types of thread spring calipers, keyhole spring calipers, straight

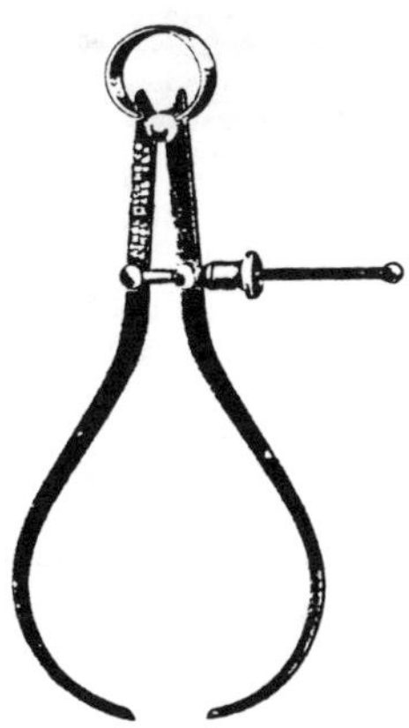

Fig. 1.—Premier Outside Spring Calipers.

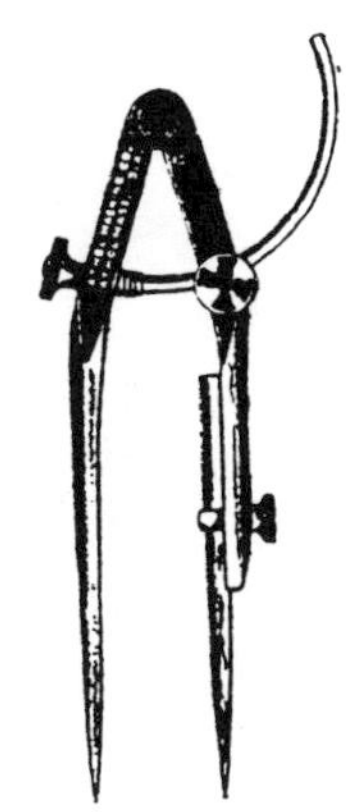

Fig. 2.—Extension Dividers.

leg spring and Hermaphrodite spring calipers. Many of them are made in six sizes, from 2½ to 8 inches, inclusive, with both quick nut and solid nut, the latter being more moderately priced. The thread calipers are made in 3, 4 and 5 inch sizes, keyhole and straight leg calipers 3 and 4 inch, and Hermaphrodite calipers are 3 to 6 inches long. The quick nut is recommended by the company for rapidity and accuracy of adjustment. Fig. 2, extension dividers, has one drop forged leg and one malleable iron leg, with tool steel points well tempered. The adjustable point can be instantly removed and a pencil inserted in place of it. The sizes are 6, 8 and 10 inches in length.

Iron Age, March 16, 1905

ATHOL LEVELS

No. 435

Fig. 36205A

No. 438

Fig. 36205B

No. 435 CARPENTERS' AND MACHINISTS' IRON LEVELS

With double plumb. Base of levels grooved for cylindrical work. The groove is exactly parallel with the edge of the base. These levels are of handsome design and accurate.

Length	inches	4	6	9	12	18	24
Price	each	1.50	1.65	1.80	2.00	2.25	2.50

With adjustable vial in level, any size except 4 inch, extra, .50 each.

No. 438 GROOVED LEVELS

Made with a deep groove running the entire length so it can be used on piping, shafting, or any cylindrical work.

Length	inches	6	9	12
Price	each	1.65	1.80	2.00

No. 439 ENGINEERS' AND PLUMBERS' LEVELS

Fig. 36205C

This level is grooved on the bottom the whole length, which any one who has occasion to level up shafting, piping, etc., will see the advantage of, as it enables the user to place the level in line with the shaft or pipe as readily as on a flat surface. These levels are so made that they can be used on a flat surface the same as ordinary levels. This level has a graduated plate on one end that can be adjusted to give the exact pitch of pipes, etc.

Length	inches	6	12
Price	each	2.15	2.50

No. 436

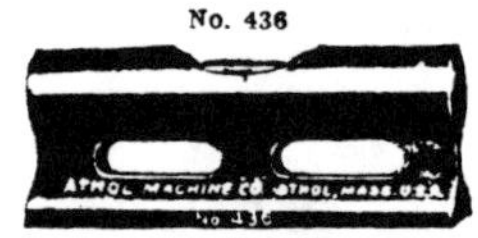

Fig. 36205D

No. 442

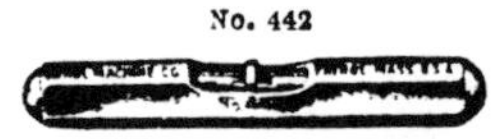

Fig. 36205E

No. 436 IRON POCKET LEVELS

Length	inches	2½	3½
Price	each	.30	.35

No. 442 NICKEL PLATED POCKET LEVELS

Length	inches	2½	3½
Price	each	.50	.60

1920

ATHOL MACHINE COMPANY

ATLAS INDICATOR WORKS Chicago, IL

Makers of "Junior" dial indicators introduced in 1921.

ATLAS DIAL INDICATOR

A small indicator reading over a range of 0.055 inch, known as the "Junior," is being sold by the Atlas Indicator Works, 160 N. Wells St., Chicago, Ill. The indicator dial

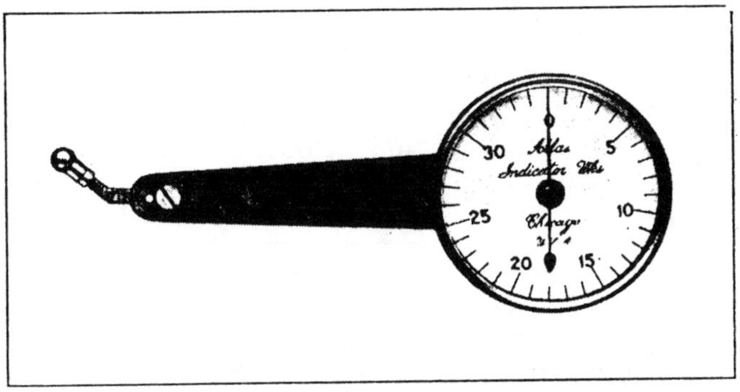

"Junior" Indicator made by the Atlas Indicator Works

is carried on a swivel mounting that offers a convenient means of attaching the instrument to suit the various jobs on which it may be employed. After adjusting the contact point on the work, the dial may be revolved so as to bring the

Machinery, December 1921

AYER, ALBERT E. East Boston, MA

Inventor and maker in 1902 of a machinist's rule which had one edge graduated in the usual manner, and the opposite edge made in a series of steps. The steps ran from 1" to ⅛" in ⅛" increments. Ayer received his patent on April 15, 1902 (No. 697701).

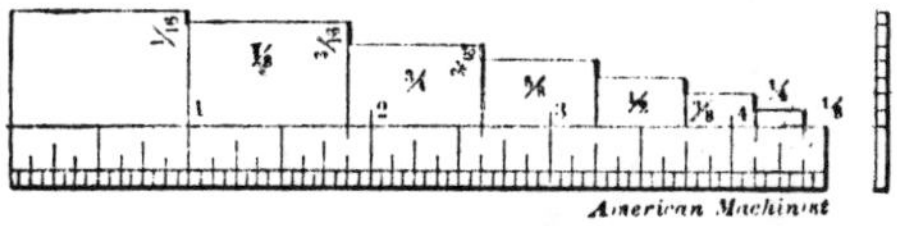

MACHINIST'S SCALE.

American Machinist, 1902

BABIN, A. E. Waterbury, CT

Operated by Arsene E. Babin who introduced, in November, 1906, a test indicator similar to one introduced earlier in 1906 by G. G. RIGGS & A. E. BABIN and based on their joint patent of October 25, 1904 (No. 773425). It appears that the Riggs and Babin partnership was shortlived.

BAKER-CASE MFG. CO. INC. Racine, WI

Maker of machinist's tool chests c1930.

BATES MFG. CO. Fitchburg, MA

Founded by J. Clayton Bates in 1913. Bates had been superintendent of SAWYER TOOL MFG. CO., Fitchburg, MA, from about 1908 until it moved to Ashburnham, MA, in 1912. The only known products, steel machinist's rules marked with the Fitchburg address, are extremely scarce.

Within the same year Bates sold the company to the UNION CALIPER CO. of Orange, MA, where he became general manager. The UNION CALIPER CO. added the line of steel rules to its line of calipers. The BATES MFG. CO. name was retained and used on a lower priced line of machinist's tools until at least 1927. Steel rules, combination squares and feeler gauges marked BATES MFG. CO., ORANGE, MA, are commonly found.

BATH, JOHN Worcester, MA

John Bath (1863-1935) received patents for calipers on June 13, 1893 (No. 499199), and indicators on May 22, 1894 (No. 520201), and October 2, 1894 (No. 526960). *(continued next page)*

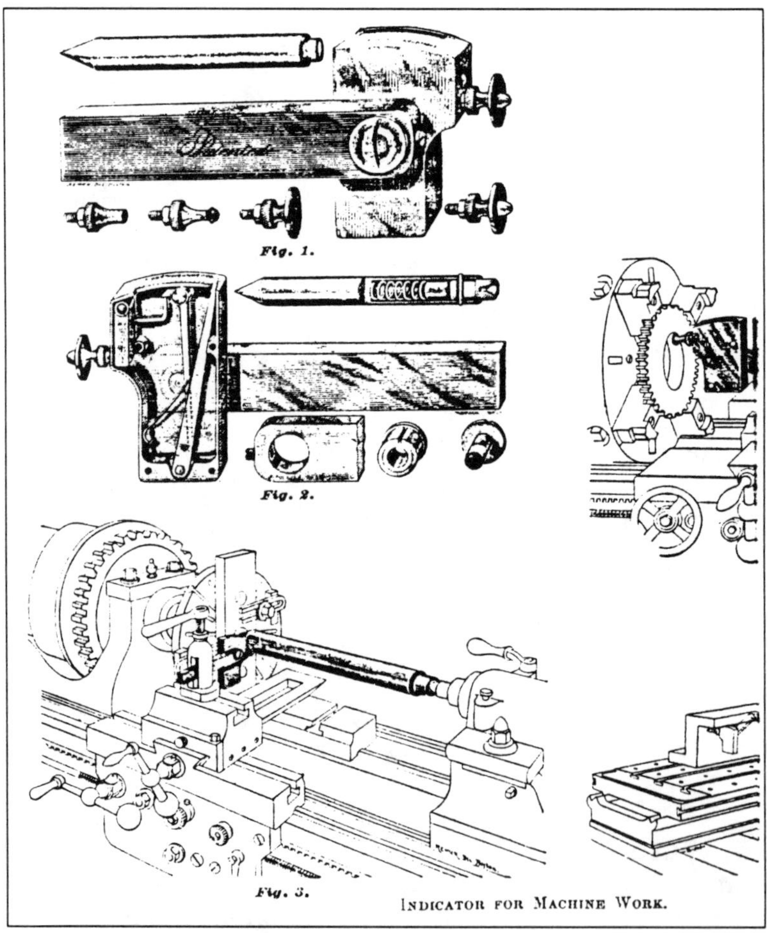

INDICATOR FOR MACHINE WORK.

American Machinist, 1895

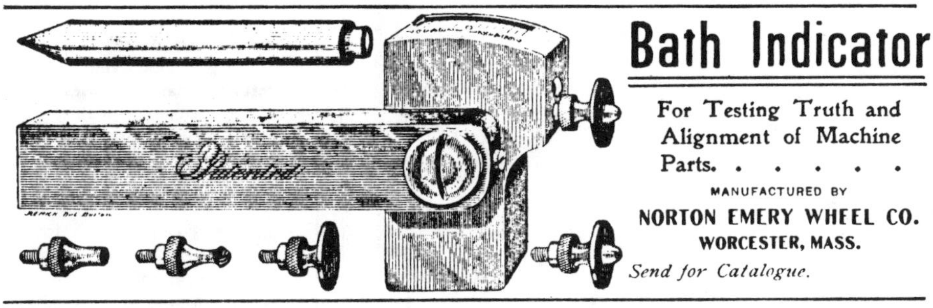

In 1895 Bath formed the BATH INDICATOR CO. to produce his new indicator in Hyde Park, MA. By 1896, however, the Bath indicators were being produced by NORTON EMERY WHEEL CO., of Worcester, MA, continuing until at least 1900. Bath may have been working for Norton during this time. By 1902, Bath was again advertising the indicators under his own name. He incorporated the company in 1910 as:

BATH & CO., JOHN Worcester, MA

The following John Bath patents were assigned to JOHN BATH & CO.:
Micrometer-calipers, October 24, 1916 (No.1202019)
Reference-gauge, February 2, 1921 (No. 1369317)
Adjustable Plug Gauge, September 12, 1922 (No. 1429183)
Ring Gauge, December 12, 1922 (No. 1438681)
Adjustable Gauge, December 11, 1923 (No. 1476681) and December 25, 1923 (No.1478541)
Micrometer-calipers, November 25, 1924 (No. 1516909)
Thread Gauge, December 27, 1927 (No.1654233)
Attachment for Internal Gauges, February 28, 1928 (No. 1660469)
Internal Micrometer, July 30, 1929 (No.1722834)
John and Stanley W. Bath's patent for a calipering mechanism of March 27, 1928 (No. 1664151) was also assigned.

In 1929, the company had capital of $100,000 with 150 employees. Officers were: John Bath, president and J. Chester Bath, son of John Bath, vice-president. The Bath company continued until c1960, making a specialty of thread gauging equipment and precision ground taps.

BEACH, JOHN Springfield, OH

Maker of a keyseat and thread-tool gauge for which he was granted a patent on August 8, 1899 (No. 630316). *(see illustration next page)*

BEARD, L. O. Lancaster, PA

Maker of an inside micrometer c1900. The tool is marked PAT. APL. FOR.

BEMIS & CALL CO. Springfield, MA

Founded by Stephen C. Bemis and Amos Call (1814-1888) in 1844. In 1855, the partnership was incorporated as:

BEMIS & CALL HARDWARE & TOOL CO. see next page

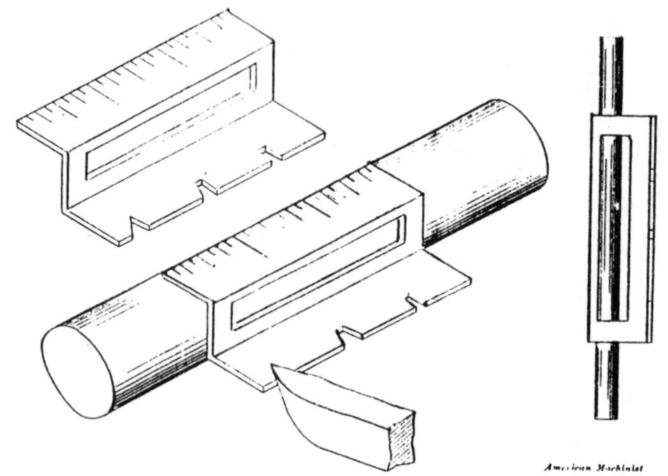

A KEY-SEAT AND THREAD-TOOL GAGE.

American Machinist, Sept. 21, 1899

A Key-seat and Thread-tool Gage.

The cut shows a little tool, recently patented, which should interest the machinist of the shop, and should be found useful for many purposes. It consists, as will be seen, of a double angle iron, the central web of which, common to both angles, is slotted, one of the outer webs being graduated in convenient divisions, while the other edge is notched as a gage for V and other tools. More notches may be inserted than here shown. The slot or slit in the central web, it will be noticed, is nearer one web than the other. The edges of this slit are used for scribing parallel lines upon shafts or other cylindrical figures. When the shaft is small, as shown in the right-hand figure, the inner edge of the slit is used to scribe by, and by using both angles and both edges of the slot four different distances from the apex of the angle are available. The tool is supposed to be correctly made, and, the ends being perfectly square, when the tool is stood upon end it may be used as a square in places where the common try square cannot be applied. Other uses of the tool will readily suggest themselves. The inventor is John Beach, Springfield, Ohio.

(see entry for John Beach, previous page)

BEMIS & CALL HARDWARE & TOOL CO. Springfield, MA

Makers of calipers and dividers, including Call's patent of May 28, 1861 (No. 32410). This patent is of special interest in that Call used the term "micrometer" adjustment. This is the earliest use of the term, applied to machinist's tools, found to date. The firm was listed as makers of calipers and small steel squares in 1905, although their primary line by that time was a wide variety of wrenches.

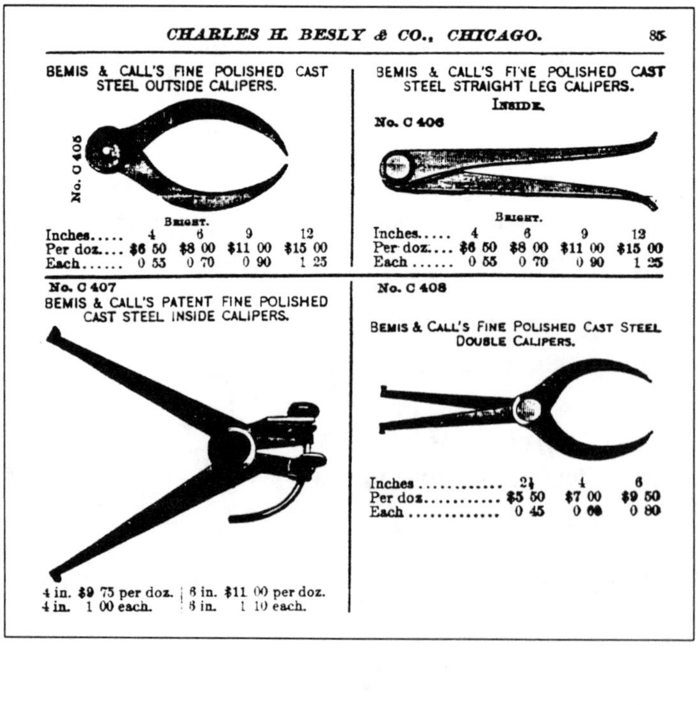

CHARLES H. BESLY & CO., CHICAGO. 85

BEMIS & CALL'S FINE POLISHED CAST STEEL OUTSIDE CALIPERS.

No. C 405

BRIGHT.

Inches.....	4	6	9	12
Per doz....	$6 50	$8 00	$11 00	$15 00
Each......	0 55	0 70	0 90	1 25

BEMIS & CALL'S FINE POLISHED CAST STEEL STRAIGHT LEG CALIPERS.

INSIDE.

No. C 406

BRIGHT.

Inches.....	4	6	9	12
Per doz....	$6 50	$8 00	$11 00	$15 00
Each......	0 55	0 70	0 90	1 25

No. C 407

BEMIS & CALL'S PATENT FINE POLISHED CAST STEEL INSIDE CALIPERS.

4 in. $9 75 per doz. | 6 in. $11 00 per doz.
4 in. 1 00 each. | 6 in. 1 10 each.

No. C 408

BEMIS & CALL'S FINE POLISHED CAST STEEL DOUBLE CALIPERS.

Inches	2½	4	6
Per doz...........	$5 50	$7 00	$9 50
Each	0 45	0 60	0 80

1889

BEST TOOL CO. Boston, MA

Maker, in 1904, of the Hjorth surface gauge patented May 31, 1904 (No. 761381) by Henry J. Hjorth. The same gauge was later made by Henry Hjorth's own company, the HJORTH LATHE & TOOL CO., Boston, MA.

Micrometer Surface Gauge.

The Best Tool Company, Boston, Mass., for whom Smith & Hemenway Company, 296 Broadway, New York, are direct representatives, have just put on the market

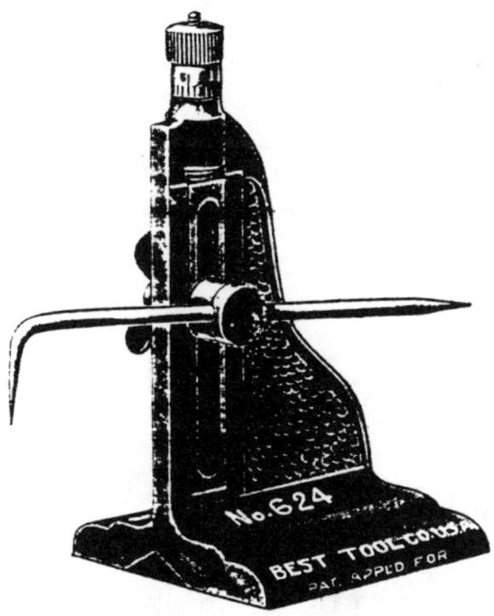

Micrometer Surface Gauge No. 624.

the micrometer surface gauge No. 624 here shown. By means of the micrometer adjustment at the top readings from 0 to 25-1000 inch are obtained at a glance, showing any differences within these limits, after the gauge arm is secured at any desired point by the thumb screw at the left. This device can be used as a surface or depth gauge, 4-inch arms being sent regularly, and extra arms in larger or smaller sizes, according to the needs of the buyer. It is suitable for use in connection with planers, shapers, lathe work or other machine shop purposes.

Iron Age, March 3, 1904

BETTS MACHINE CO. Wilmington, DE

Founded by Alfred Betts in 1878. In 1879 Betts bought control of AMERICAN STANDARD GAUGE & TOOL CO. moving it to Wilmington, DE. The company's primary products were large machine tools, but Betts also made caliper gauges, step gauges and measuring machines which had formerly been made by the AMERICAN STANDARD GAUGE & TOOL CO. The gauge line was sold in 1895 to the TAYLOR-RICE ENGINEERING CO., Wilmington, DE.

American Machinist, 1882

BICKNELL-THOMAS CO. Greenfield, MA

Introduced, in 1917, a thread lead indicator patented on April 17, 1917 (No. 1223012) by Frederick W. Whelpley and Frank A. Bicknell.

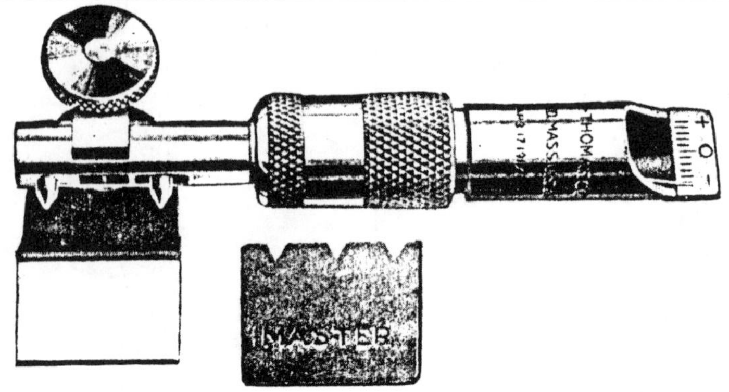

Fig. 1. Thread Lead Indicator and "Master" made by Bicknell-Thomas Co.

Machinery, 1917

BILLINGS & SPENCER CO. Hartford, CT

A partnership of Charles E. Billings (1835-1920) and Christopher Spencer (1833-1922). Spencer is best known as the inventor of the Spencer carbine used by Union forces during the Civil War. Billings was president from formation of the company as the ROPER SPORTING ARMS CO. in 1869 until his death in 1920. Spencer left the company in 1873, soon after the reorganization as the BILLINGS & SPENCER CO. The firm's primary products were contract drop forgings of all types, especially for arms makers such as Colt and Remington. As a natural outgrowth of the drop forging business, BILLINGS & SPENCER began making tools such as screw plates, ratchet drills, and wrenches about 1880. Billing's patent surface gauges were introduced in 1883 and his beam caliper in 1889. The Billing's patent magazine screwdriver was offered in 1894. Hand vises and micrometer stands were added in 1896. The company introduced a quick adjusting depth gauge and a pocket caliper gauge in 1899. Charles E. Billings was a prolific inventor, receiving a large number of patents for drop forgings and wrenches as well as the following patents for machinist's tools:

Tap wrench, May 6, 1879, (No. 215039)
Surface Gauge, August 1, 1882 (No. 261901)
Beam Calipers, May 1, 1888 (No. 382348)
Pocket Knife, March 15, 1892 (No. 47077) (the magazine screwdriver was covered by this patent)
Caliper Gauge, March 27, 1894 (No. 517208)
Hand Vise, August 20, 1894 (No. 525311)
Combination Tool, February 4, 1896 (No. 554046)
Combination Measuring Tool, August 11, 1903 (No. 735935)
Beam Calipers, November 17, 1903 (No. 744028)

The BILLINGS & SPENCER CO.'s 1911 catalog offered: Jeweler's Anvils and Bases, Caliper Gauges, Hand Vises, Combination Knives, Micrometer Holders, Steel Rules, Surface Gauges, and Tack Hammers. By 1914 they had discontinued precision machinist's tools. The catalog offered only Tap and Reamer Wrenches, Chisels, Machinist's Hammers, and Ratchet Drills. *(continued next three pages)*

Billings' Quick Adjusting Depth Gauge.

Billings & Spencer Company, Hartford, Conn., have put on the market the quick adjusting depth gauge

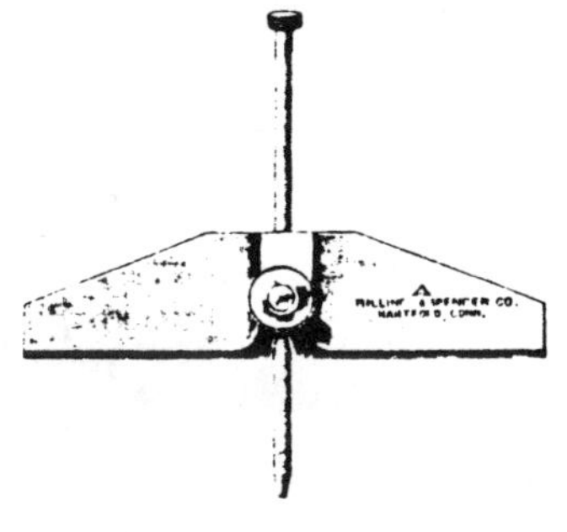

Billings' Quick Adjusting Depth Gauge.

shown in the accompanying cut. In use by making one half turn of the nut and pressing on the same the screw threaded rod is released from the half nut and can then be moved freely up and down, releasing the pressure; the half nut is locked with the screw by action of the spring inside the nut. Fine adjustment can be made by turning the screw. When adjustment is obtained the nut is screwed tight to prevent the screw turning. The gauge is designed principally for the use of die sinkers and tool makers.

Billings' Pocket Caliper Gauge.

The Billings & Spencer Company, Hartford, Conn., are offering the pocket caliper gauge shown in the accompanying cut.

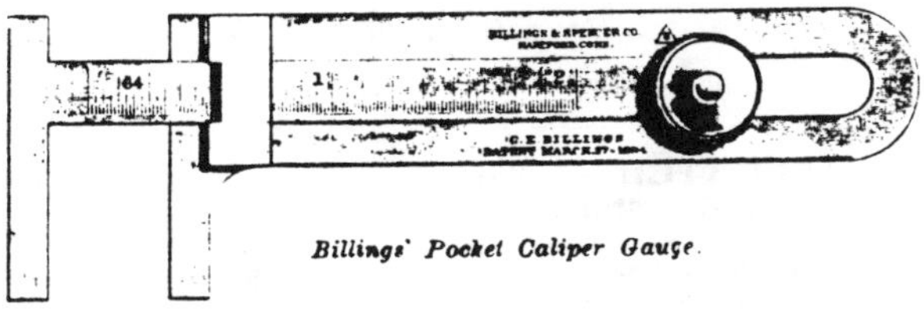

Billings' Pocket Caliper Gauge.

It is a light and convenient tool, graduated on one side to sixty-fourths and on the other to one-hundredths of an inch. The thumb nut is counterbored to receive a spiral spring which bears on the nut and washer, thus creating a moderate friction which, when the nut is loosened, facilitates the setting of the gauge to the desired measurement. The tool, it is remarked, is finished in the best possible manner and opens 2 inches.

Above top: Iron Age, January 5, 1899

Above bottom: Iron Age, November 19, 1899

Right: Iron Age, October 8, 1896

Billings Hand Vise and Clamp.

The Billings & Spencer Company, Hartford, Conn., are introducing the hand vise, also bench clamp for holding the vise, as shown herewith. All parts of the tool are drop forged of best steel for the purpose. The jaws of the vise, Fig. 1, have a positive opening and closing

Fig. 1.—Billings Hand Vise

movement in parallel lines, actuated by a right and left hand screw, moving the jaws simultaneously toward or from each other. A hole extends entirely through the handle so that the jaws will grasp and hold central round wire from 1-16 inch up to and including ¼ inch in diameter. The jaws open ¾ inch. Fig. 2 represents the clamp for holding the vise when in use as a bench vise. There is a

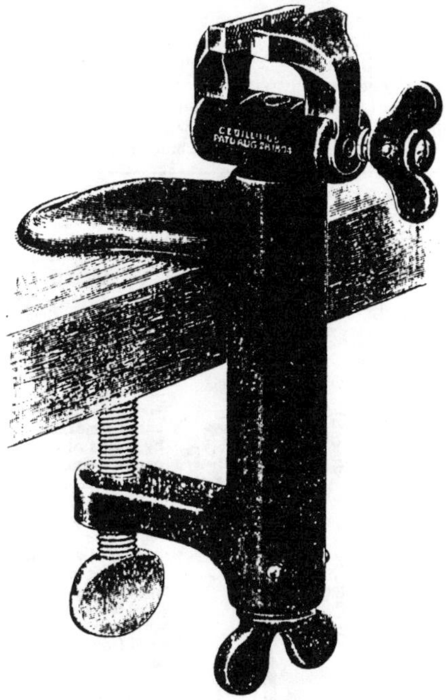

Fig. 2.—Billings Hand Vise and Clamp.

hole corresponding to that in the vise for accommodating wire, through the thumb nut, which secures the hand vise in place at the bottom of the clamp. The vise is well made and finely finished, and is designed for the use of jewelers, tool makers and machinists.

Billings' Micrometer Holder.

The Billings & Spencer Company, Hartford, Conn., is putting on the market a handy device in the shape of a micrometer holder, as shown in the accompanying illustration, which is one-half the size of tool. The holder consists of two parts, a base and a clamp. The clamp is

Billings' Micrometer Holder.

made to tilt at different angles and is furnished with parallel jaws. The piece to be measured may be held in one hand, leaving the other free to adjust the micrometer. The whole is case hardened finish, and in workmanship compares favorably with the micrometer itself. The tool is drop forged throughout and weighs 1½ pounds.

November 2, 1905

The Billings & Spencer Co.

Hartford, Conn., U. S. A.

Billings Patent Magazine Screw Driver.

Drop-forged Steel Handle.
Highly finished and full nickeled.
Made also with Rosewood Handle.

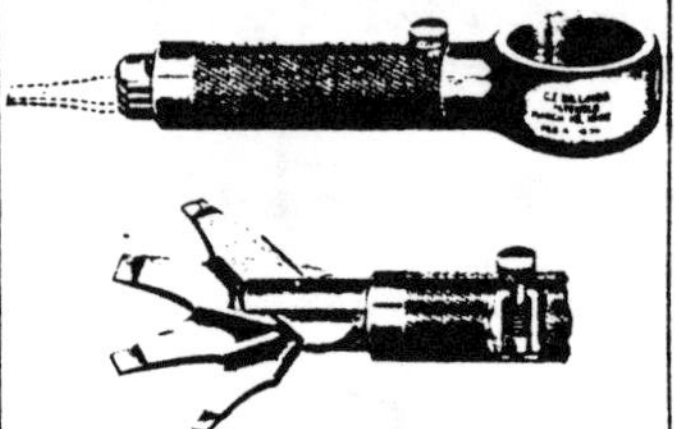

Patented March 16, 1892. Feb. 4, 1896.

A compact and convenient tool. Four sizes screw drivers contained within the handle. The blades are pivoted to the slide and cannot be lost except the whole tool is lost. By raising the locking bolt and pressing forward the slide to end of slot the screw driver blades will swing out. Select the size driver wanted, and draw the slide back into the handle and driver is ready for use.

Total length of tool when closed 3⅞ in. Width of blades ⅛ in., 3-16 in., ¼ in., 5-16 in. Weight of tool, 4 oz.

Machinery, June 1900

(Without Screw Adjustment.)

BILLINGS' IMPROVED SCRATCH GAUGE, With and Without Screw Adjustment.

The Scratch Gauge is fully as useful a tool to the mechanic as the Surface Gauge, and the designer of the tool herewith represented has overcome the objection heretofore existing in setting the Head to the exact desired distance from the Marking Point. This tool is drop-forged of the best steel for the purpose, and finished in a thorough and workmanlike manner. In addition to the sliding head there is also a sliding snug, split and held in place, in any position on the bar, by means of a thumb-screw; this snug is connected to the head by means of a screw encircled by an open spiral spring, the screw passes through a hole in the snug and is fitted with a knurled head thumb nut. A slot is cut at right angles to, and somewhat deeper than the hole that receives the bar. An eye-bolt, provided with a hole same size as that in the head, passes through a hole drilled in the bottom of the center of slot; this hole is enlarged at the bottom to receive a spiral spring that encircles shank of eye-bolt, the shank projecting above the head and being provided with winged thumb nut; the tightening of this nut draws the eye-bolt up and forces the head down upon the bar. When the nut is unscrewed, the spiral spring acting upon the eye-bolt, gives sufficient friction to prevent the head from moving too easily when setting to the desired measurement. In operation, any movement along or around the bar, of the head, with screw adjustment, will be accomplished by the head in consequence of its connection with the screw, but the head may be moved by turning the thumb nut on the screw connected to the head acting with the spiral spring, so that while the snug is held firmly in place by its binding screw, the head may be accurately adjusted to the desired measurement.

No. 1. 7 x $\frac{9}{32}$ with screw adjustment ......each, $2.00
" 2. 7 x $\frac{9}{32}$ without screw adjustment.... " 1.50

(With Screw Adjustment.)

Montgomery & Co. 1890

The Billings Hand Vise.

THE BEST VISE FOR LINEMEN.

Drop Forged from Bar Steel. Interchangeable Parts. Parallel Jaws. Hardened by Special Process.

THE BILLINGS & SPENCER CO.,
HARTFORD, CONN.

Chicago Office, 17 S. CANAL STREET.

England—CHARLES CHURCHILL & CO., 23 Cross St., Finsbury, London, E. C. France—L. ROFFO, 58 Boulevarde Richard Lenoir, Paris. Russia—J. BLOCK, Moscow.

American Machinist, June 14, 1894

DROP FORGINGS

BILLINGS & SPENCER CO. HARTFORD, CONN. U.S.A.

ALSO

BILLINGS' PATENT ADJUSTABLE POCKET WRENCHES, 4 AND 6 INCHES.

FINISHED SCREW PLATES AND DIES, GENUINE PACKER RATCHET DRILLS, BILLINGS' PATENT DOUBLE ACTION RATCHET DRILLS, TAP & REAMER WRENCHES.

CLAMP, DIE & COMMON LATHE DOGS, COMBINATION PLIERS, BEACH'S PATENT THREAD-CUTTING TOOLS, BARWICK PIPE WRENCHES.

Billings' Patent Drop Forged and Cold-Pressed Sewing Machine Shuttles.

AND ALL DESCRIPTIONS OF

STEEL AND IRON DROP FORGINGS.

American Machinist 1883

THE BILLINGS & SPENCER CO.,

HARTFORD, CONN., U. S. A.,

MANUFACTURERS OF

BILLINGS' PATENT SURFACE GAUGE.

This Surface Gauge is drop forged, of bar steel, and finished in a thorough manner, and hardened.

Send for Illustrated Catalogue and Price List of Machinists' Tools.

DROP FORGINGS OF BRONZE, COPPER, IRON AND STEEL OF EVERY DESCRIPTION.

American Machinist 1889

THE BILLINGS & SPENCER CO.,

HARTFORD, CONN., U. S. A.

MANUFACTURERS OF

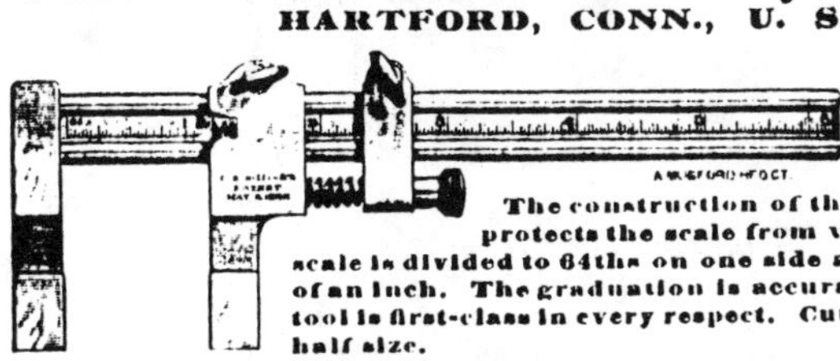

BILLINGS' PATENT BEAM CALIPER.

The construction of the Caliper is such that it protects the scale from wear and abrasion. The scale is divided to 64ths on one side and the other to 100ths of an inch. The graduation is accurate and the finish of the tool is first-class in every respect. Cut represents Caliper one-half size.

Drop Forgings of Bronze, Copper, Iron and Steel of all descriptions.

American Machinist 1889

BIRNER, C. P. Milwaukee, WI

Maker of a combination surface gauge and indicator in 1900.

BLAKE & BRADLEY Swanton, VT

Introduced a flexible fixed curve rule in 1891. The tool used a L. S. STARRETT flexible rule to which Blake & Bradley added a series of posts as shown below.

Flexible Fixed Curve Rule.

Blake & Bradley, Swanton, Vt., are putting on the market a flexible fixed curve rule, as illustrated herewith. It consists of a straight flexible steel rule, Fig. 1, with

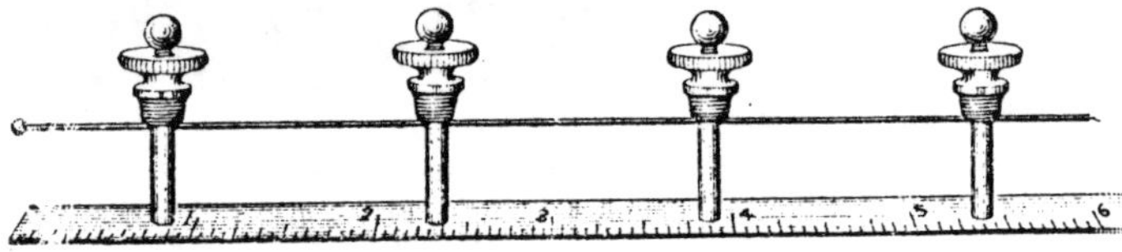

Fig. 1.—Flexible Fixed Curve Rule.

set-screw posts, through which a flexible steel rod runs. The rule is graduated both inside and outside, and can be set in any desired curve, even to form a complete circle. The rule is bent as desired, when the set screws are screwed down on to the rod, thus holding it in a fixed position, Fig. 2. It is designed for the use of

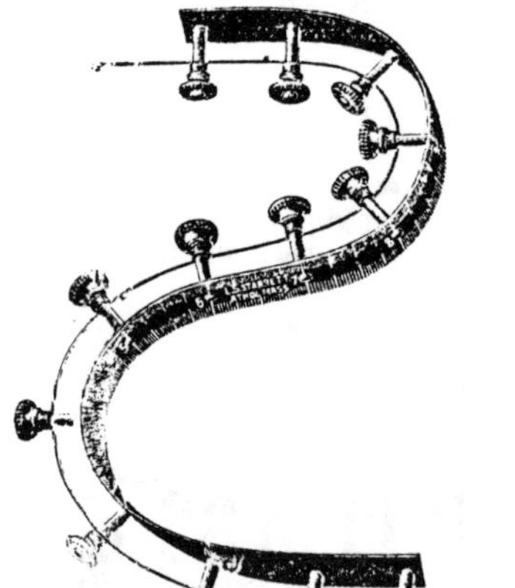

Fig. 2.—Flexible Rule Set on Curves.

draftsmen and architects in drawing, measuring and transferring regular or irregular curves; also in duplicating and measuring designs, or enlarging for working plans to a scale of inches. The rules are made in six sizes and lengths, and are put up one each in an ebonized case.

Iron Age, July 2, 1891

BLODGETT MFG. CO. Rochester, NY

Makers of screw thread gauges, thickness gauges and twist drill gauges from c1896 to c1905.

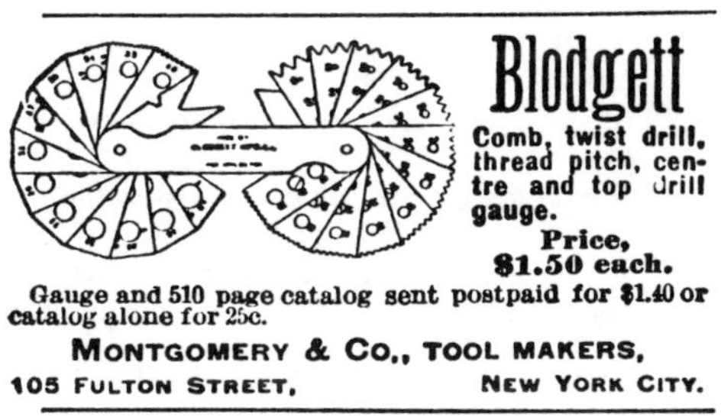

American Machinist 1898

The Blodgett Machinists' Gauge No. 7.

The accompanying cut represents one of a variety of gauges offered by the Blodgett Mfg. Company, 187 State street, Rochester, N. Y. The gauge differs from the company's No. 2 in having a machine screw gauge

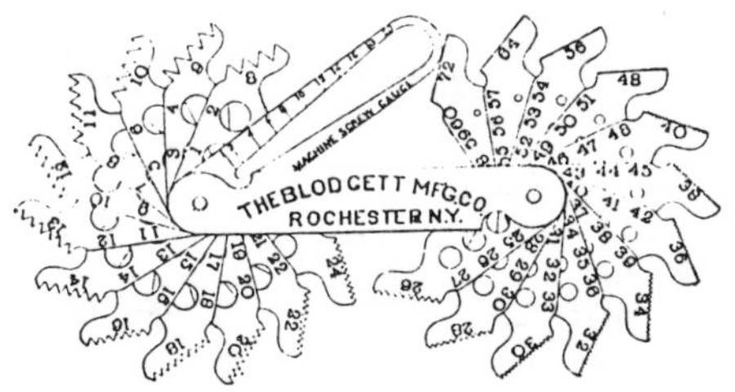

The Blodgett Machinists' Gauge No. 7.

assembled in place of a center gauge. It is referred to as making a convenient combination, weighing 2 ounces, measuring ½ inch square and 3 inches long. Machine screw gauges are manufactured separately by the company, as are also twist drill gauges.

Iron Age, January 2, 1902

BLUSH TOOL CO., A. T. Erie, PA

Introduced a "multiple micrometer" in 1920. The tool had been patented by Angus T. Blush on July 29, 1919 (No. 1311548). *(see next page for illustration)*

BOKER & CO., H. New York, NY

This company was a large importer of tools, including forged machinist's calipers. Most, if not all, appear to have been made in Germany.

BOSTON TOOL & MFG. CO. Boston, MA

Maker of the Boston Positioning Indicator. Little is known about the tool or the company.

Machinery, July, 1920

See entry for Blush Tool Co., previous page

BLUSH MULTIPLE MICROMETER

A new type of micrometer, the purpose of which is to eliminate the necessity of using two instruments when taking measurements ranging from 0 up to 2 inches, has been recently developed by the A. T. Blush Tool Co., 1145 W. 11th St., Erie, Pa. This is accomplished by the use of two screws or threads having different pitches, namely, a 20-pitch thread on the measuring screw and a 40-pitch thread for the thimble carrier. Although the mechanism embodies two screws, the arrangement is such that the accuracy depends entirely

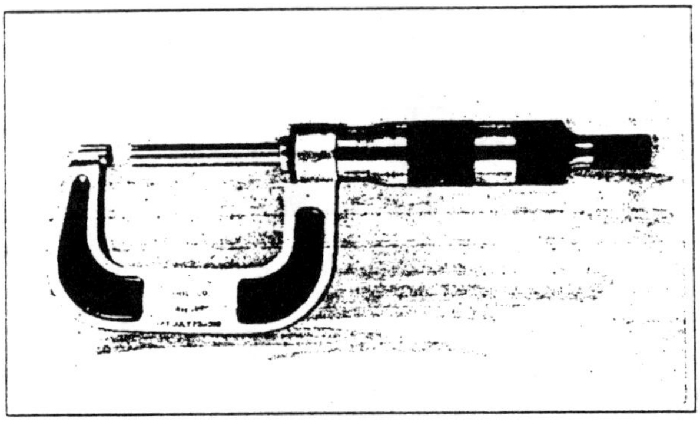

Multiple Micrometer made by the A. T. Blush Tool Co.

upon the 20-pitch thread, as the function of the 40-pitch thread is simply that of keeping the thimble in proper alignment with the graduations on the hub. The difference in the pitch and travel of these threads makes it possible to obtain a 2-inch movement of the spindle with only a 1-inch movement of the thimble.

The adjustment for wear on the anvil and spindle is made without changing the position of either the anvil or the measuring nut, which is a very desirable feature, as there is absolutely no chance for the anvil and the spindle to be turned out of parallel when making adjustments. Arrangements are being made to manufacture this micrometer in several combinations such as, 1½ to 3, 2 to 4, 3 to 5, 4 to 6, and 5 to 7 inches.

BOULET'S FINE TOOL WORKS, INC. Beverly, MA

Operated by Joseph Boulet who received patents for a Gear Tooth Caliper on February 6, 1900 (No. 643080), and Micrometer-indicators on October 2, 1900 (No. 659048), February 23, 1904 (No. 752782) and May 13, 1919 (No. 1303245). The gear tooth caliper was introduced in 1900. His first micrometer indicator was introduced in 1901 with an improved version introduced about 1910. Boulet indicators and micrometers made before he formed his own company in 1908 were made by the FALLON MACHINERY CO., Salem, MA. An article written by Joseph Boulet appeared in the October 17, 1907 issue of *American Machinist* magazine. Boulet, at that time, was chief inspector, toolmaking department, United Shoe Machinery Co., Beverly, MA. Boulet was still making indicators and other machinist's tools when he announced a move to Sebago Lake, ME, in 1913. However, no Boulet tools marked Sebago Lake have yet been observed. *(continued next page)*

Boulet's Fine Tool Works

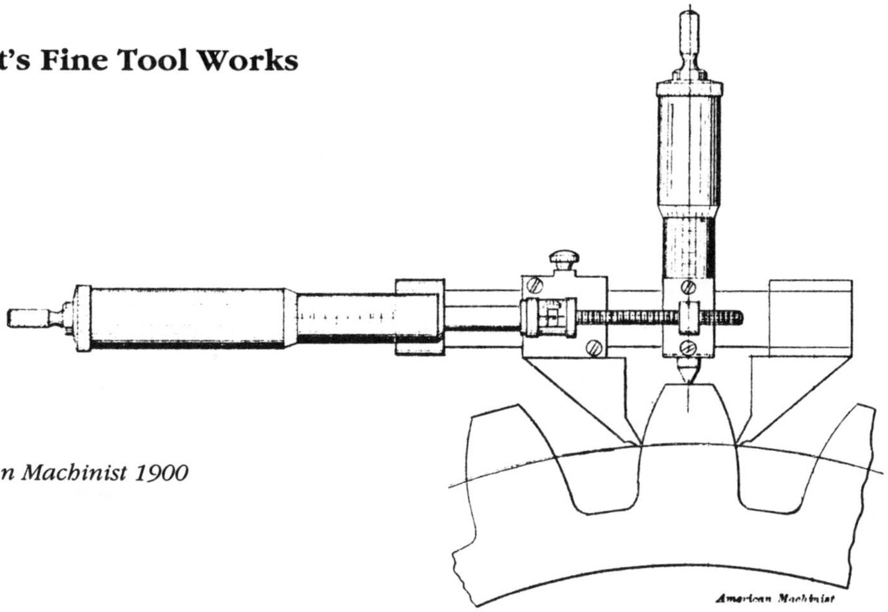

A MICROMETER GAGE FOR GEAR TEETH.

American Machinist 1900

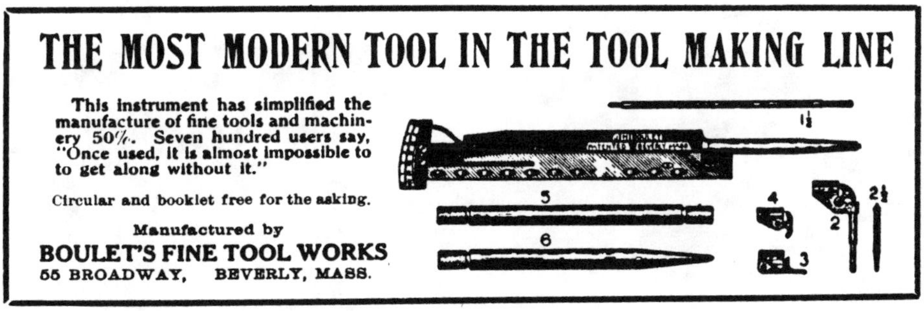

American Machinist 1910

BRAUNSDORF-MUELLER CO. Elizabeth, NJ

Established in 1888 and incorporated in 1901, the company was a maker of a variety of tools, including a speed indicator marked B-M CO.

BROWN, DAVID G. Philadelphia, PA

Inventor and maker of an inside micrometer, patented August 16, 1887 (No. 368563). This appears to be the first example of a tubular inside micrometer with screw-on extensions.

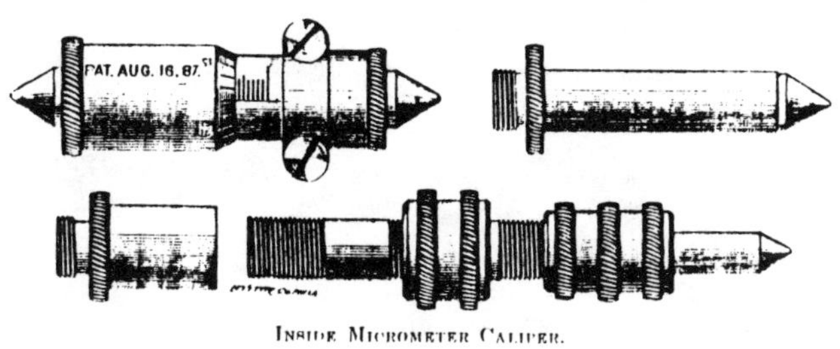

Inside Micrometer Caliper.

American Machinist 1888

BROWN & CO., R. H. New Haven, CT

Operated by R. H. Brown who advertised himself as the "Successor to Frank E. Clark". Brown made a variety of tools including the Clark patent screwdriver, Clark patent expansion bits, and bit braces. The only known machinist's tool is a micrometer-shaped slide caliper patented March 11, 1890 (No. 423284), by Morris F. Smith of New Haven, CT.

BUTT, B.B. Newark, NJ

Operated by Bozwell B. Butt who was the maker, in 1885, of a combination level, inclinometer, scribe and gauge for which he was granted a patent on August 21, 1883 (No. 283564). Butt also made a small, cast iron pocket level. The level is 3" long, of semi-circular shape and is made to fasten to a rule or straight edge. The only known example is marked, simply, PAT. JUNE 23, 85 (No. 320537).

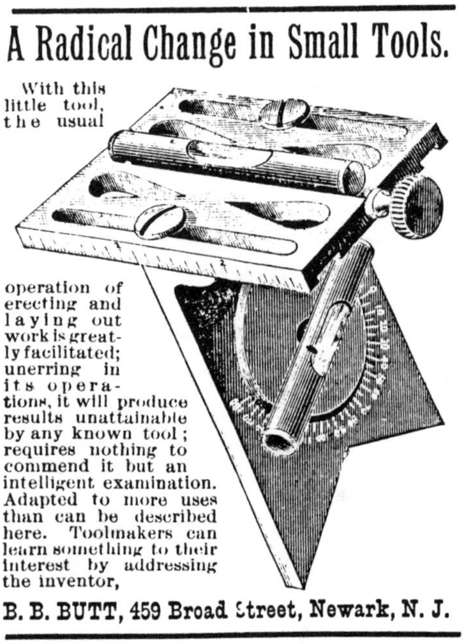

1885

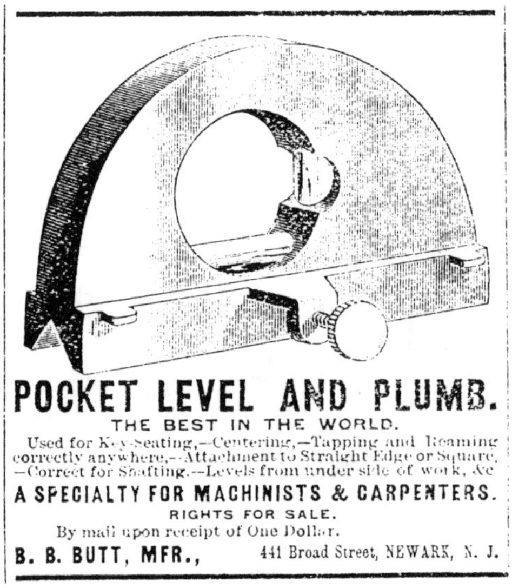

1891

both American Machinist

BYER TOOL CO. Chambersburg, PA

Maker of detachable thread points for calipers, patented April 21, 1896 (No. 558773) by David S. Byer. In 1899, the Byer Tool Co. advertised the points for 50 cents and for another 10 cents it would send a 4" steel rule. The cut of the rule shown in the ad appears to be that of a COFFIN & LEIGHTON rule even to the August 25, 1885 patent date.

American Machinist 1899

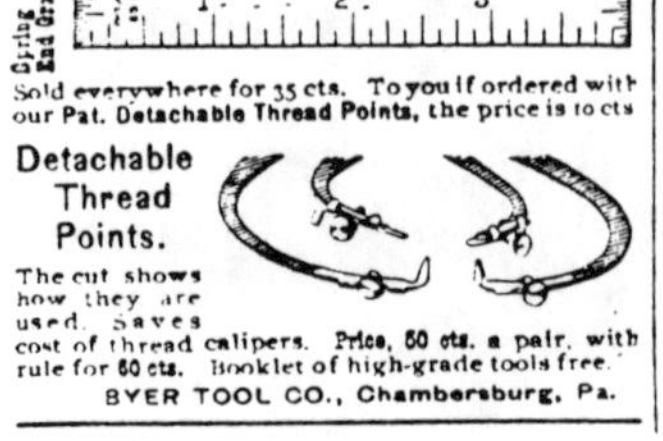

C & F Boston, MA See CHANDLER & FARQUHAR

CARR, JAMES Syracuse, NY

Advertised, in 1891, a key seat rule clamp for which he had received a patent on May 27, 1890 (No. 428890). The clamp was designed to hold together two rules or straight edges at a 90° angle. Carr also received patents for surface gauges on January 26, 1892 (No. 467732), and February 2, 1892 (No. 468093). The gauge was manufactured by the HOGGSON & PETTIS MFG. CO., New Haven, CT, beginning in 1891. Early examples are marked only PAT APL FOR without a company name. Later manufacture is marked with the two patent dates and the HOGGSON & PETTIS MFG. CO. name. The above tools, plus "Carr's improved center square" were listed in a 1902 dealer catalog only as "Carr's Patent" tools. No maker was specified. Carr may, at some time, have begun making the tools under his own name but it is more likely that all were made by the HOGGSON & PETTIS MFG. CO. Carr continued a close relationship with the HOGGSON & PETTIS MFG. CO. for many years and assigned a turning tool holder patent to the firm July 18, 1899. *(continued next page)*

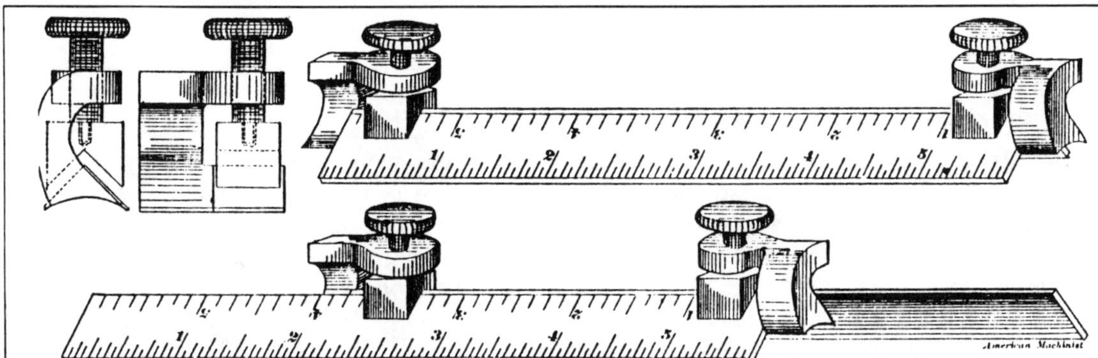

KEY-SEAT RULE CLAMP.

American Machinist 1891

CARR'S PATENT IMPROVED

COMBINATION SCRIBING BLOCK

FIG. 37.

THIS Tool fills a want for a neat and convenient Scribing Block that can be quickly adjusted to the finest work. It contains the patented fine adjustment similar to that on our surface gauge, the needle being set in any position and held perfectly rigid by the side thumb nut while using the fine adjustment, by means of the eccentric thumb washer on top. It has a V slot in the bottom of base by which work can be quickly and accurately trued to a boring bar. It also makes a useful and convenient scratch and depth gauge.

CARR'S IMPROVED

CENTRE SQUARE

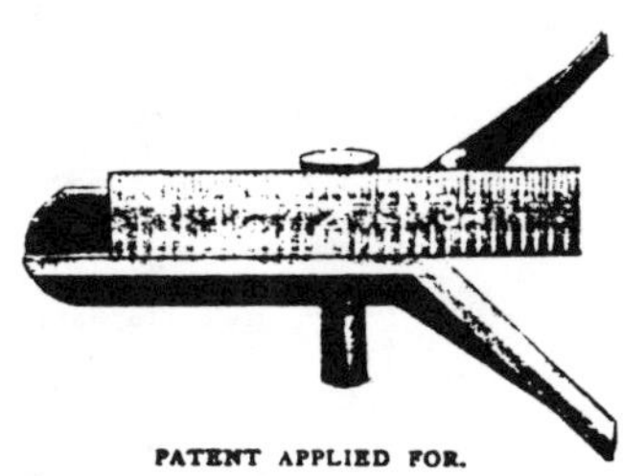

PATENT APPLIED FOR.

FOR many years machinists and patternmakers have needed a Centre Square of simple construction, which could be used with any straight-edge or scale. This square which we now offer is simple, complete, and accurate, and a straight-edge or scale of any thickness from ½ to 1¼ inches in width can be adjusted to it.

both 1902

CARR'S PATENT IMPROVED
COMBINATION SURFACE GAUGE

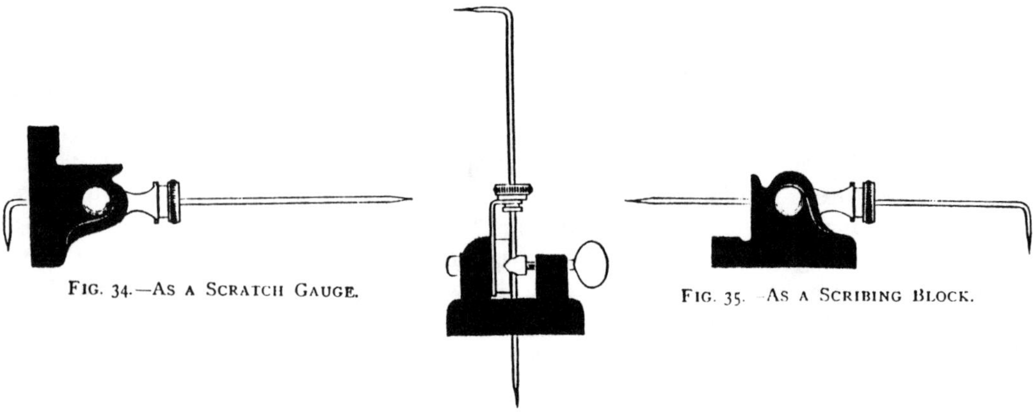

FIG. 34.—AS A SCRATCH GAUGE.

FIG. 35.—AS A SCRIBING BLOCK.

FIG. 36.—AS A DEPTH GAUGE.

1902

CENTRAL TOOL CO. Providence, RI

Organized in 1908 with $10,000 capital. Incorporators were Arthur P. Mowry, Louis S. Moulthrop and Fernando O. Jaques, Jr. The company was formed to make dental tools, micrometers and other machinist's tools. The firm may have been a reorganization of the WINDSOR & JERAULD MFG. CO., Providence, RI, to which Jaques had assigned his micrometer patents of October 27, 1908 (Nos. 902492 & 902493). The following F. O. Jaques micrometer patents were assigned to the CENTRAL TOOL CO.: September 21, 1909 (No. 934730); January 1, 1918 (No. 1252009); January 28, 1919 (No. 1292792); July 1, 1919 (1308493); May 4, 1920 (No. 1338932). The firm continued to make micrometers into the 1960's or longer. Very few, except the "Jaques" direct reading models introduced in 1919, have any collector appeal.

CENTRAL TOOL CO.'S DIRECT-READING MICROMETER

An improved type of direct-reading micrometer has recently been placed upon the market by the Central Tool Co., of Auburn, R. I. This is known as the "Jaques" easy-reading micrometer, and is sold by the Production Equipment Co., Inc., 1 Madison Ave., New York City. One of the advantages of this tool is the fact that it is so simple to read that anyone without a knowledge of the principles governing the operation of a micrometer can use it quite easily. Instead of having forty graduations on the barrel as in the case of standard micrometers, there are only ten graduations, and it is only necessary to glance at the little dial windows in order to read directly the full dimension of the work. The first decimal of the reading is taken from the barrel graduation, while the second and third decimal places are read direct from the dial. Another feature that is of value in this micrometer, is that although it is provided with a dial mechanism, giving the easy reading feature, it is very

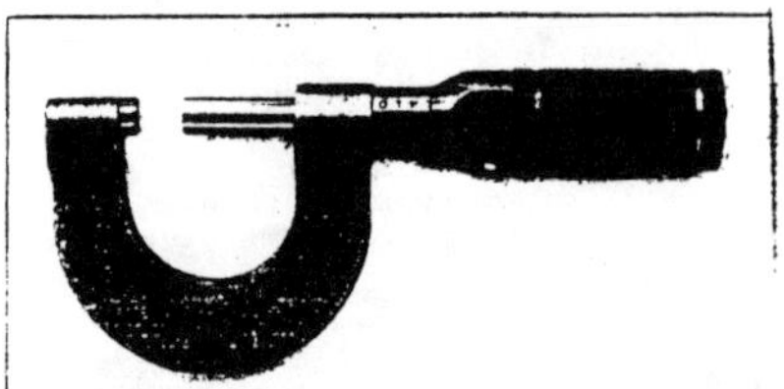

Direct-reading Micrometer made by the Central Tool Co.

little larger than a regular micrometer.

Machinery, October 1919

CHAMBERLAIN, MORRIS Bartley, NJ

Inventor and maker, in 1902, of a gauge for setting tools for planing keyways, laying out hexagons, testing twist drills, etc. The tool was also usable as a level. Chamberlain received his patent on March 18, 1902 (No. 695793).

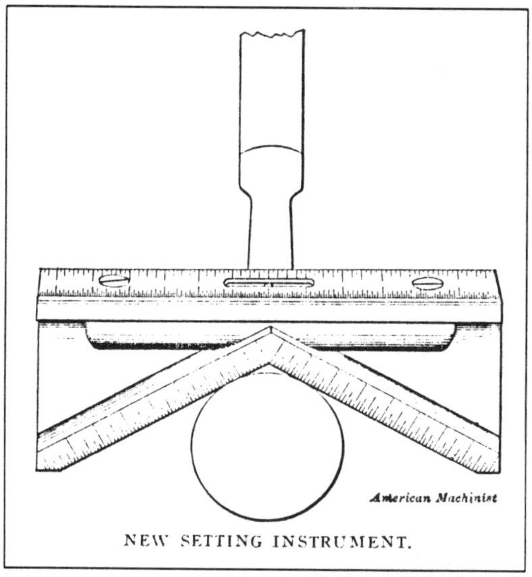

NEW SETTING INSTRUMENT.

American Machinist 1902

CHANDLER, C. E. South Boston, MA

Maker, in 1888, of a micrometer stand to convert hand micrometers to bench micrometers. A very similar unit was later offered by the L. S. STARRETT CO., Athol, MA. Chandler was also the maker of C. H. Fowler's patent speed indicator patented July 31, 1883 (No. 282067) and marked with the Chandler name. C. E. Chandler may have been associated with CHANDLER & FARQUHAR, Boston, MA, which sold the Fowler speed indicator for many years.

MICROMETER STAND.

1888

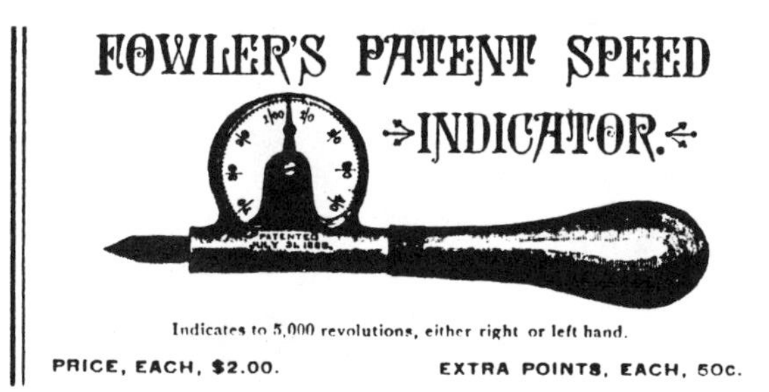

1891

CHANDLER & FARQUHAR Boston, MA

Formed in 1882 as a partnership of Frank Chandler (1852-1906) and Charles S. Farquhar who were both employees of A. J. WILKINSON before starting their own business. The firm operated as a dealer only and specialized in machinist's tools and equipment. In 1904 the partnership was incorporated as the CHANDLER & FARQUHAR CO. with Chandler as president and Robert J. Lynd as vice president. Farquhar served as president from the death of Frank Chandler in 1906 until the firm was reorganized as the LYNN-FARQUHAR CO. in 1917.

CHANDLER & CO., R. Springfield, MA

Maker of vernier calipers c1880. Little is known of the firm or the tool.

CHRISTMAN, A.B. Cleveland, OH

Inventor and maker, in 1902, of a thread gauge for gauging tools for cutting V threads, worms, or acme threads.

American Machinist 1902

A NEW THREAD GAGE.

CHURCH, GEORGE W. Roseville, NJ

Grandson of Henry F. Frasse, operator of FRASSE & CO., New York, NY. Inventor and maker of pocket and bench micrometers, patented May 22, 1883 (No. 278094). The significant part of the patent covered the first successful ratchet stop for micrometers. Church bench micrometers were advertised by FRASSE & CO. in 1882 and were shown in the 1884 Charles H. Besley & Co. catalog. Besley also listed the pocket micrometer. In 1898 Church incorporated as the:

CHURCH CO., GEORGE W. New York, NY

Directors were: George W. Church, William T. Church and W. M. Kinder. The company was incorporated to manufacture machinist's supplies. *(continued next page)*

SNAP GAUGE No. 1.

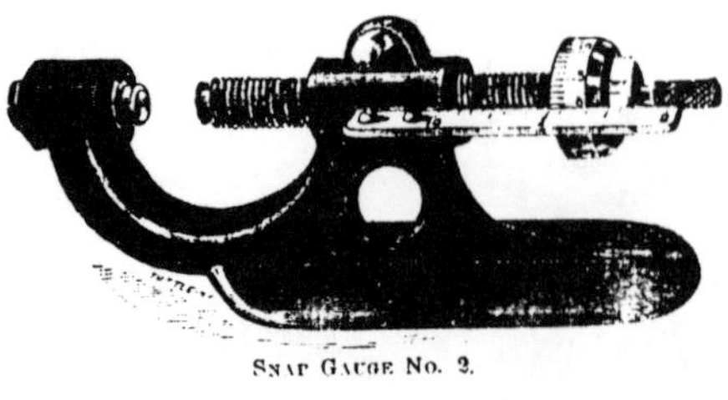

1882

SNAP GAUGE No. 2.

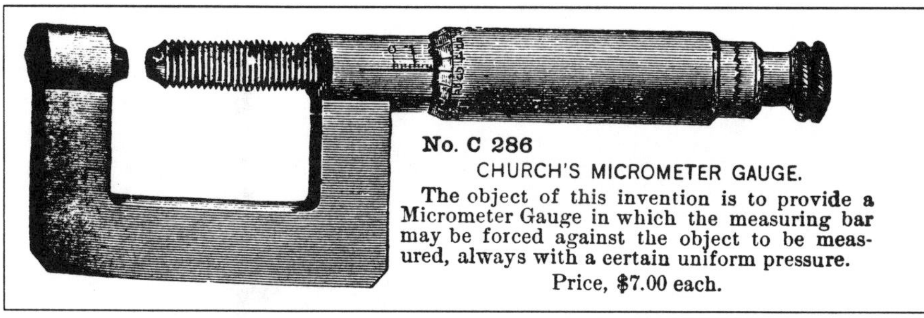

1884

CHURCH & SLEIGHT New York, NY

Advertised, in 1891, Church's speed indicator, patented by William Lang on November 28, 1882 (No. 268107), and Keenan's bevel protractor (patent applied for). This company may have been a dealer only, but there are indications that they made, or had made for them, some tools.

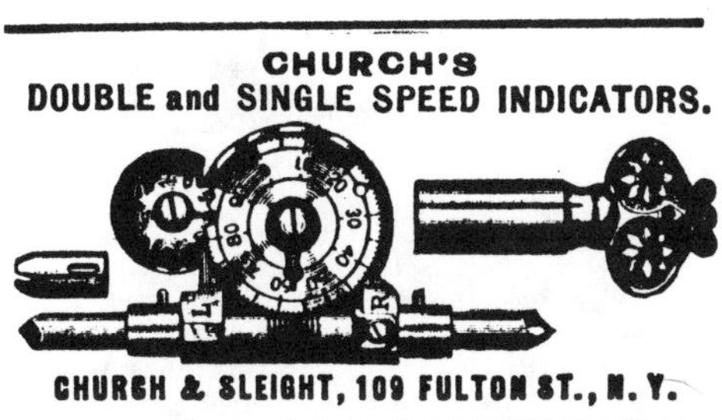

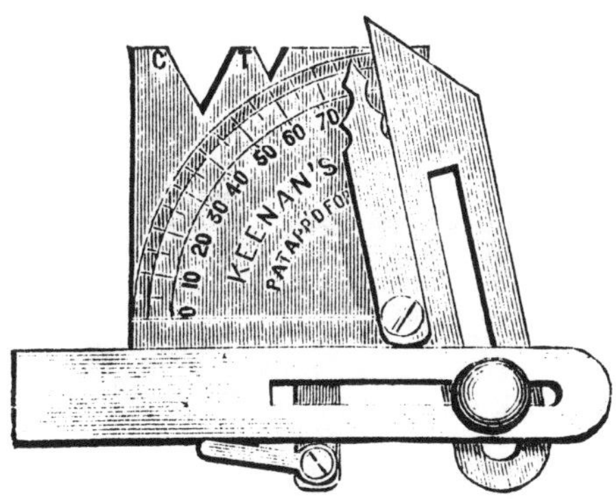

A New Bevel Protractor.

American Machinist 1891

CICHY & PESEK Chicago, IL

A partnership of Edward Cichy and Rudolf Pesek, joint patentees of an indicating caliper patented November 14, 1922 (No. 1435630). The only known example of their caliper is completely unmarked.

CINCINNATI TOOL CO. Cincinnati, OH

This company was primarily a maker of woodworking tools. In 1892, however, they began offering a surface gauge which appears to be based on the H. Avery patent of November 21, 1882. *(see illustration next page)*

CLARK, C.D. Torrington, CT

Inventor and maker of a micrometer height gauge patented September 12, 1905 (No. 799533). *(see illustration next page)*

CLEMENS & FARREL Ansonia, CT

Makers, in 1883, of an adjustable bevel for machinists. The tool was made under a patent granted December 27, 1881 (No. 251524), to E. V. Clemens, Ansonia, CT, who assigned one-half to Franklin Farrel, also of Ansonia. Franklin Farrel was the owner of the FARREL FOUNDRY & MACHINE CO., Ansonia, CT, and was probably a financial backer for Clemens.

Improved Surface Gauge.

The Cincinnati Tool Company, Cincinnati, Ohio, are offering this article, as shown in the accompanying illustration. The special adjustment is a stud fitted to the base of the tool with a fine thread at the upper end and a round knurled nut fitting the screw. A steel spring is also fitted to the stud in the base, holding the nut firmly against the top of the base, and by turning the nut as close an adjustment as is desired can be obtained. The method of adjusting the scriber or needle on the staff and clamping it at the same time is referred to as making this a very desirable tool. This part consists of a loose lug with a hole in it to fit the staff. The lug fits into another piece also on the staff, and two washers with half a hole in each the staff, and at the same time clamps the washers tightly on the needle, holding, it is stated, all perfectly rigid and tight. An extra long staff is furnished when required. The staff is made tapering and fits in a taper hole in the stud at the bottom, a tap of a hammer being sufficient to drive out the staff or to fasten it firmly in place. The tool is designed especially for both close and rapid work, and also for its durability. It can be shifted by hand closely to the work, and when very close adjustment is required the nut at the bottom of the staff is used. The hight of the tool over all is 14½ inches, length of staff 12 inches, length of needle 11 inches. The special adjustment is ⅜ inch, scratches sideways 10 inches and scratches in hight 23 inches. The tools are finished with japanned and nickeled bases.

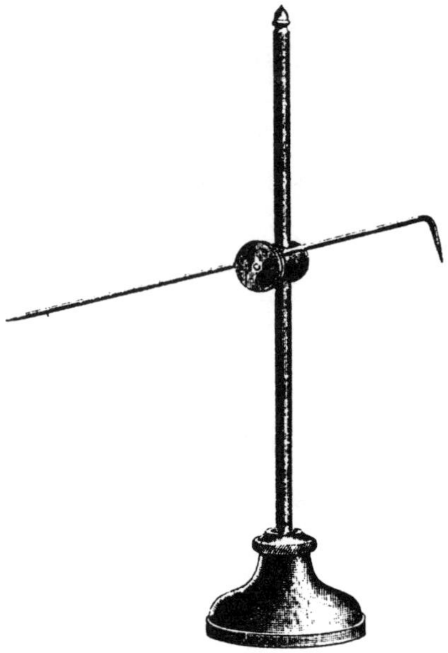

Improved Surface Gauge.

Iron Age, July 14, 1892

see entry for Cincinnati Tool, previous page

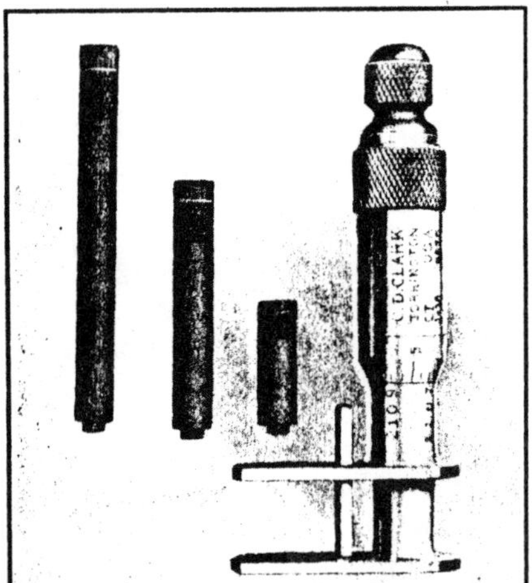

FIG. I. MICROMETER HIGHT GAGE.

Machinery 1905

See entry for C.D. Clark, previous page

CLEVELAND TWIST DRILL CO. Cleveland, OH

A large manufacturer of cutting tools, founded in 1874 by Joseph Cox. The company began offering for sale, in 1893, an adjustable snap gauge which it had been making for its own use. The gauge is similar to a micrometer without calibration numbers and with a spindle lock. The company was listed, in 1905, as makers of snap gauges, universal screw cutting gauges and drill gauges as well as the primary line of cutting tools.

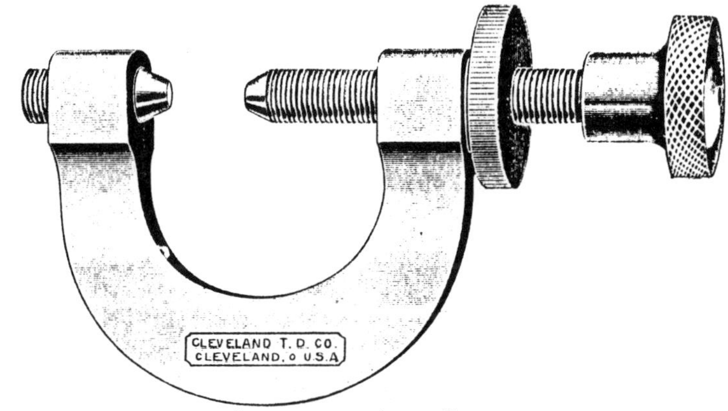

ADJUSTABLE SNAP GAUGE.

COFFIN & LEIGHTON Syracuse, NY

A partnership of John Coffin and Herbert J. Leighton formed in 1885. Their first products were steel rules, patented jointly on August 25, 1885 (No. 325096). The patent covered end graduations, the first to be offered. The partners introduced, in 1895, an adjustable notch centre gauge, and by 1897 a bevel protractor, patented by Herbert J. Leighton on January 17, 1893 (No. 490121). Its 1897 catalog of "improved machinist's tools" offered: End Graduated Rules, Light Tempered, Semi-flexible, Flexible and Narrow Tempered Rules, Adjustable Notch Center Gauge, Improved Center Gauge, Machinist's Protractor and Draftsmen's Protractor. On October 24, 1901, the company was bought by the GOODELL-PRATT CO. and moved to Greenfield, MA. The same line of rules were then made under the MASSACHUSETTS TOOL CO. name. The BROWN & SHARPE MFG. CO., in its 1904 catalog, offered rules with end graduations and marked with the Coffin & Leighton August 25, 1885, patent date. Evidently, at some point, B&S bought either the patent or a license.

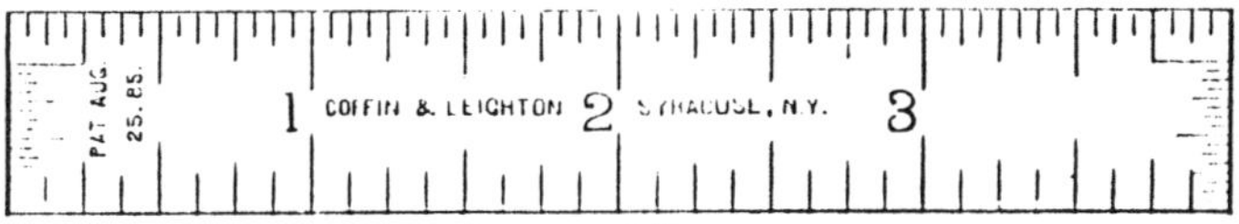

MACHINISTS' SCALES.

Machinists' Improved Steel Scales.

The engraving represents an improved steel scale, made by Coffin & Leighton, Syracuse, N. Y. The improvement consists in graduating the ends, as plainly shown in the cut. Both ends are graduated, to provide for taking either right or left-hand measurements. By means of this improvement, measurements can readily be taken in grooves, recesses and places where it would be impossible to use the long end of the scale. It will be found very convenient for measuring shoulders on turned work without taking it from the lathe, where the center prevents the use of the ordinary scale.

In other respects, the graduating lines on these scales are clearly and accurately cut, and the general finish is excellent.

American Machinist 1885

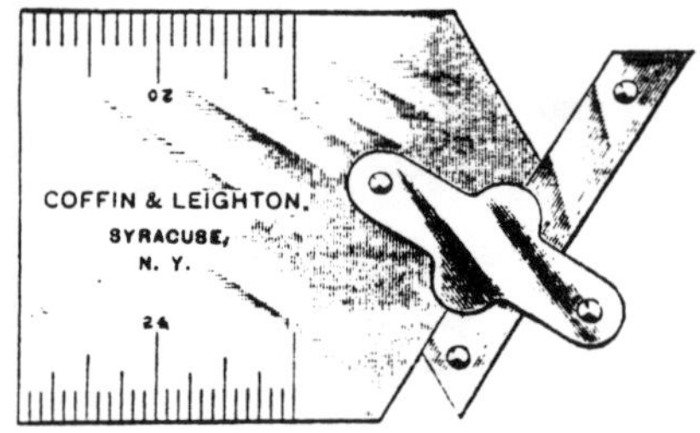

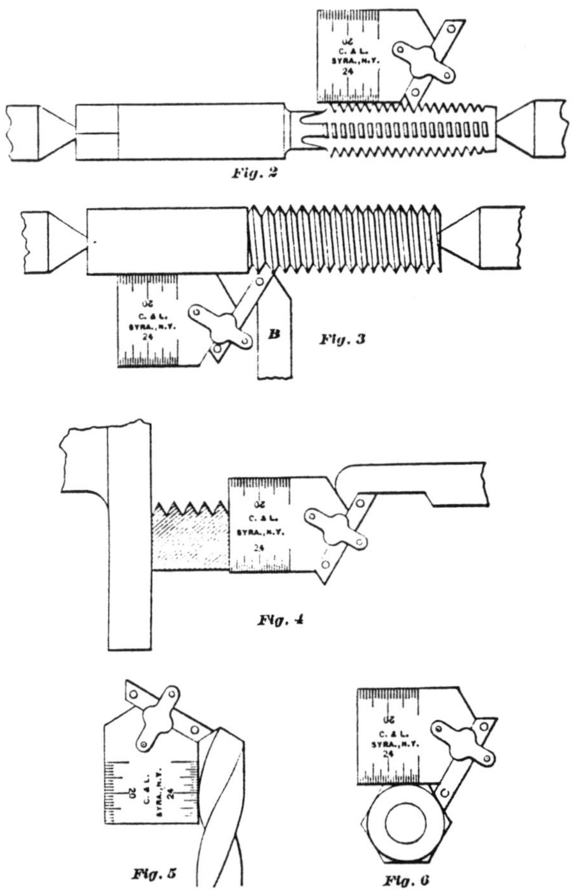

American Machinist

A NEW THREAD AND CENTER GAUGE.

American Machinist 1895

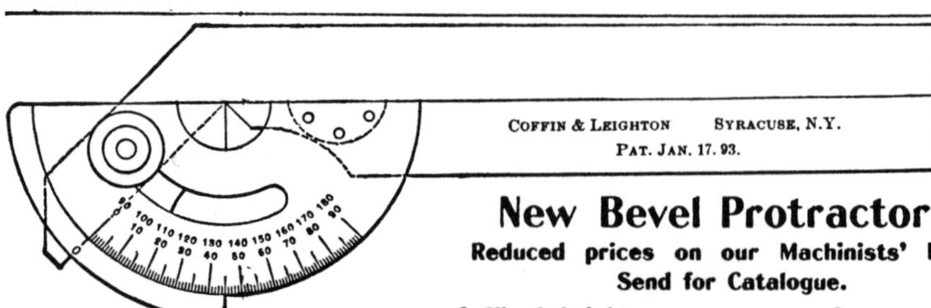

New Bevel Protractors.

Reduced prices on our Machinists' Rules.

Send for Catalogue.

Coffin & Leighton, Syracuse, N. Y.

American Machinist 1897

COLUMBUS DIE, TOOL & MCH. CO. Columbus, OH

Maker of surface gauges in 1929.

CONN. CUTLERY CO. Naugatuck, CT

Maker of the Taylor's patent, double dial speed indicator from about 1879 until at least 1890.

Double Dial Speed Indicator.

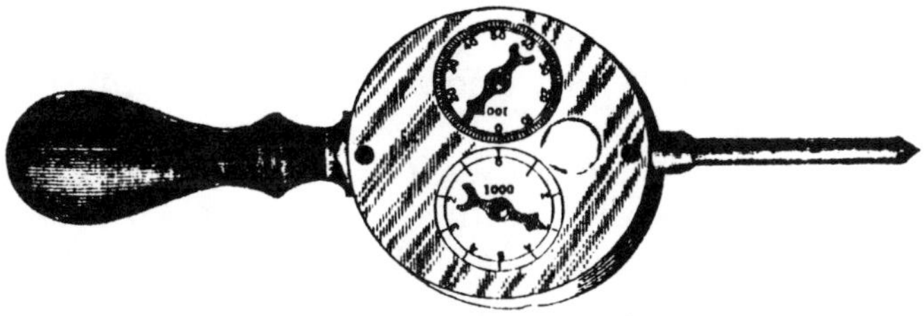

Registering to 10,000 revolutions.

Each $6 00

1889

CONSOLIDATED TOOL WORKS, INC. New York, NY

Introduced the "Simplex" direct reading micrometer in 1920.

CONSOLIDATED TOOL WORKS DIRECT-READING MICROMETER

The Consolidated Tool Works, Inc., 261 Broadway, New York City, have recently brought out a new line of direct-reading micrometers. The design of these tools is essentially the same as that of standard types, but they have the

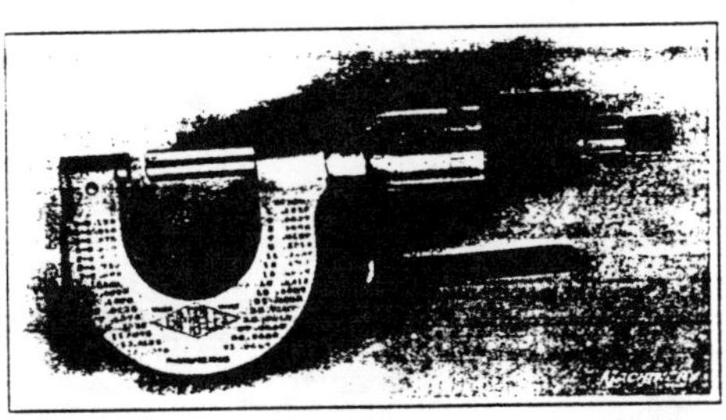

"Simplex" Direct-reading Micrometer made by the Consolidated Tool Works

advantage of direct-reading numerals, which give the exact reading at all times through small apertures in the barrel. This feature not only prevents mistakes being made by unskilled operatives in reading lines, but also avoids errors in computing measurements. There are only two working parts in the direct-reading unit of these micrometers, and both of these units are composed of hardened steel members which are of substantial design. Nine sizes of micrometers are made for the English system of measurement, with a range for taking measurements from 1 inch to 6 inches. The 1-, 2-, and 3-inch sizes can be obtained with graduations reading either to 0.001 or 0.0001 inch, while the 4-, 5-, and 6-inch sizes read to 0.001 inch. These micrometers are also made in six sizes for taking measurements by the metric system. They have a range for measuring to 0.01 millimeter, from 25 up to 150 millimeters. Both the English and the metric system micrometers can be furnished with ratchet stop, lock-nut, or with both ratchet stop and lock-nut.

Machinery 1920

COOK, Albert A. Milford, MA. See COPELAND & CHAMBERLIN

COOK, J.H. Millers Falls, MA

Maker of "J. H. Cook New Micrometer Surface Gauge". The gauge is marked "Pat Applied For" but no patent has been found. This may be the same J. H. Cook that worked at both L. S. STARRETT CO. and SAWYER TOOL MFG. CO. and who received the patent for the calipers made by SAWYER TOOL MFG. CO., but this is not certain.

COPELAND & CHAMBERLIN Worcester, MA

A partnership of Salem Copeland (1815-1891) and Charles W. Chamberlin formed about 1872 to make Cooke's Patent Extension Dividers and Calipers. The tools were based on a patent granted December 12, 1871 (No. 121850), to Albert A. Cook of Milford, MA. Copeland patented his own version of extension dividers on May 24, 1887 (No. 363600)5. He died in 1891 and, in 1901, the firm was reorganized as the COPELAND HARDWARE MFG. CO., Worcester, MA. *(continued next page)*

New Calipers and Dividers.

Neale & Goulding, Worcester, Mass., are introducing to the trade the Copeland Patent Divider and Caliper, as shown in the accompanying illustrations, Figs. 1, 2 and 3. Fig 1 represents the No. 1 divider, which is made in 4, 5, 6, 7, 8, 9 and 10 inch sizes. A small pencil can be used in 4 and 5 inch and a common pencil in other sizes in place of movable point. The 4 inch will scribe an 11-inch circle, the 5 inch a 13-inch circle, the 6 inch a 17-inch circle, the 7 inch a 20-inch circle, the 8 inch a 22-inch circle, the 9 inch a 27-inch circle and the 10 inch a 30-inch circle. The 4 and 5 inch are especially adapted for draftsmen and students. Fig. 2 represents the No. 2 divider, made in 6, 7, 8, 10 and 15 inch, in which a common pencil can be used in place of movable points. The 6 inch by adjustment of points becomes 7 inch, the 7 and 8 inch become 8 and 9 inch and will scribe same size circle as No. 1 of corresponding size. The 10 inch by adjustment becomes 12 inch and will scribe a 35-inch circle and the 15 inch will scribe a 52-inch circle. Fig. 3 represents the outside and inside caliper, which are used with the 6, 7, 8 and 10-inch divider body. With points at right angles to body they are enabled to caliper as small as ¾ inch inside diameter. Adjustment as per dotted lines enables the 6 and 7 inch to caliper 8 inch outside diameter, and by turning the points over they will caliper 9 inch inside diameter. The 8 inch will caliper from ¾ to 13 inch inside diameter and 12 inch outside diameter; the 10 inch will caliper from ¾ to 15 inch inside and 14 inch outside diameter; the 15 inch will caliper 28 inch inside and 28 inch outside diameter. The above described tool is nicely finished, quickly and easily adjusted, and expressly designed, it is stated, for durability and usefulness. The points are steel and properly hardened to mark wood or iron. The combination embodied in this tool will be appreciated by the mechanic, as he has in it a complete divider and an outside and inside caliper. We are advised that great care is used to make this tool a superior article in every way, and it is suggested that a trial order will suffice to show its utility and excellence.

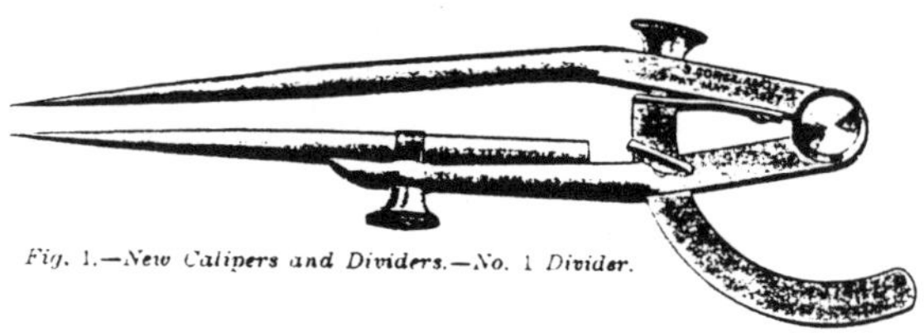

Fig. 1.—New Calipers and Dividers.—No. 1 Divider.

Fig. 2.—No. 2 Divider.

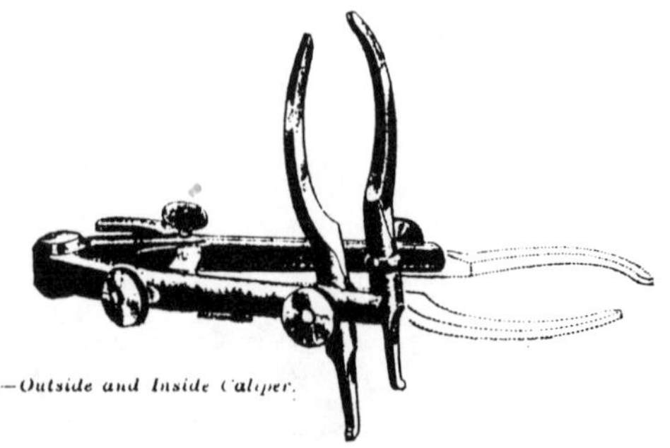

Fig. 3.—Outside and Inside Caliper.

COOKE'S EXTENSION DIVIDERS AND CALIPERS.

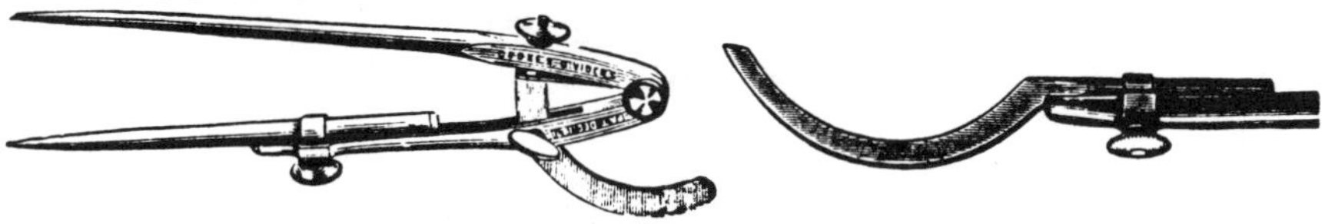

Furnished with either one or both legs movable. The 5 inch are made of brass with steel points, the larger sizes of iron with steel point. A pencil can be inserted instead of the movable point when desired.

ONE EXTENSION LEG.		BOTH LEGS EXTENSION.		CALIPER POINTS.	
5 inch, each	$1 00	6 inch, each	$1 00	For 6, 7, 8 and 10 inch Dividers, per pair	$0 60
6 "	1 00	7 "	1 25	For 15 inch Dividers per pair	1 00
7 "	1 00	8 "	1 50		
8 "	1 25	10 "	1 75		
10 "	1 50	15 "	2 00		

CORNELL TOOL CO. Detroit, MI

Maker, in 1919, of dial test indicators made in a "No. 1 size adapted for use in the tool room" and "No. 2 instrument suitable for meeting the requirements of general production."

CORNELL DIAL TEST INDICATOR

A dial test indicator which has recently been placed on the market by the Cornell Tool Co., Detroit, Mich., is shown in the accompanying illustration. Indicators of this type are made in two sizes, namely, a No. 1 instrument which is adapted for use in the tool-room and a No. 2 instrument which is suitable for meeting the requirements of general production work. Both sizes are regularly furnished with a complete set of attachments. It is stated that these indicators are of simple design and that they are so constructed that there is no chance for them to lose their original degree of accuracy. An important feature of the design is that the dial is mounted on a universal joint so that the operator is always able to have the dial located in a position where it may be easily seen. This construction makes it unnecessary to resort to the use of a small mirror in order to gain a view of the dial when indicating parts where the dial would otherwise be located face downward or away from the operator. Attachments are furnished with these indicators which enable them to be employed with a height gage or to indicate internal work down to 3/8 inch and up to 8 inches in depth. The graduations on the dials are spaced at a sufficient distance apart so that they can be easily read, even in places where the light is bad. The No. 1 instrument will indicate 0.135 inch and the No. 2 size will indicate 0.235 inch. The F. Conley Co., 502 Book Bldg., Detroit, Mich., is the sole distributor of these indicators.

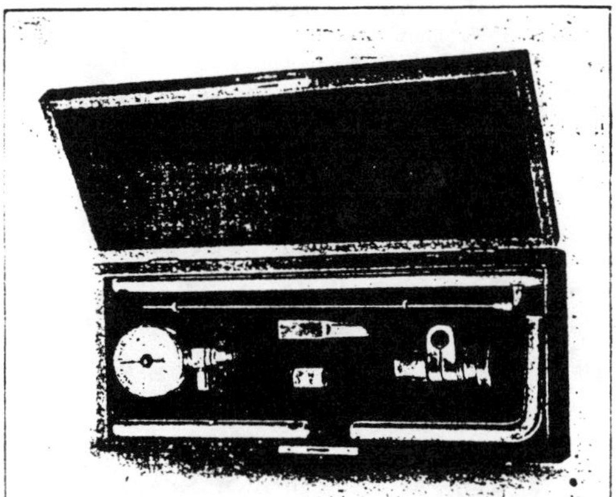

Cornell Dial Test Indicator distributed by the F. Conley Co.

Machinery, April 1919

COSTELLO, HENRY O. Providence, RI

Costello patented, on September 15, 1908, a quick adjusting micrometer which he introduced in October of 1908. It is not certain that the tool was actually made commercially.

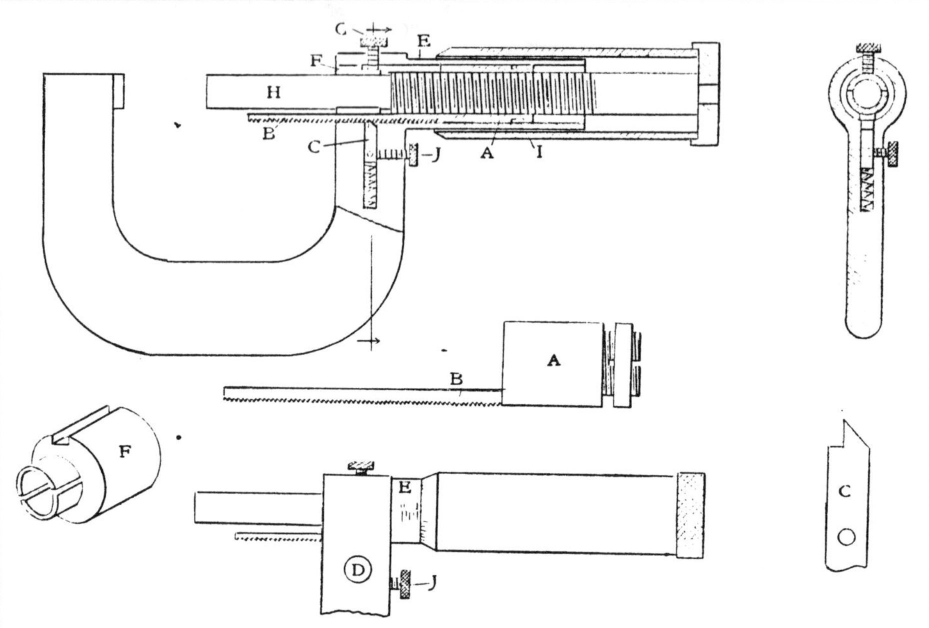

QUICK ADJUSTING MICROMETER

Machinery 1908

CRESCENT TOOL WORKS Bridgeport, CT. See KENNELLY & CAIN

CROSBY, G.S. Brooklyn, NY

Maker of precision levels in 1913. He offered two sizes, 5 and 8 inches, both with one-minute accuracy.

D. & B. Bangor, ME See DARLING & BAILEY

D. & S. Bangor, ME See DARLING & SCHWARTZ

DARLING & BAILEY Bangor, ME

A partnership of Samuel Darling (1815-1896) and Edward H. Bailey, formed in 1852 to make steel machinist's rules. The partners used a dividing machine developed and built by Darling. They may also have made machinist's try squares. All products, thus far examined, are marked D. & B., Bangor, ME. Darling was granted a patent for a grinding machine on August 30, 1853 (No. 9976). This is generally considered to be the first surface grinder and was, no doubt, one of the reasons for the high quality of the firm's rules. In 1853, Darling bought out Bailey and took on a new partner, Michael Schwartz, a Bangor saw maker and hardware dealer. The business was continued under the name: *(see next page)*

DARLING & SCHWARTZ Bangor, ME

The new partnership continued to make the same steel machinist's rules and try squares. The squares were made under Darling's patent of October 6, 1857 (No. 18327), and many are so marked. All products, thus far examined, are marked D. & S., Bangor, ME. On May 24, 1866, Darling & Schwartz merged with the rule making part of its major competitor, J. R. BROWN & SHARPE of Providence, RI. Samuel Darling, along with his graduating machine, grinder, workmen and stock moved to Providence to become part of the newly formed company, DARLING, BROWN & SHARPE. Schwartz was not associated with the new firm.

DAVIS, Jr., DANIEL See entry on page 169

DAVIS & COOK See entry on page 168

DAVIS & THOMPSON Milwaukee, WI

Primarily a maker of production machine tools, the company also made tubular frame micrometers which had been patented August 15, 1905 (No. 797409), by Frank M. Davis, president and treasurer of DAVIS & THOMPSON. The same micrometers had been made by F. M. DAVIS & CO. c1904-1905 and by A.J. MACHEK & CO. which bought the line in 1905.

DAVIS & CO., F.M. Milwaukee, WI

Operated by Frank M. Davis who was the inventor and maker of tubular frame micrometers patented August 15, 1905 (No. 797409). He was bought out by A. J. MACHEK & CO., Milwaukee, WI, in 1905. Davis was also president and treasurer of a machine tool builder, the DAVIS & THOMPSON CO., Milwaukee, WI, when it was formed in 1917.

DAVIS CO., L.L. Springfield, MA

Formed by Leonard Leroy Davis (1838-1867) in 1867 to make iron levels under his patent of September 17, 1867 (No. 68961). Davis also received two more level patents: on March 17, 1868 (No. 75533) and one on November 21, 1871 (No. 75534). In 1875, the name of the company was changed to:

DAVIS LEVEL & TOOL CO. Springfield, MA

Its 1880 catalog lists: machinist's levels patented May 29, 1877 (No. 191389) by F. T. Ward and T. Bedworth, spring calipers and dividers, screw calipers, patent helical coil spring calipers and dividers, firm joint calipers, Hathaway's surface gauge, machinist's scratch awls and screwdrivers, hack saws including double blade type, and thread gauges. The spring calipers and dividers, screw calipers, patent helical coil spring calipers and dividers, firm joint calipers, and Hathaway's surface gauge were products of J. STEVENS & CO. of Chicopee Falls MA. Some others were made by the BILLINGS & SPENCER CO., Hartford, CT. It appears that Davis operated a wholesale tool business as well as making tools. By 1880, the company was advertising the Marshall patent square (January 2, 1877, No. 185830). It appears that it was the maker of the Marshall square but it may have only been the sales agent. In 1883, the company was included in a list of machine tool builders as a maker of planers. Davis levels are listed in dealer catalogs as late as 1911. The cuts show, however, a M.W.R. CO. marking instead of the D.L.T. CO. mark. The M.W. ROBINSON CO., New York, NY, bought the DAVIS LEVEL & TOOL CO. in 1893. *(continued next two pages)*

Davis Level & Tool Co's Patent Tools.

UNEQUALED IN ACCURACY AND FULLY WARRANTED.

MACHINISTS' PLUMB AND INCLINOMETER.

Showing style of Nos. 1, 2, 3, 4.

No. 1. 6 inches long. Price, each...$2 00
No. 2. 12 inches long. " 2 50
No. 3, 18 inches long. " 3 00
No. 4, 24 inches long. " 3 50

DOUBLE PLUMB AND LEVEL.

No. 6, 6 inches long. Each..............................$2 00

DOUBLE PLUMB AND LEVEL.

No. 7, 12 inches long. Each.....$2 25
No. 8, 18 inches long. " 2 50
No. 9, 24 inches long. " 3 00

No. 09. Carpenters' and Machinists' adjustable Plumb, Level and Inclinometer combined, 24 inches long. Price, each......................................$6 00

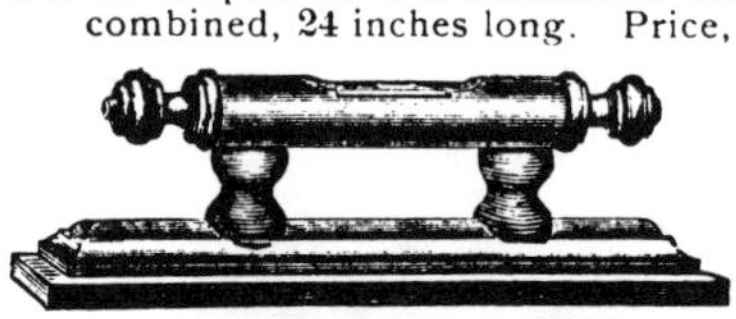

ADJUSTABLE BENCH LEVEL.

No. 10, 4 inches long. Each $0 50
No. 12, 5 inches long " 75

BENCH LEVEL.

FOR SQUARE OR STRAIGHT EDGE.

No. 11, 4 inches long. Each......$0 75

POCKET LEVEL.

No. 14, 3 inches long. Each, $0 25
No. 15, 5 inches long. " 35

POCKET LEVEL FOR SQUARE OR STRAIGHT EDGE.

No. 17, 3½ inches long. Each..$0 30

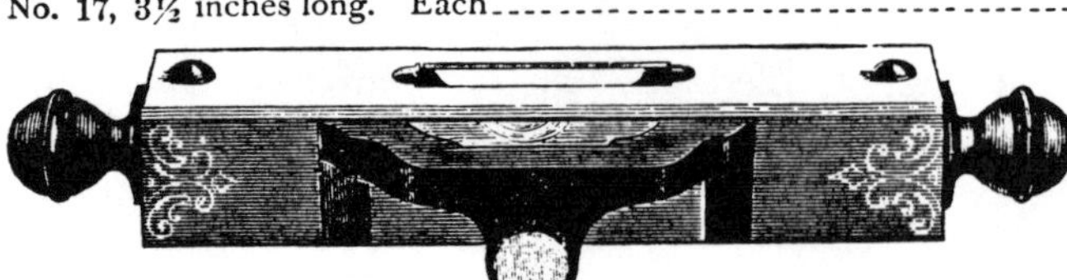

This Level can be attached to a Square or Straight Edge the same as No. 11. It will be found accurate, useful, durable and cheap.

LEVEL GLASSES.

Inch	1 to 1¾	2	2½	3	3½	4
Price, per doz	$0 95	$1 00	$1 00	$1 25	$1 35	$1 50
Price, each	10	10	10	15	15	25

1891

Marshall's Patent Universal Square

Price, each, $3.50.

This tool combines, in the most convenient form, seven different instruments, viz.: The **Try-Square,** the **Miter,** the **T-Square,** the **Bevel,** the **Center Square,** the **Depth Gauge,** the **Graduated Rule or Machinists' Scale.**

The scale furnished with this instrument is of Brown & Sharp Mfg. Co. make, 6 inches in length, graduated on both sides in 64, 32, 16 and 8 parts to the inch. The scale can be removed from the frame instantly for other use; it is adjustable, and can be used at any length and angle with the frame, making it applicable to a great variety of uses, and is highly appreciated by Machinists, Pattern Makers and, in fact, by mechanics of every class. Each tool is accurately made and put up in a neat pasteboard box.

December 1881

DAVIS' HEXAGON LEVELS.

BRASS, NICKEL PLATED.

PRICE LIST.

No. 0.	2½ inches long............................	per dozen, $3 00.	Each, $0 30
No. 00.	3 " "	" 3 60.	" 35
No. 000.	3½ " "	" 4 20.	" 40

Iron Pocket Level, for Square or Straight Edge.

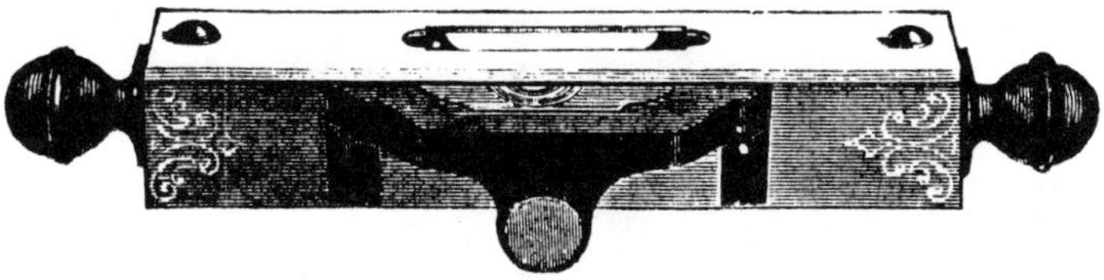

This level is so arranged that it may be attached to a square or straight edge. It will be found accurate, useful, durable, and cheap.

Length, 3½ inches..........................List, per dozen, $3 00. Each, 25c.

Adjustable Iron Spirit Level, Plumb, and Inclinometer.

No. 1, 6 inches...each, $2 00

DAVIS & COOK Watertown, NY

A partnership of George Davis and Gustavus Cook, formed in 1886 to make levels under Cook's patent of December 7, 1886 (No. 353846). The partnership lasted until Davis' death in 1906. The company continued until about a year after Cook's death in 1909. The levels were not intended for precision machinist's work, but no doubt the cast iron models found a good deal of use as machinist's bench levels.

Cook's Patent Levels.

Davis & Cook, Watertown, N. Y., are manufacturing Cook's patent iron plumb

Fig. 1.—Iron Plumb and Level, No. 25.

and levels in the new patterns shown in Figs. 1 and 2. Fig. 1 represents level with a cast iron frame 18 inches long on the longest edge, $2\frac{3}{4}$ inches wide and $1\frac{5}{8}$

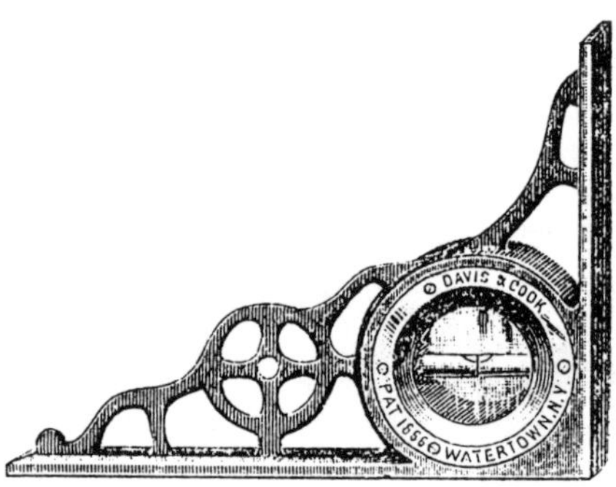

Fig. 2.—Iron Plumb, Level and Square, No. 20.

inches thick on the edge, weighing about 3 pounds. They also make an iron plumb, level and square, Fig. 2. The plumb and level glasses are fastened in a nickel plated metal ring, protected by glass, this inner ring being set in an outside ring, which is fastened to the frame by screws. Countersunk screws in the face of the frame are used to adjust the level glasses. The inside ring must revolve on the outside ring in adjusting, and we are advised there is no up and down or side motion whereby the level can get out of adjustment.

Iron Age, July 10, 1890

DAVIS, Jr., DANIEL Worcester, MA

Joint patentee, with Edward Wright, of a speed indicator (September 12, 1876, No. 182177). The first production of the tool was by Davis, operating as DANIEL DAVIS, Jr. In 1884, R. WOODMAN, Boston, MA, bought the patent and took over production. The illustration below is a copy of a sheet packed in the box with the tool.

Improved Speed Indicator.

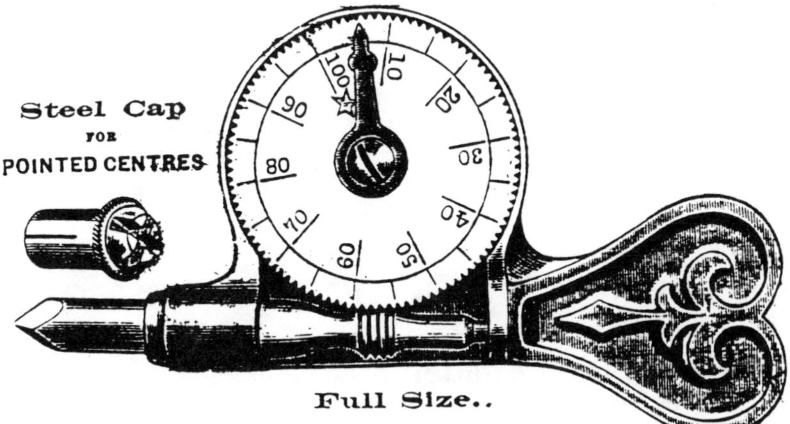

The attention of DEALERS and others using Machinery, is invited to the qualities which combine to make this INDICATOR preferable to any in use.

It is adapted to *pointed* or *hollow centres*,

IS ACCURATE, RELIABLE, OF CONVENIENT SIZE,

and suited for carrying in the pocket.

For STYLE, FINISH and DURABILITY it is without a rival, while the PRICE is within the reach of ALL.

PRICE, per Dozen, - - - - - - - - - **$30.00.**

Liberal trade discount.

RETAIL PRICE, Each, - - - - - - - **2.50.**

SAMPLES of One Dozen or less sent by mail, post paid, as above.

Address,

DANIEL DAVIS, Jr.

WORCESTER, MASS.

DEMING, Charles M. Hartford, CT

Inventor and maker of the Deming machinist's indicator beginning about 1903. The unit appears to be very fragile and the underside of the box lid is taken up with a warning notice not to try to repair or adjust the tool as further damage would result. In 1918, Deming announced that the SHEFFIELD MACHINE & TOOL CO., Dayton, OH, would take over the manufacture and sale of the indicator.

DEMING INDICATOR CO. Dayton, OH

Maker of a machinist's test indicator beginning in May, 1920. The company announced that it "has taken over the manufacture and sale of the Deming indicator, which has been made for the last two years by the SHEFFIELD MACHINE & TOOL CO., Dayton, OH, and for fifteen years previous by the the inventor, Charles M. Deming, under whose personal supervision the indicator will now be made."

DISSELL, W.F. Cincinnati, OH

Introduced, in 1909, a combination square "consisting of two steel rules, provided with suitable surfaces and clamps for engaging each other firmly in the various positions required for a tool of this kind. It may be set up for use as a depth gauge, try square, hook rule, T square, vernier caliper, etc."

DISSTON & MORSS Phildelphia, PA See H. DISSTON & SONS

DISSTON & SONS, H. Philadelphia, PA

The Disston company was, of course, primarily a maker of saws and saw related tools. From the mid-1880's to about 1910, however, they made a wide variety of machinist's tools. It appears that most of the tools are of the type which would use the Disston company's expertise in working with thin, flat pieces of steel. Early examples are usually marked DISSTON & MORSS, which was a "company within a company" at H. DISSTON & SONS. Machinist's tools offered in the 1885 catalog were: machinist's marking gauge, combination steel gauge, center square, machinist's levels, machinist's straight edges, steel squares and try squares. The catalog also offered steel draftsman's T-squares and triangles. Offerings in the 1906 catalog were: steel rules, machinist's marking gauge, combination steel gauge, center gauge, machinist's levels, steel squares with or without a center finder, thin steel squares, shafting levels and speed indicators. The speed indicator is obviously a L.S. STARRET CO., product, but the other tools appear to have been made by Disston. *(continued next five pages; all illustrations are from the 1906 catalog)*

SPEED INDICATOR.

Full Size
Weight, 5 Ounces.

Correct and Reliable. Any speed can be accurately counted.

Price, complete, ... $1.35 each.

DISSTON

MACHINISTS' TOOLS.

STEEL SQUARE.

Graduated to 1-32 of an inch.

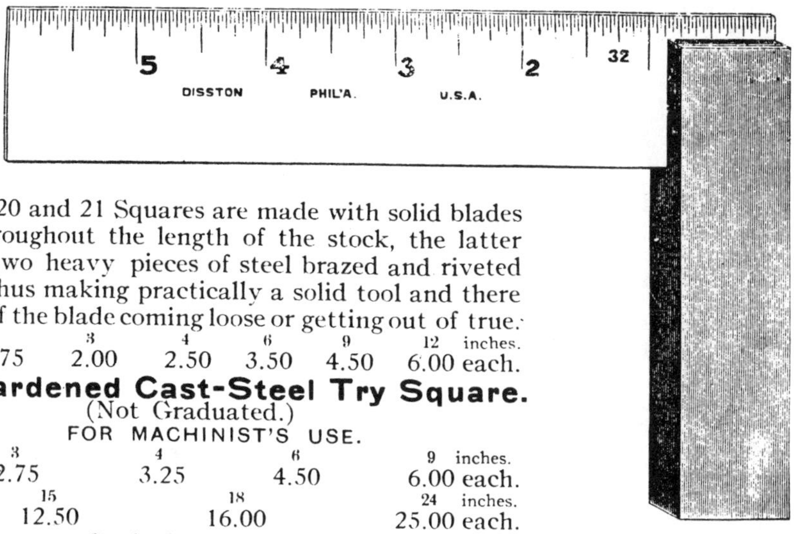

The Nos. 20 and 21 Squares are made with solid blades continuous throughout the length of the stock, the latter consisting of two heavy pieces of steel brazed and riveted to the blade, thus making practically a solid tool and there is no liability of the blade coming loose or getting out of true.

	2	3	4	6	9	12 inches.
No. 20.	$1.75	2.00	2.50	3.50	4.50	6.00 each.

No. 21. Hardened Cast-Steel Try Square.

(Not Graduated.)

FOR MACHINIST'S USE.

2	3	4	6	9 inches.
$2.00	2.75	3.25	4.50	6.00 each.

12	15	18	24 inches.
$8.50	12.50	16.00	25.00 each.

Packed one in box.

PATENT STEEL CENTRE SQUARE, No. 22.

FOR MACHINISTS' USE.

Graduated to 1–32 of an inch.

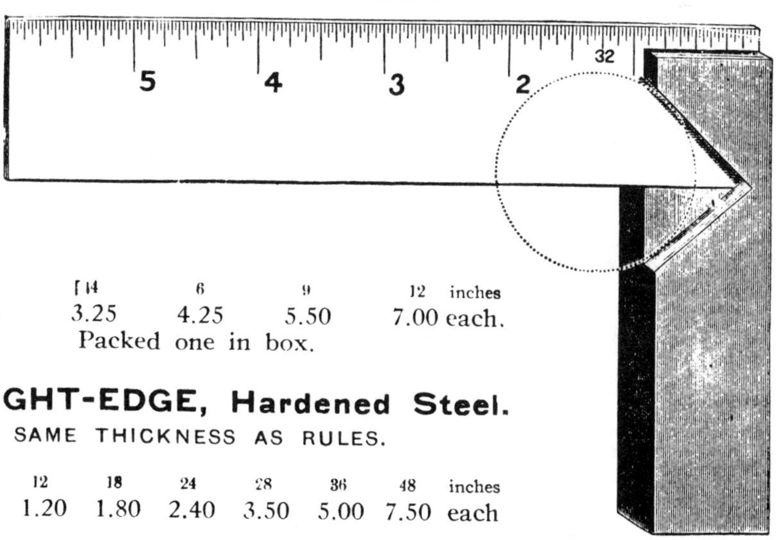

2	3	4	6	9	12 inches
$2.25	2.75	3.25	4.25	5.50	7.00 each.

Packed one in box.

STRAIGHT-EDGE, Hardened Steel.

SAME THICKNESS AS RULES.

3	6	9	12	18	24	28	36	48 inches
$0.40	.60	.90	1.20	1.80	2.40	3.50	5.00	7.50 each

DISSTON
MACHINISTS' TOOLS.

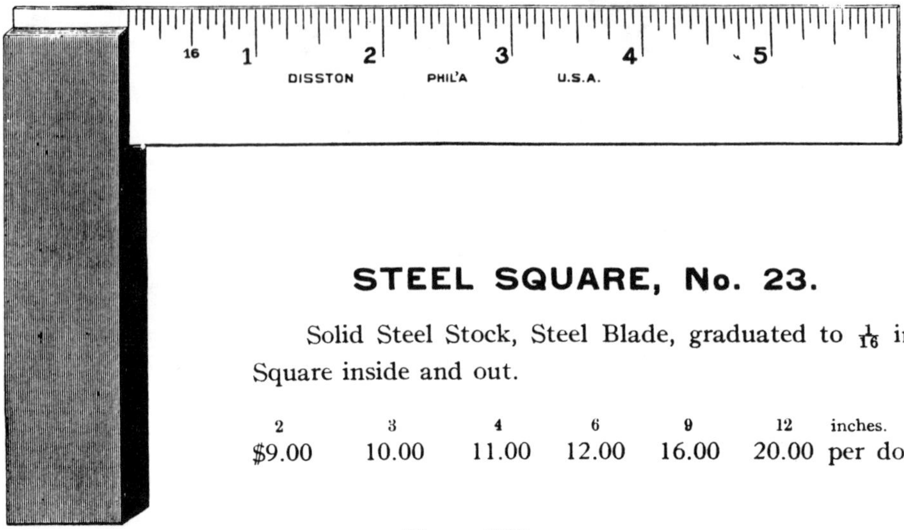

STEEL SQUARE, No. 23.

Solid Steel Stock, Steel Blade, graduated to $\frac{1}{16}$ inch. Square inside and out.

2	3	4	6	9	12	inches.
$9.00	10.00	11.00	12.00	16.00	20.00	per dozen.

No. 30.

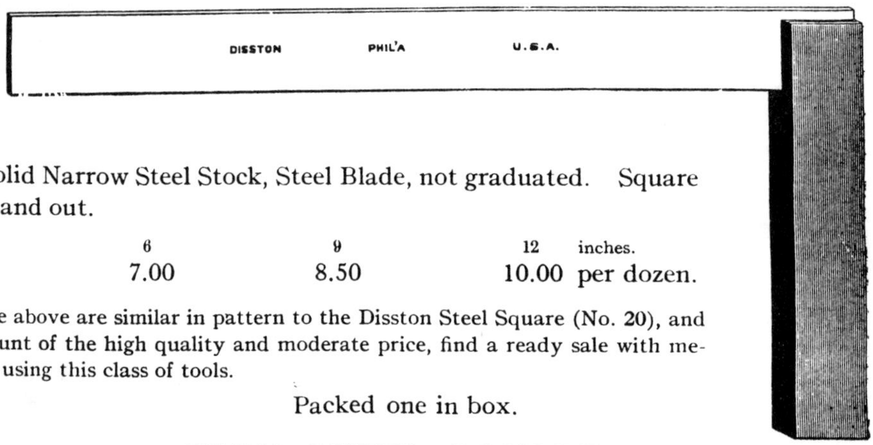

Solid Narrow Steel Stock, Steel Blade, not graduated. Square inside and out.

4	6	9	12	inches.
$6.50	7.00	8.50	10.00	per dozen.

The above are similar in pattern to the Disston Steel Square (No. 20), and on account of the high quality and moderate price, find a ready sale with mechanics using this class of tools.

Packed one in box.

THIN STEEL SQUARE.

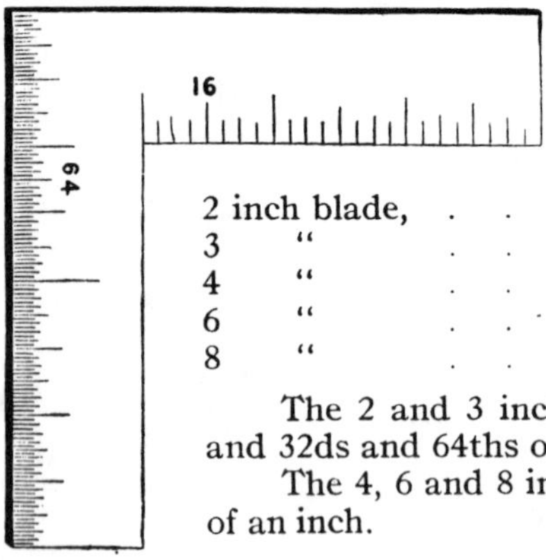

2 inch blade,	. . . .	$1.00 each.
3 "	. . . .	1.50 "
4 "	. . . .	2.00 "
6 "	. . . .	2.85 "
8 "	. . . .	3.75 "

The 2 and 3 inch are graduated to 16ths and 64ths on one side, and 32ds and 64ths on the other.

The 4, 6 and 8 inch are graduated on both sides to 16ths and 32ds of an inch.

DISSTON
MACHINISTS' MARKING GAUGE.

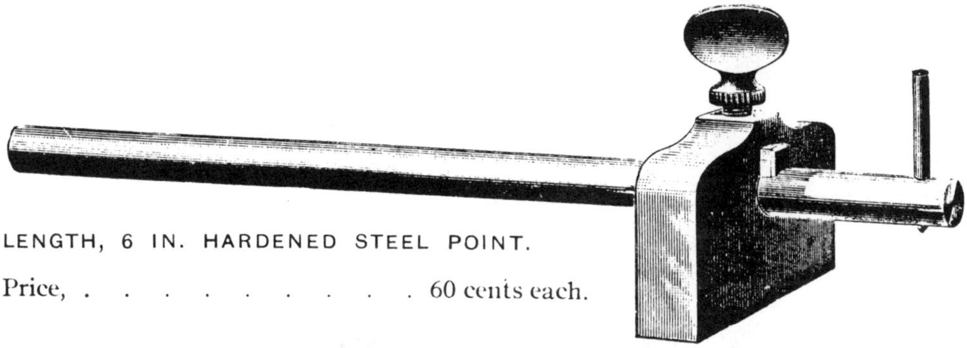

LENGTH, 6 IN. HARDENED STEEL POINT.

Price, ... 60 cents each.

PATENT COMBINATION STEEL GAUGE.

The most useful and convenient gauge in the market. It comprises five different gauges; a hexagon, or 120 degrees; a centre, or 60 degrees; a square, or 90 degrees; and a mitre, or 45 degrees; with scale graduated to 32ds of an inch.

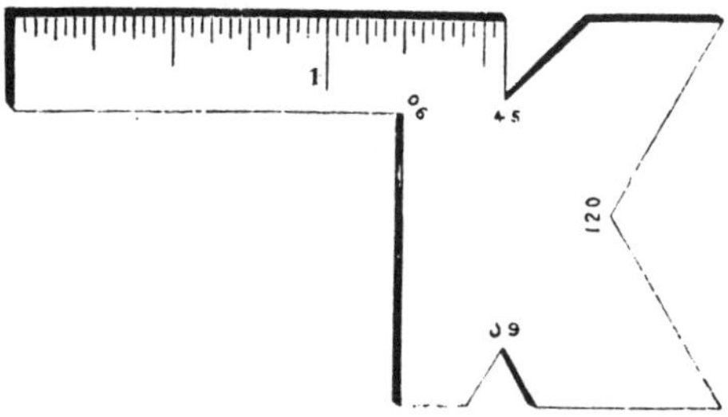

Price, ... $1.00 each.

CENTRE GAUGE.

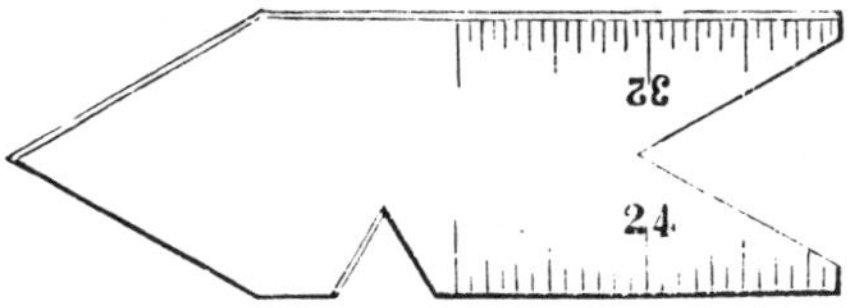

Price, ... 30 cents each.

The above cut represents our Centre Gauge and gauge for grinding and setting screw tools.

MACHINISTS' LEVEL.

IRON FRAME, WITH STEEL TOP.

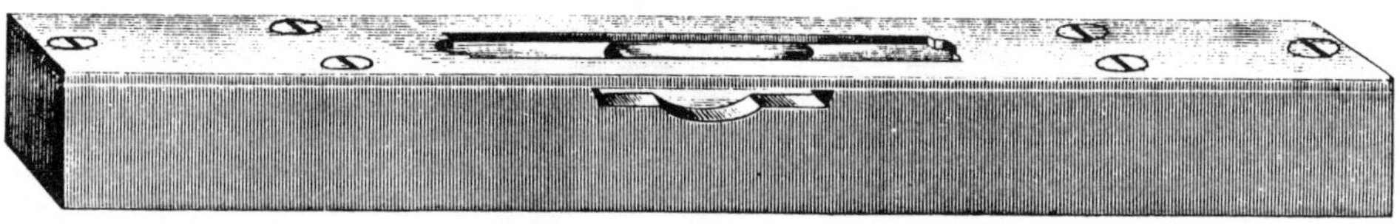

☞ These levels are warranted true, the spirit glass being equal to the ground; great care is taken in testing them; the bubble in spirit glass is large, which makes the level more sensitive.

6	8	10	12 inches.
$1.00	1.25	1.50	1.75 each.

We can furnish machinists' levels as above, with ground bottles, at an advance of $3.35 each, over above prices.

DISSTON

MACHINISTS' TOOLS.

STANDARD STEEL RULE.

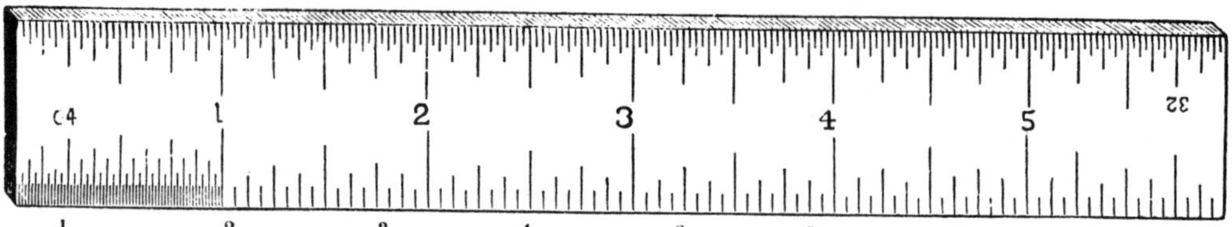

1	2	3	4	6	9	12	18 inches.
$0.20	.30	.40	.50	.75	1.12	1.50	2.25 each.

24	36	48 inches.
$3.00	$7.00	10.00 each.

The Rules in this list are divided as follows:

	No. 1 Graduations.	No. 2 Graduations.	No. 3 Graduations.
First edge	8 of an inch, whole length.	10, 20, 50, 100.	32 of an inch, whole length.
Second "	16 " "	12, 24, 48,	48 " "
Third "	32 " "	16, 32, 64,	50 " "
Fourth "	64 " "	14, 28,	64 " "

Metric graduations marked on one or both sides when specially ordered.

NARROW STEEL RULE.

3 inches, $\frac{3}{8}$ wide, graduated, 8, 16, 32, 64 of an inch, whole length . . $0.40 each.
4 " $\frac{3}{8}$ " " 8, 16, 32, 64 " " . .50 "
6 " $\frac{1}{2}$ " " 8, 16, 32, 64 " " . .75 "

These Rules are adapted to measuring depths of holes, key seats, etc., and are in other respects the same as the Standard Rule.

SQUARE STEEL RULE.

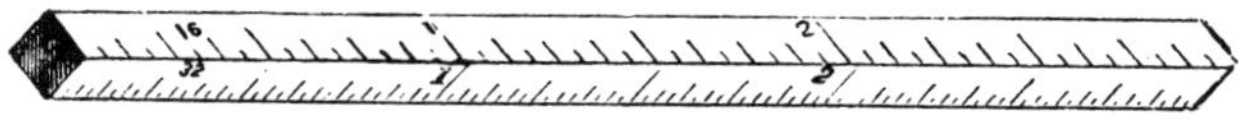

For determining the depth of holes.

3	4	6 inches.
$0.40	.60	.85 each.

Graduations, No. 1, ... 8, 16, 32, 64 of an inch, whole length.
" " 2, ... 16, 32, 64, 100 to the inch.
" " 3, ... 16, 64, 50, 100 " "

TRIANGULAR STEEL RULE.

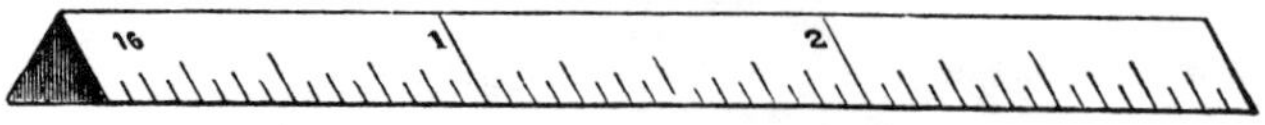

3	4	6	12 inches.
$0.50	.65	1.00	2.50 each.

Graduations, No. 1, ... 16, 32, 64 whole length.
" " 2, ... 16, 64, 100 to the inch.
" " 3, ... 20, 50, 100, 12, 24, 48, 16, 32, 64 to the inch.

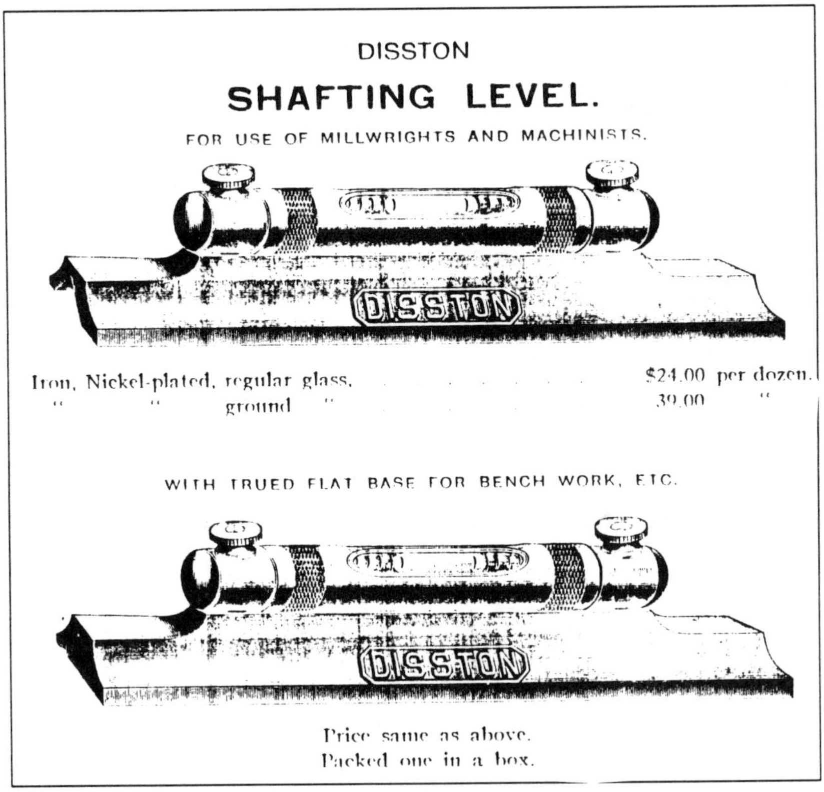

DURAND & CO.

Maker of inside micrometers (they called them "rod gauges") circa 1906. The location of the firm is unknown.

DYKE, HERBERT New Castle, PA

Maker, in 1906, of a combination gauge "intended to provide a universal tool by which the gauge, bevel or angle of any piece of work may quickly and accurately be determined". The tool was patented by James D. Cann on July 17, 1906 (No. 826311).

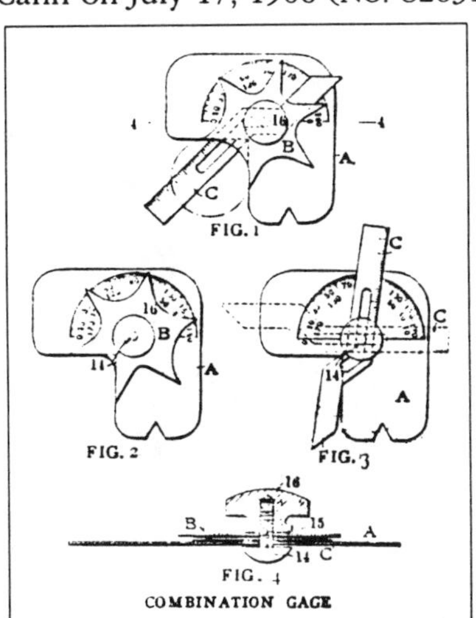

1906

E. & T. MFG. CO. See MODERN CALIPER & TOOL WORKS

EATON & CO., GEORGE H. Boston, MA

Maker of a surface gauge patented May 7, 1889 (No. 402763), by W.H.H. Norcross.

ECONOMY MFG. CO. Philadelphia, PA

Maker of a trammel point which they introduced in 1908.

New Trammel Point.

We illustrate herewith a new trammel point, which is being placed on the market by the Economy Mfg. Company, 506 Commerce street, Philadelphia. The tool which

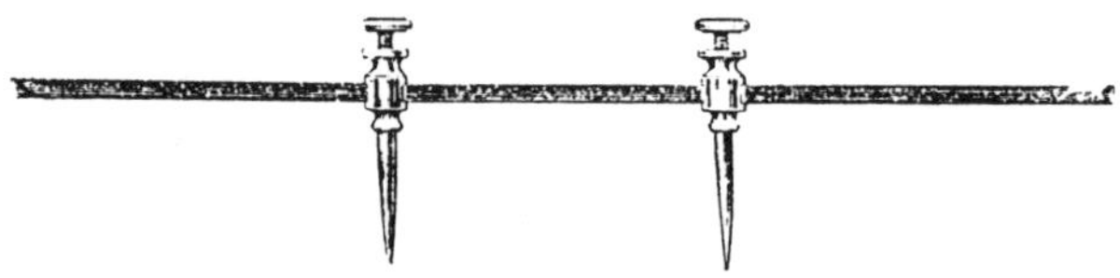

New Trammel Point.

is particularly adapted for pattern makers, sheet metal workers, cabinet makers and machinists, has a 3-16 in. square steel bar, 12 in. long, which is a unique feature and prevents the turning of the point, which is possible when a round bar is used. The bodies of the adjusting screws are of cast brass, nickel plated, and the points of tempered steel. The tool is well made and finished, and is being offered to the trade at a price to meet the demand for a high grade, moderate priced tool of this character.

Iron Age
May 14, 1908

EDDY ELECTRIC MFG. CO. Hartford, CT

Maker of Heath's patent self-timing speed indicator in 1892.

SELF-TIMING AND SELF-REGISTERING SPEED INDICATORS.

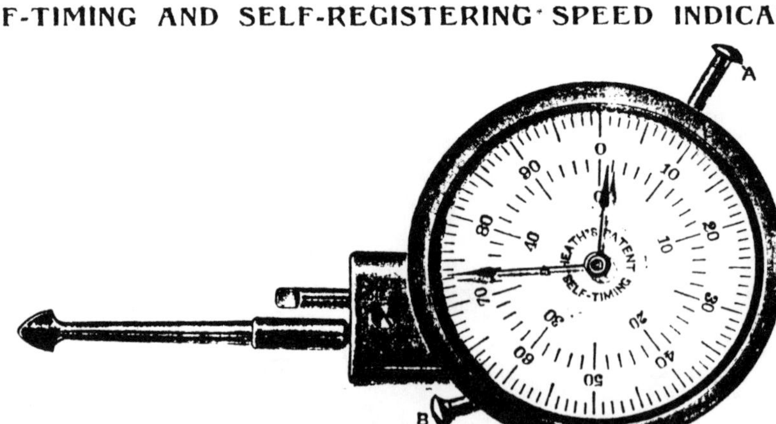

Specially adapted to Electric Motors and High Speed Engines. Just what is wanted for accurate speed taking.

Each $20.00

ELLRICH HARDWARE MFG. CO. Plantsville, CT

Operated by Robert C. Ellrich of Southington, CT, c1883-1895. Makers of a variety of small tools including a machinist's depth gauge offered in the 1891 catalog of Montgomery & Co., New York, NY.

ELLRICH DEPTH GAUGES.

In this tool the needle is kept from slipping when the thumb-screw is loosened to set it by a friction plug, when the depth wished is measured, it is fastened in position by the thumb-screw ; it makes a very convenient and handy tool for machinists.

No. 1 DG,.................. per dozen, $7.20

ELMER, D.E. Springfield, MA

Inventor and maker of a micrometer caliper (he called it an index gauge), patented October 1, 1867 (No. 69418). The only known example may be, in fact, the patent model.

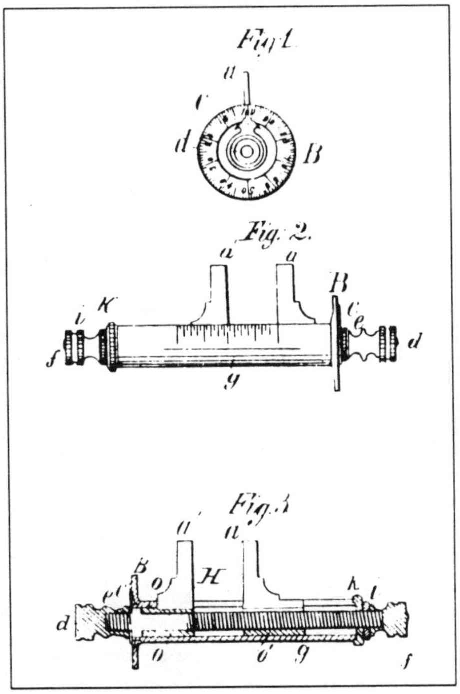

1867 patent drawing

EMERSON APPARATUS CO. Boston, MA

Makers of an inside micrometer introduced in 1910. The micrometer was covered by the Henry S. Plant patent of September 21, 1909 (No. 934692). *(continued next page)*

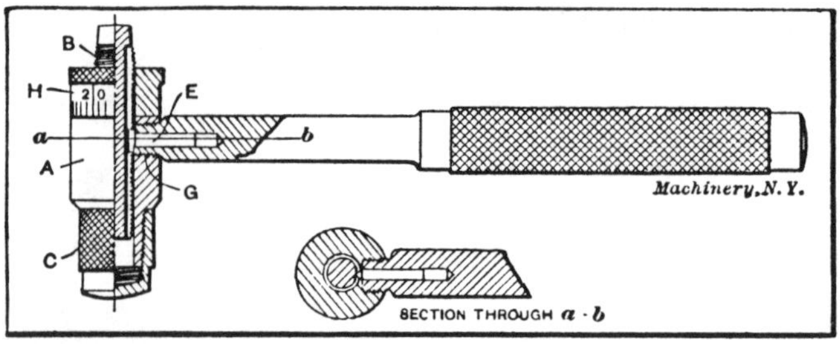

Fig. 2. Sectional View of the Plant Caliper

Machinery 1910

EMPIRE MACHINE CO. Pittsburg, PA

Makers of the Empire speed indicator c1905.

ENNIS, BENJAMIN F. Syracuse, NY

Maker, in 1909, of a machinist's test indicator.

ERB MFG. CO., E.M. Jersey City, NJ

Makers, in 1911, of a machinist's test indicator patented October 24, 1911 (No. 1006924), by E.M. Erb.

ERMENTROUT, W.H. Reading, PA

Maker of a surface gauge about 1890. The tool does not appear to have been patented.

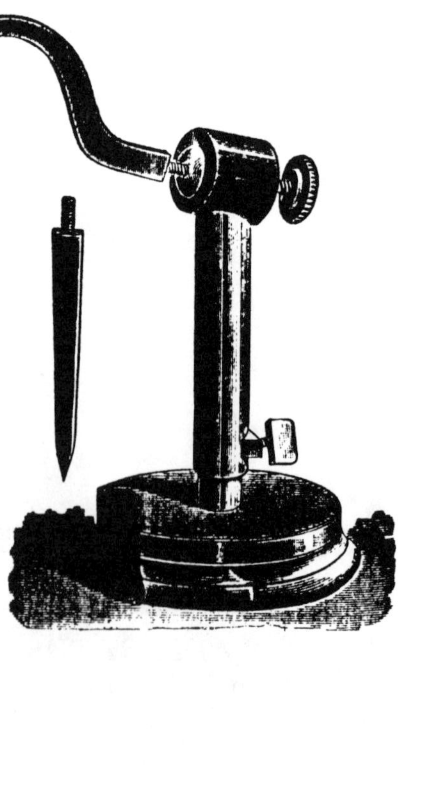

SURFACE GAUGES.

Any machinist using a Surface Gauge knows the inconvenience of a Gauge that requires a hammer or something else to move the points. This sketch shows one which is easily and quickly adjusted, and is permanent when once set. For convenience, there is none on the market to equal it. It can be used to scribe or follow a circle, and used at any angle sideways, top or bottom. It can be brought close to your work, has two points, is highly polished, and made in the best possible manner. It can be raised on stand 3 inches. Heighth from bottom to the highest point is 10 inches.

Each....................................$3.00

FALLON MACHINERY CO. Salem, MA

See BOULET FINE TOOL WORKS.

FAY & CO., CHARLES P. Springfield, MA

Fay began his career at J. STEVENS & CO. where his father was vice president and part owner. While there, he developed a screw caliper, patented December 20, 1881 (No. 251200), and a surface gauge, patented September 23, 1884 (No. 305515). The surface gauges were produced by J. STEVENS & CO. for many years. The calipers were never produced by Stevens. Fay sold the patent to a Mr. Hyde of Springfield, MA, who produced them as the Hyde patent screw caliper.

In 1883 Fay began working for L.S. STARRETT, Athol, MA, but left in 1884 to start his own company. He developed and patented, on June 2, 1885 (No. 319215), a line of calipers and dividers which are still in production today as the Starrett "Fay" calipers. He also received a patent for a rapidly adjusting nut for calipers and dividers on January 26, 1886 (No. 334764), and James Bullard assigned his related caliper patent of February 9, 1886 (No. 335740), to Fay. The Samuel B. Dover patent for a rapid adjusting nut, November 7, 1882 (No. 267166), was also used by Fay.

During the summer of 1886, Fay introduced his "Yankee" line of flat leg calipers. These were cheaper to produce and were meant to compete with the cheap calipers then imported from Germany. These calipers are also still in production as the Starrett "Yankee" calipers. In 1887, Laroy Starrett bought out Fay's patents, machinery and stock and induced Fay to rejoin L.S. STARRETT. Fay remained at L.S. STARRETT from 1887 to 1896, receiving the following patents:

Leg Joint for Calipers, December 27, 1887 (No. 375703, assigned to L.S. STARRETT);
Micrometer-gauge, September 24, 1889 (No. 411536, joint with Starrett);
Calipers and dividers, May 21, 1895 (No. 539759, joint with Starrett);
Surface gauge, March 17, 1896 (No. 556703, joint with Starrett);
Cutting-pliers, September 22, 1896 (No. 568242 assigned to L.S. STARRETT).

In 1896 he acquired the interest of his late father, W.P. Fay, in the J. STEVENS ARMS & TOOL CO., Chicopee Falls, MA, and returned there as vice president and superintendent. Patents received by Fay after rejoining J. STEVENS ARMS & TOOL CO. were:

Calipers, October 5, 1897 (No. 591348)
Center indicator, January 10, 1899 (No. 617501)
Protractor, January 1, 1901 (No. 665316).

All were assigned to J. STEVENS ARMS & TOOL CO. After 1901 he received a number of patents for firearms but no further tool patents. *(see illustration next page)*

FISK, C.A., New York, NY.

Operated by Carl A. Fisk, inventor of a rapidly adjustable micrometer patented June 26, 1906 (No. 824420). Introduced in early 1905, the tool operated: "in the ordinary way for the final adjustment of a reading and for slight changes in setting may be screwed out or in by rolling the knurled knob. Radical changes in setting are made by twisting the smaller knurled knob while holding the knurled handle. This rotates a fluted rod, allowing the threads on the screw to recede from the nut, and leaves the plunger free to be moved instantly to any other position." *(see illustration next page)*

FITCHBURG LEVEL CO., Fitchburg, MA

Makers of cast iron machinist's levels. In 1888 the company moved to Leominster, MA, and changed the name to the Leominster Novelty and Spirit Level Works.

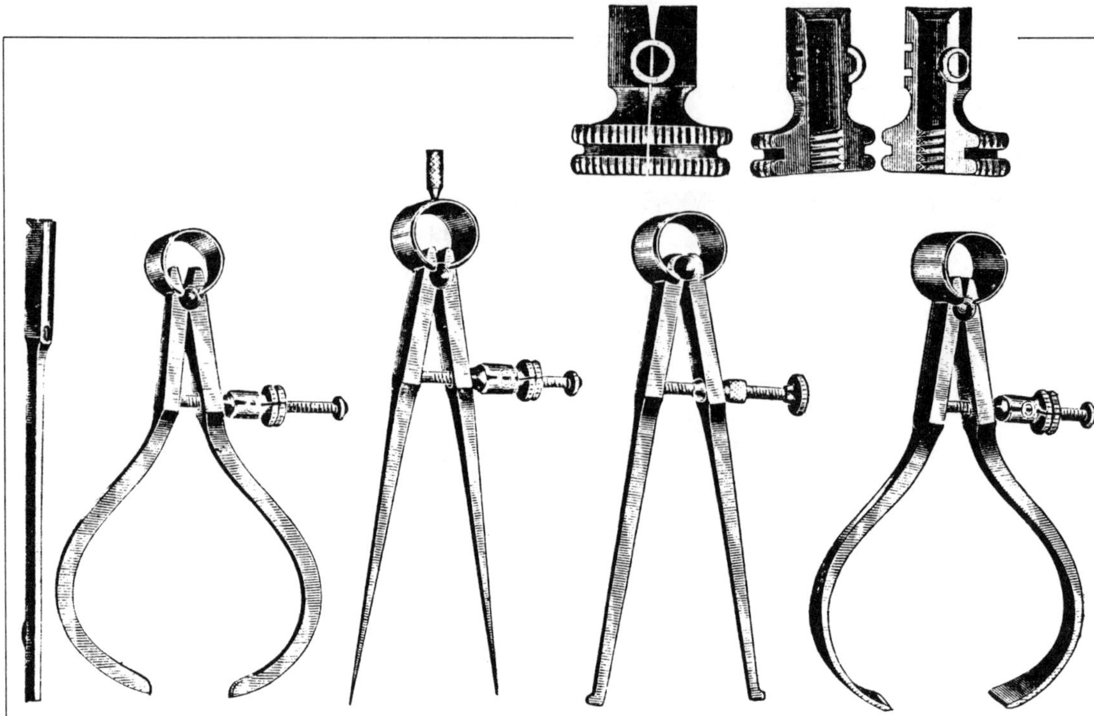

IMPROVED MACHINISTS' TOOLS.

See entry for Fay & Co., previous page *American Machinist 1885*

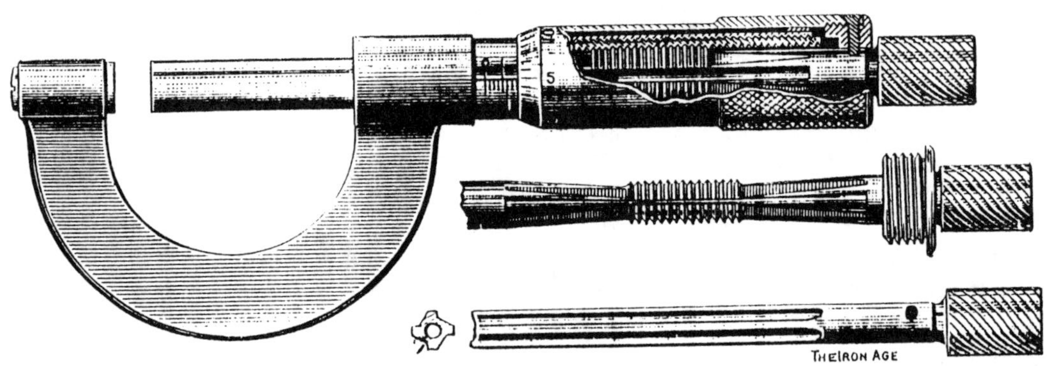

A Quick Change Micrometer Calliper.

See entry for C.A. Fisk, previous page *Iron Age, April 6, 1905*

FLEMING MFG. CO., INC., Worcester, MA

Formed in 1929 when the Fleming Machine Co., Worcester, MA, organized its small tool division into a separate company. Officers were: G.W. Fleming, President; S.B. Wilson, Vice-president and General Manager; and Charles F. Davis, Treasurer. The firm made Fleming precision tools and screw drivers. The micrometers may have been made for them by REED SMALL TOOL WORKS.

FOOT, JAMES D. New York, NY

Advertised surface gauges, pin vises, tap wrenches, pocket wrenches, machinist's marking gauges, and machinist's scratch awls in 1880. Foot, however, was probably a dealer rather than a manufacturer.

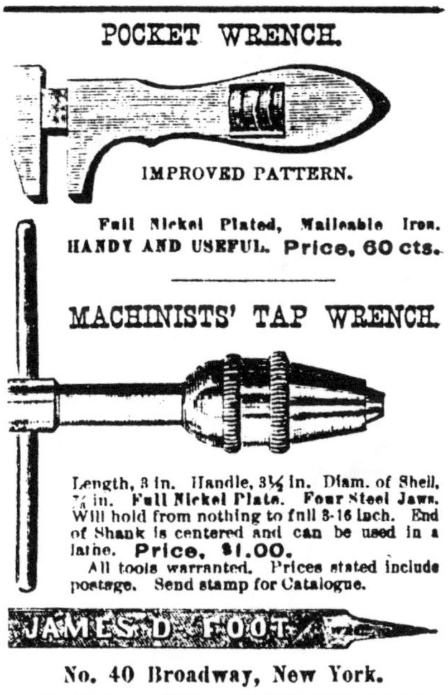

FORBES, W.D. New London, CT

Announced, in 1915, "A surface gauge of simple construction which can be made to sell at a low price." No examples are yet known.

FORREST, B.L. Fitchburg, MA

Maker of steel machinist's rules. Examples, which are fairly common, appear to date from the first quarter of this century.

FOWLER, CHARLES H. Boston, MA

Inventor of a speed indicator patented July 31, 1883 (No. 282067). The indicator was advertised by GOODNOW & WIGHTMAN in 1884 and CHANDLER & FARQUHAR in 1890. The tools are marked MF'D BY C.E. CHANDLER, BOSTON, MASS., USA. Fowler also patented, on December 31, 1889 (No. 418608), a speed indicator with a bell that rang after every 100 revolutions. *(see illustration next page)*

FRASSE & CO., PETER A. New York, NY

A large dealer in machinist's tools, it may have made, or had made for it, some tools. In 1893, it advertised a speed indicator which rang a bell after 300 revolutions and at each 100 thereafter. The maker is not specified, but it appears to have been made under the CHARLES H. FOWLER patent of December 31, 1889 (No. 418608).

The company was founded in 1816 by Henri Frasse who was a dealer in watch and clockmaker's supplies. After his death in 1849, the business was expanded by his son, Peter,

AN IMPROVED SPEED INDICATOR.

The accompanying cut represents a very convenient little instrument for registering the speed of any revolving piece of machinery. The advantage claimed for it over the ordinary indicator is that it will register up to 2,500 revolutions, those usually sold only counting up to 100. This is accomplished by having two dials, arranged back to back, one of which counts up to 100 and the other up to 2,500, the two being connected by suitable gearing. It is used by pressing the hardened steel point of the spindle of the instrument against the center of the revolving shaft of which it is desired to obtain the speed; one dial shows the number of revolutions up to 100, and the second shows the number of revolutions the first has made, or the number of hundreds.

This indicator has been patented by Mr. C. H. Fowler,

FOWLER'S IMPROVED SPEED INDICATOR.

and is now being introduced by Messrs. Goodnow & Wightman, of Boston, Mass., to whom communications should be addressed.

1884.

See entry for Charles Fowler, previous page

to include machinist's tools and supplies. Peter Frasse reorganized the company as PETER A. FRASSE & CO., when he was joined by James M. Montgomery in 1868.

FREEMAN, T.

Maker of a precision, triangular, machinist's rule marked, T. FREEMAN, MAKER. It appears to have been made in the 1860's.

FUCHS MFG. CO., LEON Dayton, OH

Maker of the Fuchs universal test indicator, c1915. Leon Fuchs had also developed a micrometer protractor for which he was granted a patent on May 9, 1911 (No. 991839). The tool was manufactured by the BROWN & SHARPE MFG. CO., Providence, RI. *(see illustration next page)*

FULTON MACHINE WORKS Brooklyn, NY

Maker of the "Fulton Single Speed" and the "Fulton Improved Double Speed" speed indicators, c1900. The speed indicators were covered by patents issued to William Lang on November 28, 1882 (No. 268107), and October 9, 1894 (No. 527207). The earlier patent had been used by CHURCH & SLEIGHT, New York, NY, for speed indicators made in 1891.

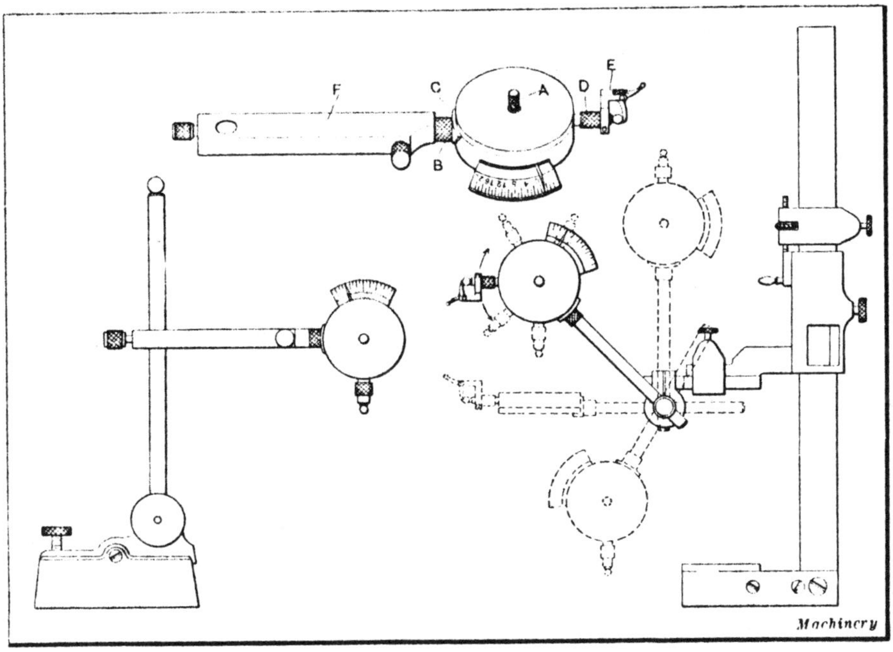

Fuchs Universal Test Indicator and Two Methods of mounting it

Machinery 1915

see entry for Fuchs Mfg. Co., previous page

GALLAGHER MFG. CO. Lynn, MA

Maker of "Gallagher's Improved Speed Indicator" c1887-1889. The firm also made emery grinders, ratchet drills and breast drills.

Improved Speed Indicator.

Nickel Plated.

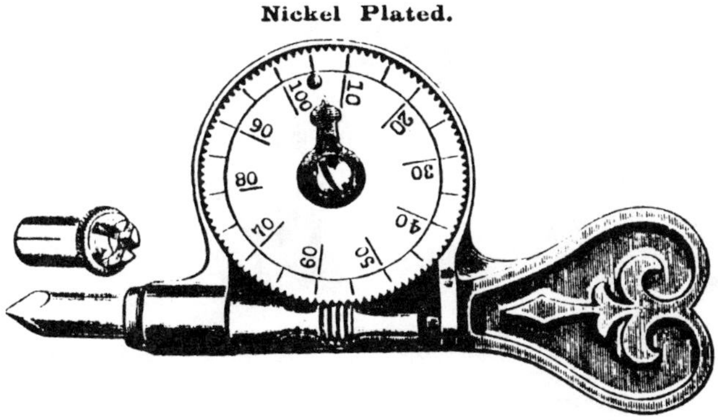

It is adapted to pointed or hollow centres, and has a raised sight on the dial so as to count by touch.

With split cap....................................each, $1 25
" plain cap.................................. " 1 00

GARDINER BROS. Danielson, CT

Operated by brothers Andrew H. Gardiner and Walter E. Gardiner. The only known product is a surface gauge patented by Andrew H. Gardiner, May 9, 1893 (No. 497071). He assigned one-half of the patent to his brother Walter. By 1897 the surface gauges were in production and were being advertised by the brothers.

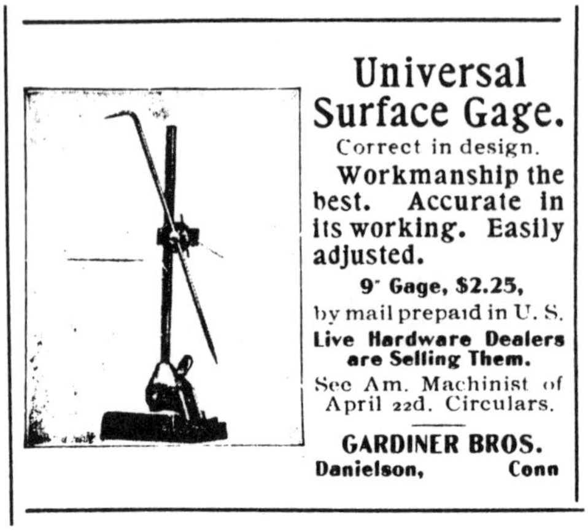

American Machinist 1897

GATES, B.L. Chicago, IL

Maker, in 1918, of a dial indicator designed so "that the dial is always visible to the workman." The dial was graduated to .001", with a range of 0.130".

GATES DIAL INDICATOR

The dial indicator which is made by B. L. Gates, 125 S. Wells St., Chicago, Ill., is intended for setting up work on machines and for the other purposes which indicators of this class are generally used for. This indicator is so arranged that the dial is always visible to the workman, so that it is unnecessary to walk around a machine when setting a jig

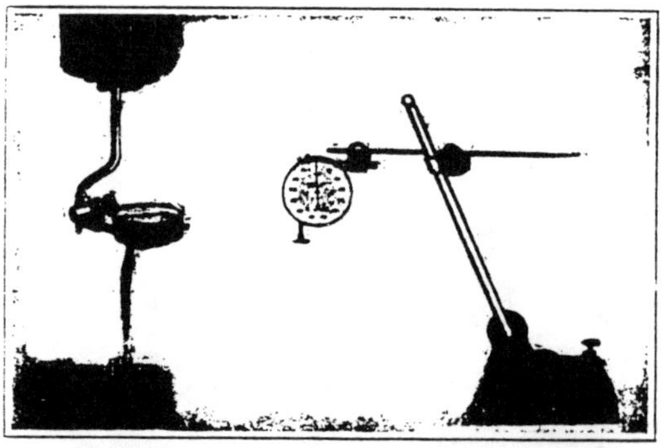

Gates Dial Indicator applied to Machine Spindle and to Surface Gage

button or other part in alignment with a spindle to which the indicator is attached. The view to the left in the accompanying illustration shows the position of the dial while it is being used for truing a button. As the dial is in a horizontal position, it remains visible as the spindle revolves and a looking glass is not required to show variations in the movement of the indicating hand. The dial is graduated to read to thousandths inch, the graduation figures giving a direct reading up to 0.130 inch. This indicator may, of course, be applied to a surface gage as the illustration shows, or it may be used in conjunction with limit gages and other forms of measuring tools.

Machinery, March 1918

GERSTNER & SONS, H. Dayton, OH

Maker of machinist's tool chests from c1909 until today. The firm introduced a line of portable tool chests for machinists in 1909. This style had the patented front lid which stored under the bottom drawer when the case was in use, a feature which later became the standard for nearly all machinist's tool chests.

1910

GILBERT, HARRIS & CO. Chicago, IL

Maker, in 1910, of a combination inside/outside micrometer caliper covered by the Herman Sauter patent of August 9, 1910 (No. 966773). Sauter had assigned his patent to the company.

GILLARD, WM. J. Minneapolis, MN

Maker, in 1909, of a Combination Depth Gauge and Square. "The tool consists of a pair of telescoping scales and a clamp, so arranged that the scales may be used either as a depth gauge or as a square." No examples of this tool have been observed.

GODDARD, A.H. Worcester, MA

Maker, in 1891 of an "improved marking gauge for woodworkers, machinists and workers of metal."

Improved Marking Gauge.

A. H. Goddard, Worcester, Mass., is introducing improved marking gauges, all illustrated in Figs. 1 and 2. The manufacturer claims that the gauge shown in Fig. 1 can be used for any and every kind of measurement required by woodworkers, machinists and workers of metal. The point is made that it will do circular and irregular work, both inside and outside,

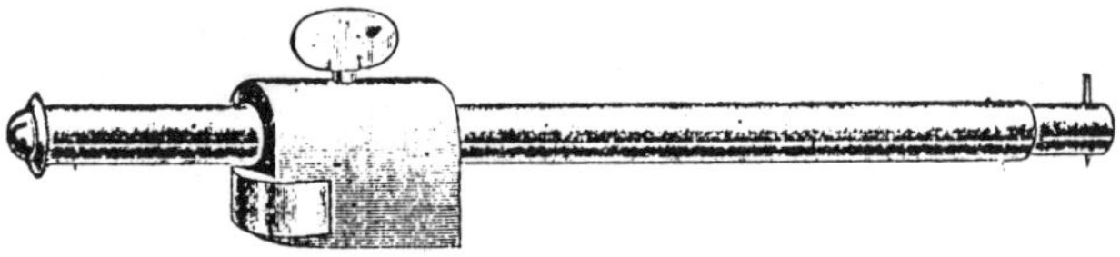

Fig. 1.—Improved Marking Gauge.

by the use of the roller guide. It has a milled-edge marking wheel, made of steel, which it is claimed will not follow the grain of the wood. The one thumb screw is designed to hold both tubes firmly in place. This gauge is referred to as being quickly and easily adjusted, and as expressly designed for durability and usefullness, and will doubtless be appreciated by mechanics and workers of metal. Fig. 2 represents a mortise gauge. These are made wholly of metal nickel-plated. It is stated that the material and make-up are first-class.

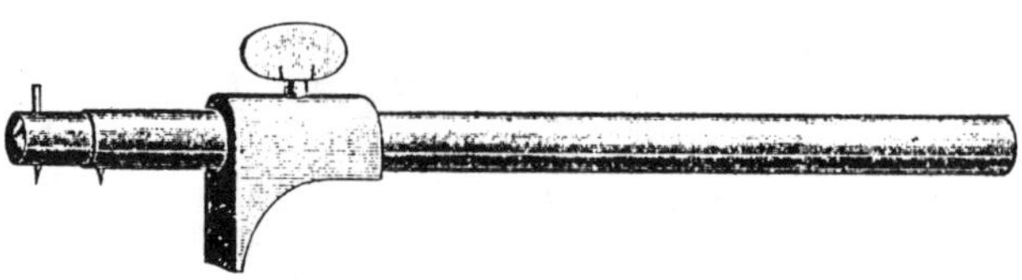

Fig. 2.—Improved Mortise Gauge.

Iron Age, April 9, 1891

GOODELL-PRATT CO. Greenfield, MA

Formed by William M. Pratt (1867-1946) when he bought control of GOODELL BROS. CO. in 1899. The Goodell firm had made only carpenter's and mechanic's tools and Pratt apparently decided to add a line of machinist's tools. In 1900, Pratt formed the MASSACHUSETTS TOOL CO. as a precision tool department of the GOODELL-PRATT CO. Although technically a separate company, it shared manufacturing facilities with Goodell-Pratt and the products were sold exclusively by Goodell-Pratt. Pratt bought the COFFIN & LEIGHTON CO., Syracuse, NY, for its line of machinist's rules and the LAVIGNE MICROMETER CO., New Haven, CT, owner of a large number of micrometer patents.

Goodell-Pratt bought a line of cast iron levels from C.F. RICHARDSON & SONS, Athol, MA, about 1904. This line was later supplemented by the purchase of the STRATTON LEVEL CO., Greenfield, MA, in 1913.

The GOODELL-PRATT CO. 1905 catalog offered, in addition to the line of MASSACUSSETS TOOL CO. machinist's tools, Richardson's pocket and bench iron levels, and combination squares. Some other machinist's tools were added, including a caliper patented by Martin Ball, February 17, 1903 (No. 720773); centering device patented July 3, 1906 (No. 824975) by Edward L. Cook; and a revolution counter patented February 22, 1916 (No. 1172793), by Oscar Hapgood. All these patents were assigned to the GOODELL-PRATT CO.

At some time prior to 1920, the MASSACHUSETTS TOOL CO. operation was shut down and the machinist's tools were thereafter made and sold under the Goodell-Pratt name. Catalog illustrations still showed the MASSACHUSETTS TOOL CO. name, but since GOODELL-PRATT CO. marked tools are common and those marked MASSACHUSETTS TOOL CO. are relatively scarce, it would appear that the actual marking was changed.

An early victim of the depression, the company was merged with the MILLERS FALLS CO. in 1931. The Goodell-Pratt name continued to appear on the tools until 1949 when all tools, including those few machinist's tools which were still offered, were marked with the Millers Falls Co. name. The following illustrations are from the GOODELL-PRATT CO. catalog number 17, published in 1930. They are limited to tools of special interest to collectors and do not include the very wide line of rules, squares, gauges, etc. which were much like those made by others. *(continued on next 9 pages)*

Inside Micrometer Gauges

With One Inch Run

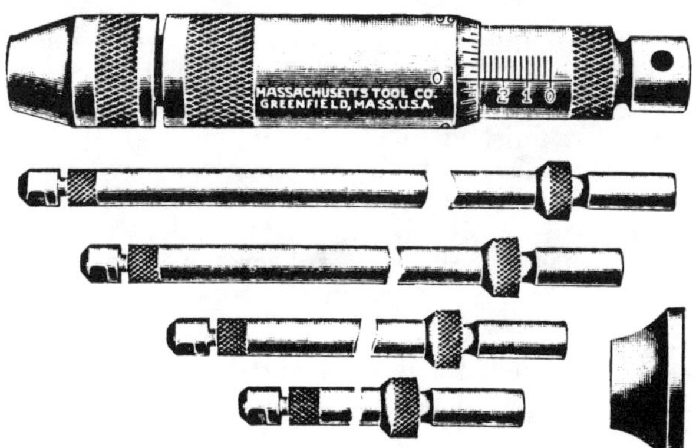

Patented May 8, 1894

Goodell-Pratt

No. 55 Universal Surface Gauge

With Micrometer Adjustment

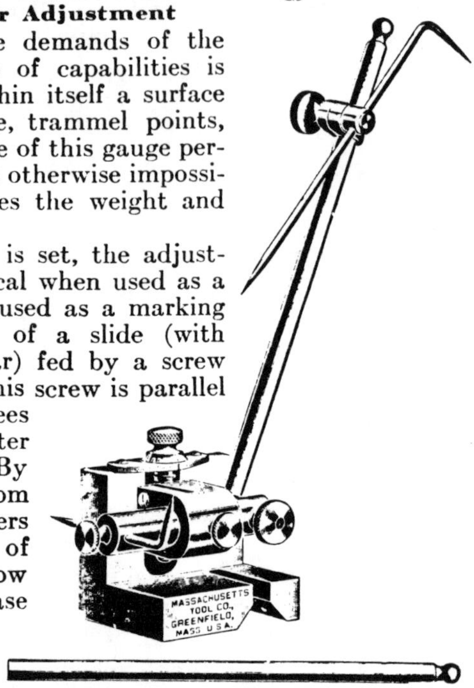

Designed especially to meet the demands of the most critical mechanics, its range of capabilities is almost limitless. It is at once within itself a surface gauge, depth gauge, marking gauge, trammel points, set or height gauge. The small base of this gauge permits accurate work at close quarters otherwise impossible and at the same time decreases the weight and space occupied.

At whatever angle the standard is set, the adjustment of the scriber is always vertical when used as a surface gauge, or horizontal when used as a marking gauge. Adjustment is by means of a slide (with compensating take-up for any wear) fed by a screw graduated to read to .001 inch. This screw is parallel with one base face and at 90 degrees with the other, making a micrometer surface, depth, or height gauge. By removing the standard and spindle from the base, and using the two scribers with them, a most convenient set of trammel points is arranged. For low work remove the standard from base and use scriber in slide spindle. Has V-slot in one base for cylindrical work. Extra length standards (jointed for folding) can be furnished at small cost, so that circles of almost any diameter may be described to a nicety by means of the micrometer adjustment. Furnished as shown above, with two standards, 5 inches and 10 inches long, and two scribers. Net weight, $2\frac{1}{8}$ pounds.

Price, each . . . . (YAGUC) $11.00

Packed one in a carton, $10\frac{3}{4}$ x 3 x $2\frac{1}{2}$ inches. Weight, $2\frac{3}{8}$ pounds.

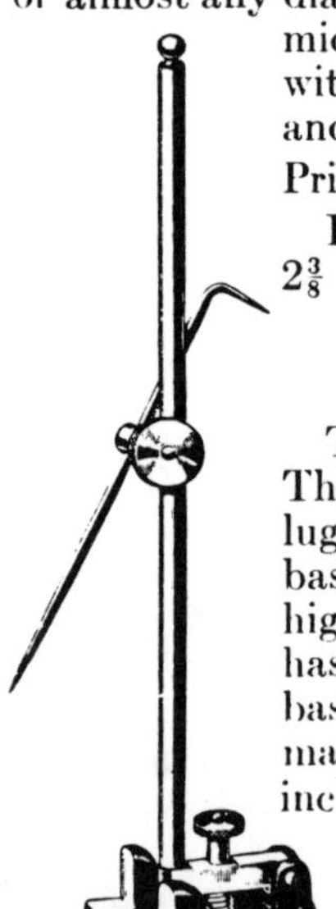

No. 56 Surface Gauge

This gauge is simple in construction, but accurate. The face of the base and the angles formed by the two lugs in front are milled and finished. The rest of the base is finished in black enamel. The standard is highly polished steel, and the scriber best drill rod. It has a fine adjustment by means of the knurled nut and base screw. It can be used as a depth gauge and, for many cases, makes a useful scratch gauge. Spindle is 8 inches long. Net weight, $\frac{3}{4}$ pound.

Price, each . . . . (YAGYE) $2.75

Packed one in a carton, $10\frac{3}{4}$ x 3 x $2\frac{1}{2}$ inches. Weight, 1 pound.

Goodell-Pratt

Surface Gauge

No. 57

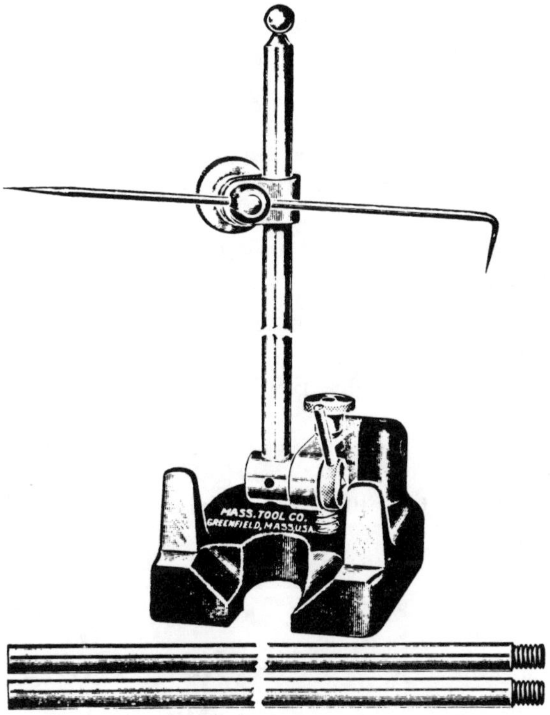

This large surface gauge with a solid base is one of the best ever offered to mechanics for large work or heavy duty, both on account of its range and its practical uses.

The spindle has a movement of 180 degrees, with a fine adjustment. After tightening the slide on the spindle, close adjustment is made by turning the knurled head nut on the screw through the supporting arm.

The tool can also be used as a depth gauge. By removing the spindle and inserting the scriber in the clamping stud it makes a satisfactory scratch gauge.

The angle milled on top of the base is of great convenience in working against a surface plate or planer bed. The base is 4 x 5 inches, all finished in black enamel except the bearing surfaces, which are polished.

Three 12-inch jointed standards, that can be screwed together for large work, are furnished with each tool. Net weight, 6½ pounds.

Price, each..(YAHEZ) $14.60

Packed one in a carton, 13½ x 4¾ x 4 inches. Weight, 7 pounds.

Goodell-Pratt

Surface Gauge
No. 115

This is a very useful and efficient surface gauge. The base is solid and stands square on the work. The spindle has a fine adjustment operated by turning the knurled headed nut on top of the base; after setting, this can be locked firmly by means of the tightening screw shown in the illustration. The scriber is made of carefully tempered tool steel, 4 inches long.

The base is $2\frac{1}{2}$ inches in diameter, finished in black enamel, with polished bearing surfaces. Height, 9 inches. Net weight, 1 pound.

Price, each. . . . . (YEADZ) $3.35

Packed one in a carton, $10\frac{1}{4}$ x $3\frac{1}{4}$ x 3 inches. Weight, $1\frac{1}{8}$ pounds.

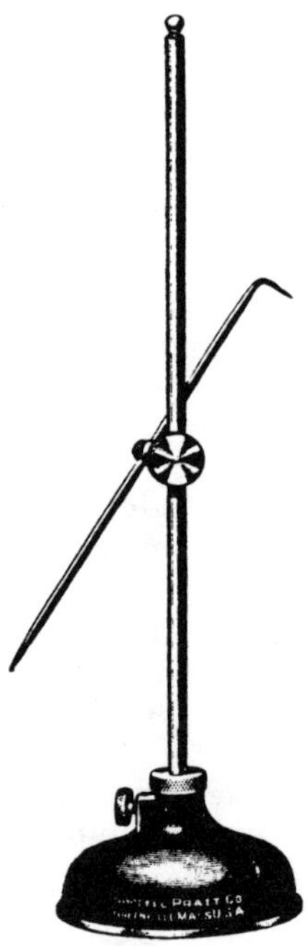

Surface Gauge
No. 116

This surface gauge is similar to the No. 115 shown above, except that it is larger and has a heavier base.

The gauge has a fine adjustment of the spindle and can be locked by the tightening screw. The base is $3\frac{1}{2}$ inches in diameter, finished in black enamel, with polished bearing surfaces. The scriber is made of tempered tool steel, $7\frac{1}{2}$ inches long. Height, 12 inches. Net weight, $1\frac{3}{4}$ pounds.

Price, each. . . . . (YEAJF) $4.60

Packed one in a carton, $13\frac{1}{2}$ x $4\frac{1}{4}$ x 4 inches. Weight, $2\frac{1}{8}$ pounds.

Goodell-Pratt

Adjustable Notch Center Gauge 60°

No. 44

(Tempered)

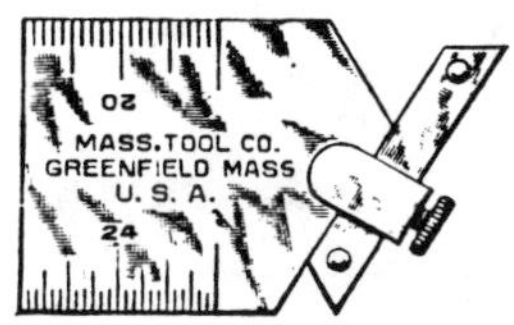

These center gauges are made of tempered crucible steel, and all angles are accurately ground. The notch, being made of separate pieces, insures a perfect angle to the extreme point. By tightening thumbscrew the sliding blade is held firmly in any position desired. It is the only center gauge that will fit any size inside threading tool. The sliding blade, together with the size of the tool, makes it very useful in many other ways. Graduated one corner each in 32ds, 24ths, 20ths, and 14ths.

Price, each..(YAECT) $1.25

Packed one half dozen in a carton.

Adjustable Notch Center Gauge 55°

No. 45

(Tempered)

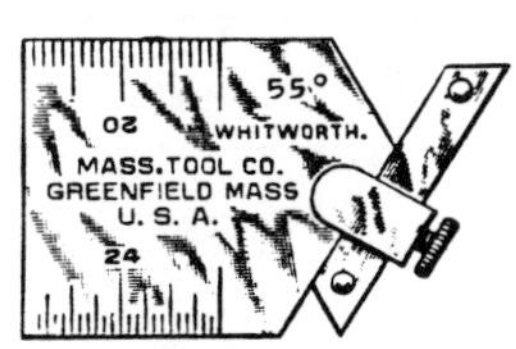

Same as the No. 44 above except that the angles are 55 degrees for the Whitworth English Standard.

Some of the uses to which this versatile little gauge can be put are illustrated on the opposite page.

Graduated one corner each in 32ds, 24ths, 20ths, and 14ths.

Price, each..(YAELD) $1.25

Packed one half dozen in a carton.

Goodell-Pratt

Adjustable Notch Center Gauge 60°

No. 46

(Tempered)

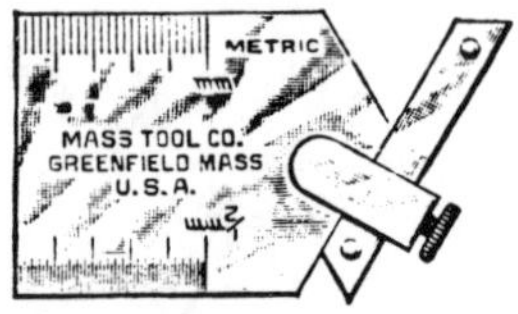

This gauge is exactly the same as No. 44 on the page opposite but with the four corners graduated in Metric as follows: one corner in ½ millimeters and the other three corners in millimeters.

Angles are 60 degrees.

Price, each. . . . . (YAENG) **$1.25**

Packed one half dozen in a carton.

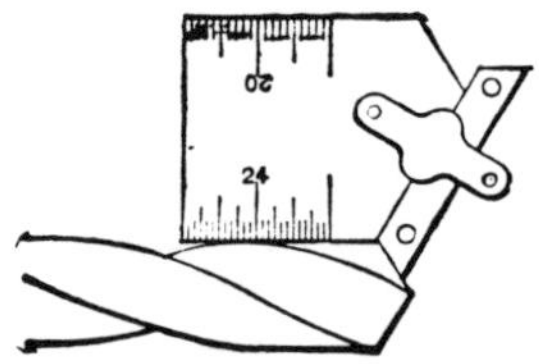

GRINDING TWIST DRILLS

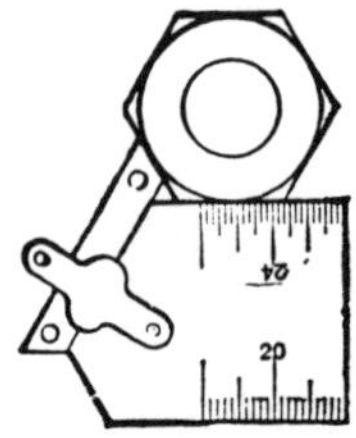

TESTING ANGLES

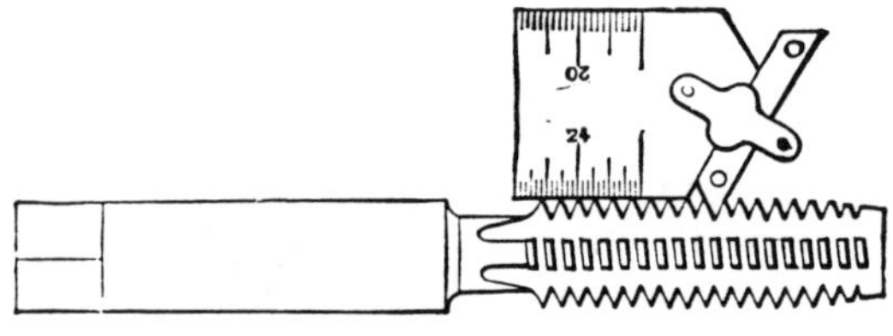

TESTING ANGLES OF THREAD

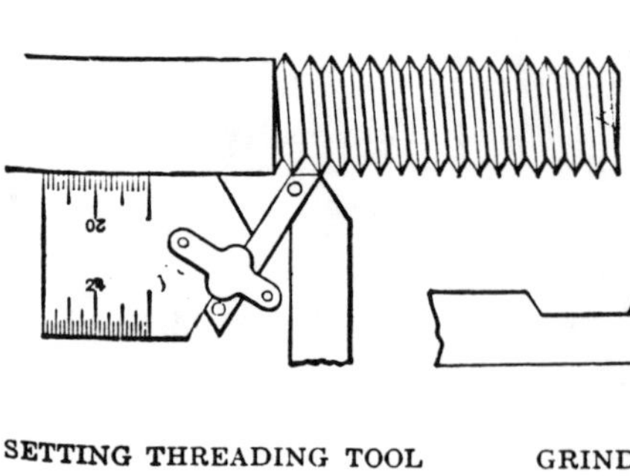

SETTING THREADING TOOL

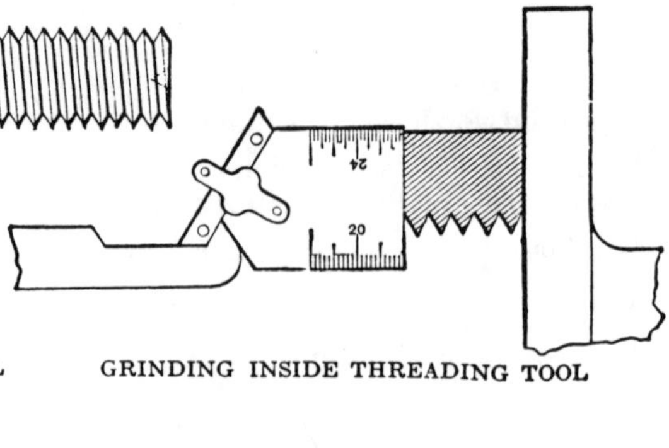

GRINDING INSIDE THREADING TOOL

Goodell-Pratt

Outside Transfer Spring Calipers

Patented February 17, 1903

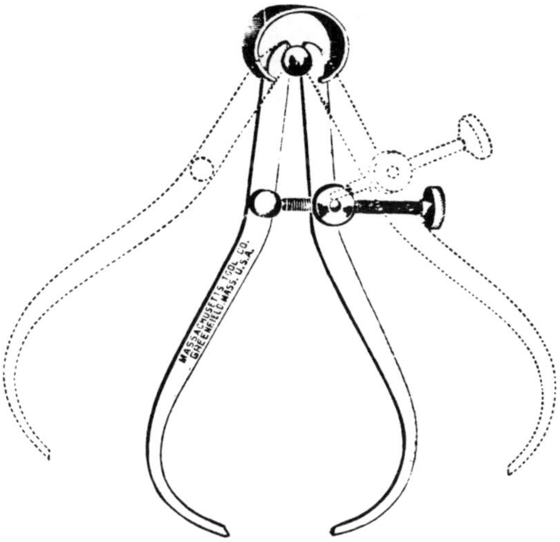

Indispensable for calipering back of shoulders, etc. The reading is taken in the usual manner and the screw then locked by tightening the thumb screw provided for the purpose. The legs are then spread sufficiently to draw them over the shoulder. The stiff tempered spring then returns the legs to exactly the same position they were in when the setting was made.

			Price, Each
No. 554.	Size 4 inches	(YUHWO)	$1.70
No. 556.	Size 6 inches	(YUIFS)	1.85
No. 558.	Size 8 inches	(YUIRG)	2.00

Packed one fourth dozen in a carton.

Inside Transfer Spring Calipers

Patented February 17, 1903

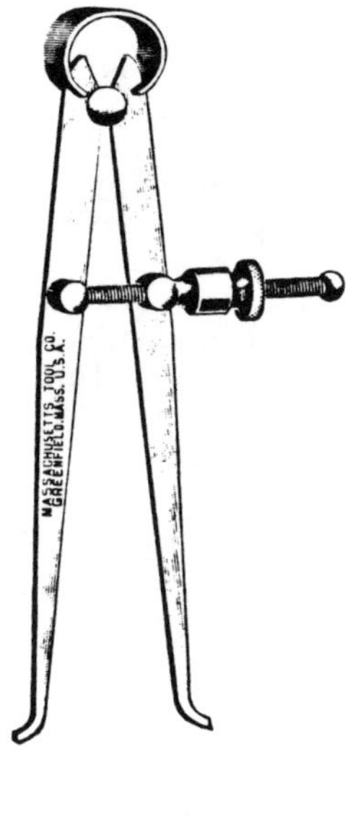

For accurate inside calipering where the legs must be compressed before the tool can be removed. The nut has an internal friction to prevent accidental change of the setting.

			Price, Each
No. 544.	Size 4 inches	(YUGIT)	$1.70
No. 546.	Size 6 inches	(YUGRA)	1.85
No. 548.	Size 8 inches	(YUGVO)	2.00

Packed one fourth dozen in a carton.

Goodell-Pratt

Iron Bench Levels

Made of smooth, well-seasoned gray iron castings. The base and top are accurately ground, the edges are polished and the body finished in glossy black enamel.

The vials are accurately set in the frame.

	Length	Weight		Price, Each
No. 503	4 inches	5 ounces	(YOWKY)	$0.85
No. 504	6 inches	6 ounces	(YOWUJ)	1.00

Packed one in a carton.

Adjustable Bench Levels

With Ground and Graduated Vials

These levels will meet the requirements of the most particular users. All the vials used are accurately ground and graduated, and each one is inspected before and after being set. The adjustment is close and positive.

The handles not only add to the convenience and attractiveness of the tools, but also form efficient protectors for the vials. On all the larger sizes, the supports for the handles are fastened directly to the base, so that the tube containing the vial is not disturbed in handling. Where accurate work is essential, this feature is particularly valuable, as it enables the operator to handle the level without danger of affecting its accuracy by the heat of the hand.

The bases of these levels are accurately ground and are finished in black enamel. All the tubes are polished and nickel plated.

	Length	Weight		Price, Each
No. 719	4 inches	$\frac{3}{8}$ pound	(ZASAF)	$3.20
No. 720	6 inches	$\frac{3}{4}$ pound	(ZASFA)	3.85
No. 721	8 inches	$1\frac{1}{8}$ pounds	(ZASGE)	4.50
No. 722	12 inches	3 pounds	(ZASJO)	5.95
No. 723	18 inches	$5\frac{1}{2}$ pounds	(ZASOJ)	7.70

Goodell-Pratt

No. 3 Micrometer Caliper

Range 0 to 1″ With Depth Gauge Attachment

Patented February 20, 1894

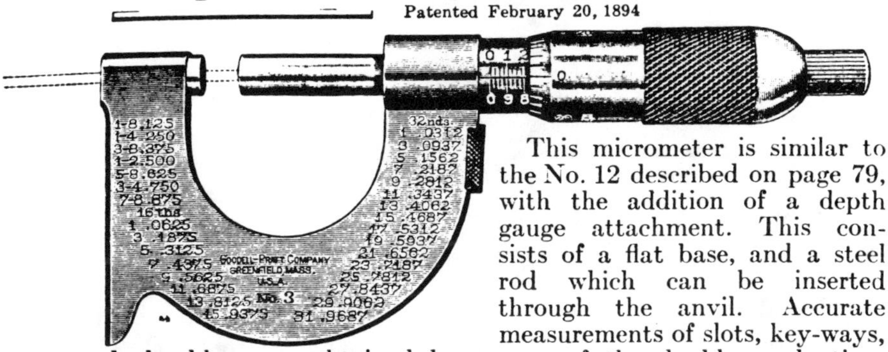

This micrometer is similar to the No. 12 described on page 79, with the addition of a depth gauge attachment. This consists of a flat base, and a steel rod which can be inserted through the anvil. Accurate measurements of slots, key-ways, and shoulders are obtained by means of the double graduation. When the rod is used, read the lower row of figures on the barrel and the outer row on the thimble.

Price, each . . . . (WYEGH) $13.00
Price of leather case . . . . (WYDDA) 1.25

Metric

No. 3 M. For measurements by 1/100 mm. from 0 to 25 mm.

Price, each . . . . (WYELM) $13.00

No. 6 Beam Micrometer Caliper

Range 0 to 6″ by Thousandths of an Inch

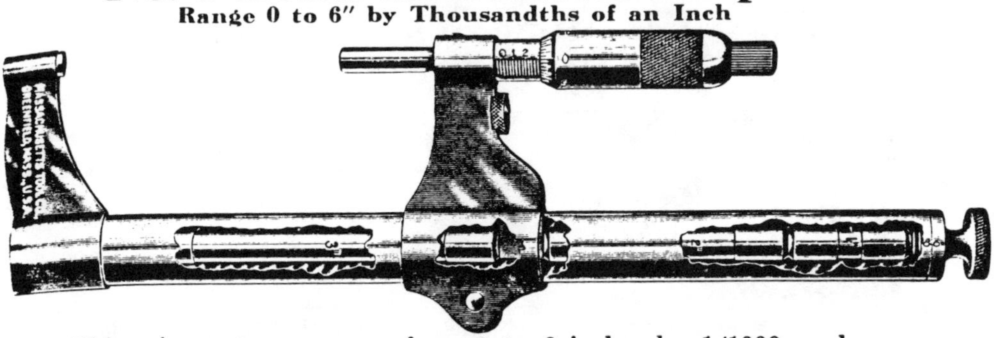

This micrometer measures from 0 to 6 inches by 1/1000, and diameters up to $2\frac{3}{4}$ inches. Three standard plugs—one, two and three inch—are placed in the tubular beam of the frame, which is slotted to allow a key fastened to the traveling head to come in contact with the ends of the standards. This enables setting the traveler, which carries the micrometer head, at any even inch from 0 to 6. The standards not in use are placed to the right of the key, as illustrated. The test screw at the extreme right is turned down tight so that the 0 mark on it and the frame coincide.

The micrometer head, anvil and locking device are exactly the same as in our other micrometers.

Price, each . . . . (WYGKO) $45.90
Price of leather case . . . . (WYGUL) 4.00

Metric

No. 6 M. For all measurements from 0 to 15 cm.

Price, each . . . . (WYGOK) $45.90

Goodell-Pratt

Steel Center Square

No. 78

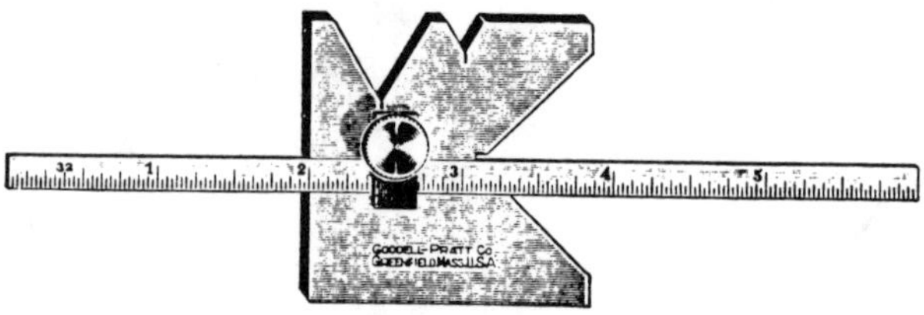

This all-steel tool combines the most, in the smallest compass and lightest weight, that has ever been offered to tool maker or machinist. It is a center square, T-square, depth gauge, center gauge, and steel rule. The narrow tempered steel rule furnished with the tool is 6 inches in length.

Price, each.......................................(YAPAG) $3.35

Furnished in No. 10, No. 11, or Metric graduation, as specified.

No. 50 Draftsman's Protractor

Stainless Steel

Patented January 17, 1893

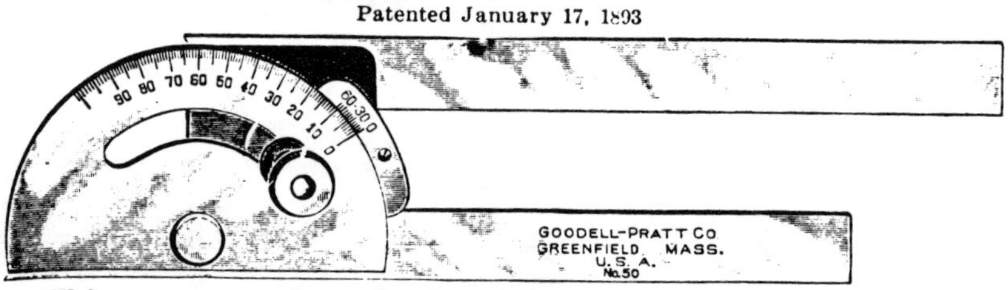

This protractor is made of stainless steel with blades 9 inches long. The arc is 4 inches in diameter, graduated in degrees, with a vernier reading to five minutes. It has a binding screw on one side that securely holds the blades at any angle and enables it to be picked up and moved about readily. The blades are fastened into the arc in such a manner as to make all parts come flush on the under side, thus making a perfectly flat surface for resting on the table. Either blade can be used in contact with a T-square, giving any angle and its complement from 0 to 90 degrees. It forms a perfect adjustable triangle. Finely polished.

	Weight		Price, Each
Price, in carton	$\frac{1}{2}$ pound	(YAFIX)	$9.20
Price, in polished mahogany case	$\frac{7}{8}$ pound	(YAFOZ)	10.50

Blades of extra length furnished to order.

GOODNOW & WRIGHTMAN Boston, MA

In its 1882 catalog it listed itself as "Importers, Manufacturers and Dealers in tools." It appears, however, that it did little or no manufacturing. Some tools were sold under its own name but these were probably made by others. The 1882 catalog includes a number of English calipers and dividers made by P.S. Stubbs, Call's patent calipers (J.H. Call patent of May 28, 1861, No. 32410), a variety of tools from DARLING, BROWN & SHARPE, VICTOR micrometers, and a number of German- and American-made calipers and dividers without brand names. It is interesting to note that the German-made tools were about one third the price of the equivalent P.S. Stubbs English tools and about one half the price of the unbranded American tools. In 1883, it offered machinist's tool chests. *(see illustration next page)*

GRANT MFG & MACHINE CO. Bridgeport, CT

Buyer, in 1907, of the TABOR MFG. CO.'s speed indicator business.

GRAY INSTRUMENT CO. Philadelphia, PA

Maker of "Queen" triangular levels advertised in 1930.

GREBLE, TURNER & CO. Hamilton, OH, and Orange, MA

Maker of machinist's try squares, c1900-1902. It sold the squares in sets of 2", 3", 4", 6", and 8" sizes in a wooden case. Its advertisments state that its factory is located in Orange, MA. This may mean that there is some relationship with the HOWELL TOOL CO., which was making machinist's try squares in Orange, MA, at about the same period.

Machinery, May 1900

Fig. 1.—Machinists' Tool Chest.

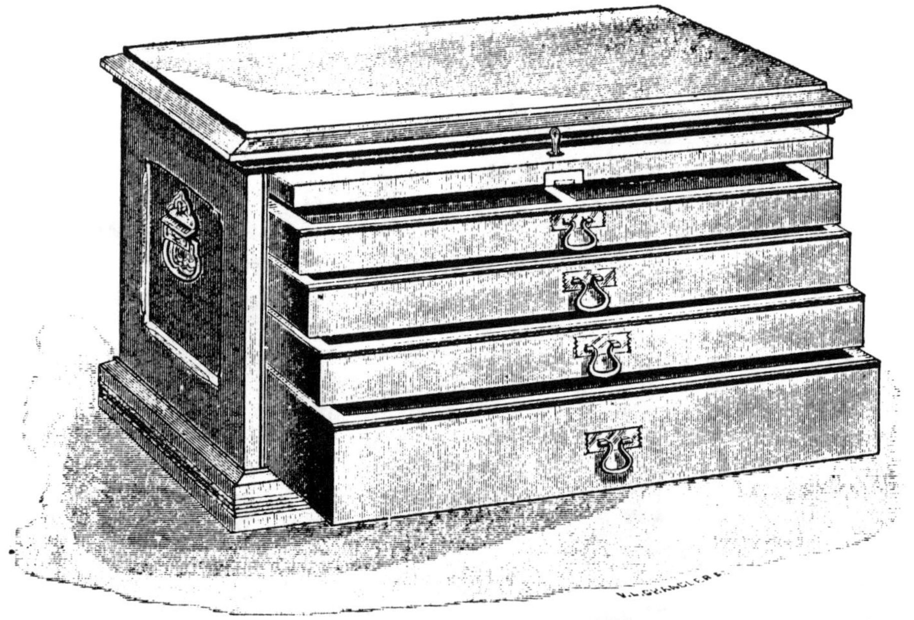

Fig. 2.—Machinists' Tool Chest.

Machinists' Tool Chest or Cabinet.

We present with this, engravings representing a tool chest or cabinet, which although more especially designed for machinists' use, is equally available for watchmakers, jewellers, amateurs, photographers, &c. Fig. 1 shows the cabinet closed, and Fig. 2 shows it opened. The dimensions are, outside,

These cabinets are made of black walnut, highly polished. The drawer handles are nickel-plated, and the end handles are bronzed. They are provided with a "Yale" lock with two keys. They are manufactured by Goodnow & Wightman, 176 Washington street, Boston, Mass.

See entry for Goodnow & Wrightman, previous page

American Machinist 1883

GREELEY & CO., E.S. New York, NY.

Maker, c1884-1888, of the Pratt Speed Indicator patented June 3, 1884 (No. 299677), by Robert J. Pratt.

GROVE MFG. CO. Dayton, OH

Maker of outside calipers with a .001" indicator on one leg, marked "Pat. Pending". Thomas E. Grove received the patent on January 18, 1921 (No. 1365779). In use, the calipers "are first set with a micrometer by tapping against the solid leg as in the case of an ordinary pair of calipers, and particular attention is paid to where the indicator registers when at its smallest point. The gauge is then tried in the hole or slot, and if it registers the same as for the micrometer, assurance is obtained that an accurate measurment has been secured." By 1926 the same caliper was advertised by J.M. WATERSON of Detroit, MI.

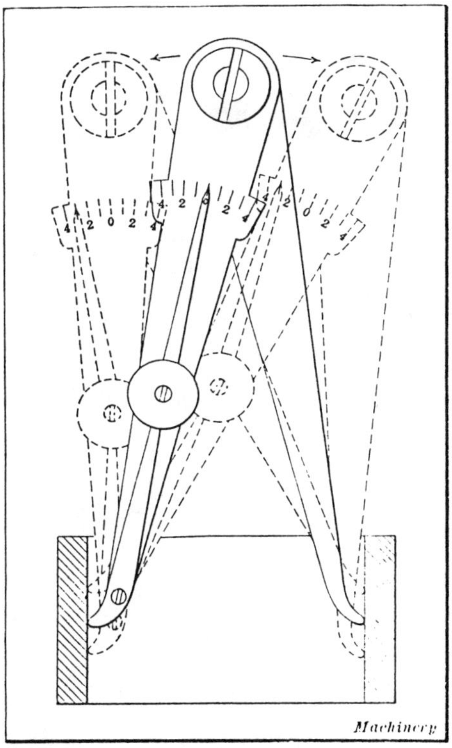

Indicating Gage made by Grove Mfg. Co.

Machinery September 1918

HADDOW, WM. Newton, MA

Patentee, on April 10, 1888 (No. 380797), of a surface and planer gauge. It is not now known if the tool was actually manufactured. *(continued next page)*

A New Measuring Instrument.

The accompanying illustration shows a tool which combines a micrometer caliper both outside and inside, a surface gauge, scratch block and depth gauge. It may be used to set a planer tool to plane any given size, and also for measuring a planed piece without removing it from the planer bed.

It consists of a micrometer screw mounted in the main casting and carrying a vertically movable nut. This nut has fore and aft projections, to the former of which is pivoted

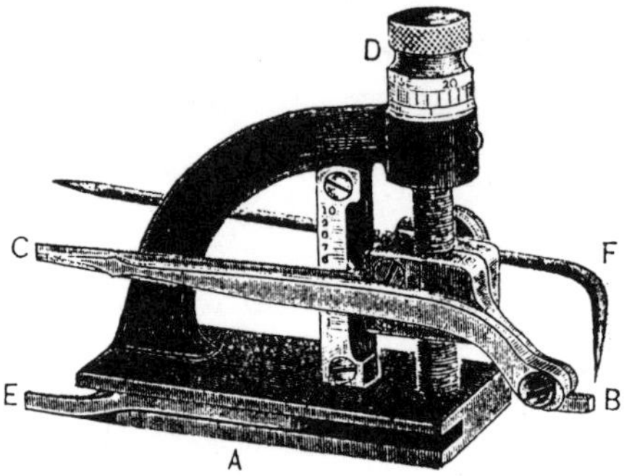

a lever *B*, whose longer and pointed end stands against the projection of nut at *C*, and is normally held by a light, double-acting spring at the central line, as shown.

To set a planer tool, adjust the end *B* to proper height and bring the tool down until the point *C* will indicate, by a slight motion, the contact. To measure the piece after it is planed, the part *B* is brought down onto the work until the point moves slightly, when, by reading the screw in connection with the vertical graduated bar, the thickness may be determined. Of course allowance must be made for thickness of projection *B*, which is $\frac{1}{8}$ inch. By swinging the bar *E* around under *B*, the two may represent the legs of a caliper, the distance between them being read from the bottom of nut where it crosses the graduated bar, and in using the tool as an inside caliper read from top of nut, which equals the combined thickness of *B* and *E*. Extensions of different lengths are made to attach to *B* when used as a depth gauge, and also to increase the capacity of tool in either direction.

The scriber, which may be placed on back of nut, as shown, makes a convenient and easily adjustable surface gauge, and as a "scratch block," the screw furnishes an accurate means of marking two or more parallel lines, the distance between which may be as desired, within a minute fraction.

The maker is Wm. Haddow, Newton, Mass.

American Machinist 1888

HALLSTRONG, CHARLES Somerville, MA

Introduced finely adjustable trammel points and a sliding step block type of height gauge in 1907.

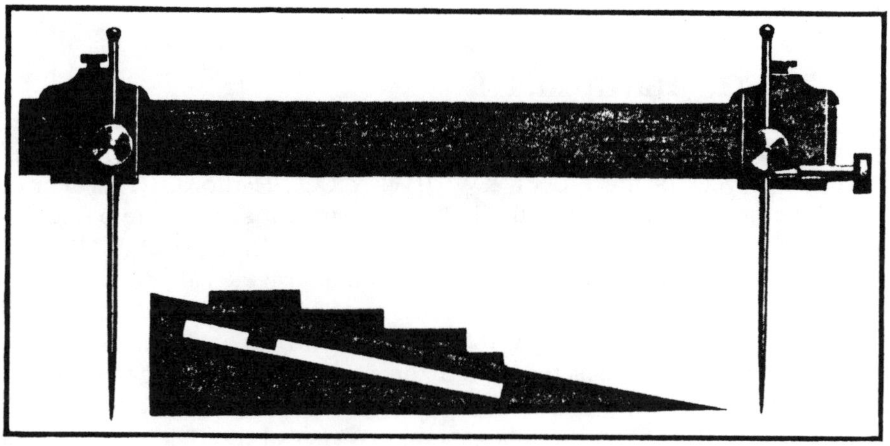

Machinery 1907

TRAM POINTS AND HIGHT GAGE

HAND MFG. CO., S. ASHTON Toughkenamon, PA

Maker, in 1885, of a "piston packing" indicator designed to measure for the center of a cylinder.

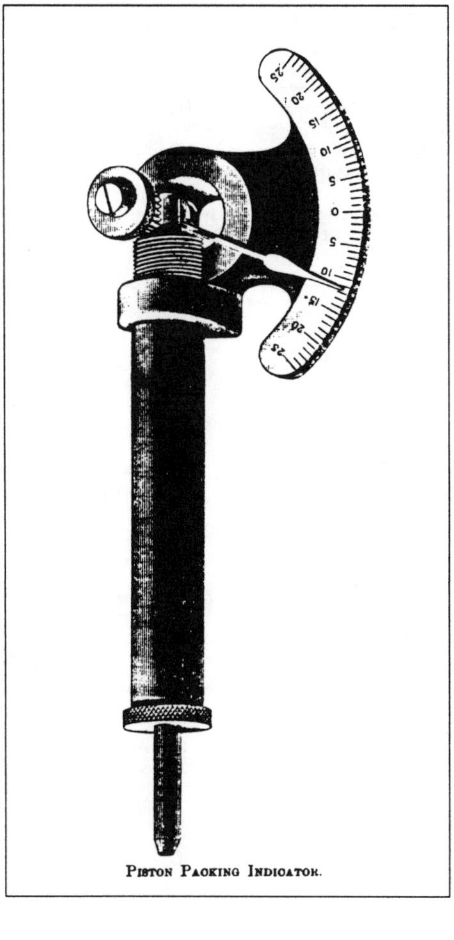

PISTON PACKING INDICATOR.

American Machinist 1885

HARRIS, WILLIAM H. Laurium, MI

Maker of center calipers for which he was granted a patent September 20, 1910 (No. 970817). The calipers were designed with "legs which are linked to a central bar containing a scriber that automatically maintains a central position."

HARTFORD TOOL CO. Hartford, CT

Formed in 1883 by E.G. Parkhurst, M.D. Pratt, J.E. Woodbridge, and F.N. Gardner to make knife edge straight edges developed by Gardner, and the Woodbridge lathe and planer tool. The partners were all employees of the PRATT & WHITNEY CO., Hartford, CT, which may actually have owned the company. After 1886, the knife edge straight edges were being made by the PRATT & WHITNEY CO., and some years later, by the BROWN & SHARPE MFG. CO., Providence, RI. The partners also made, beginning in 1885, an adjustable caliper gauge which is much like an inside micrometer but without graduations, and Kidd's finely adjustable divider. F.N. Gardner was granted a patent September 28, 1886 (No. 349840), for a variation of the Kidd Divider which added a fine adjustment screw. Both the adjustable caliper gauge and the Kidd/Gardner divider were later made by PRATT & WHITNEY CO. The firm closed in 1890 when Gardner left to join C.H. BESLEY & CO., as general manager of the Beloit, WI, works.

HARTFORD ADJUSTABLE CALIPER CAUCE.

Holder full Nickel Plate. Diameter of Holder, ¼ inch. Diameter of Steel Wire, 5-32 inch.

This little Tool will be found a convenient and economical substitute for the wire, usually made after repeated filing and "drawing" to get the exact size. Two "chucks" are arranged, one on each end of a hollow tube. The hole in one end of tube is smooth, so that by loosening the chuck the wire may be quickly slid out or in. In the other end a fine thread is tapped, and a knurled wire is threaded to fit. In use, in taking a size, the smooth wire is moved out to give nearly the diameter required, and fastened; then the fine adjustment is made by the threaded wire. Ends of wire are hardened. Any large diameter may be taken by changing the plain wire for a longer one. This Tool will be found convenient in taking distances between uprights, etc., being positive.

No. 1, Shortest 3 inch, Longest 6¾ inch, 3 Wires $0 90
No. 2, Shortest 6 inch, Longest 18 inch, 3 Wires 1 00

Wires of any length will be furnished at 1 cent per inch.

1890

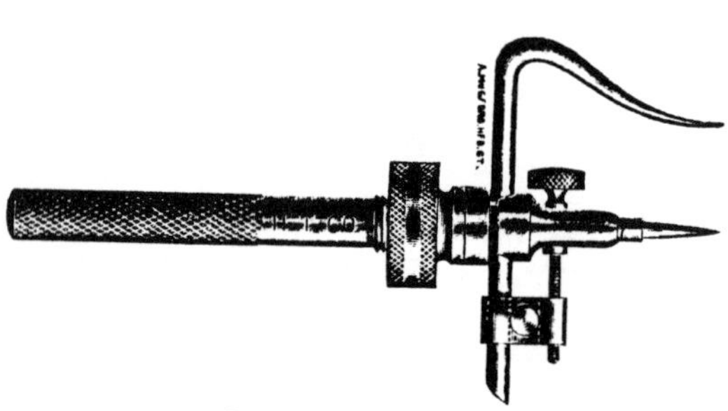

IMPROVED DIVIDER.

American Machinist 1885

KIDD'S IMPROVED DIVIDER

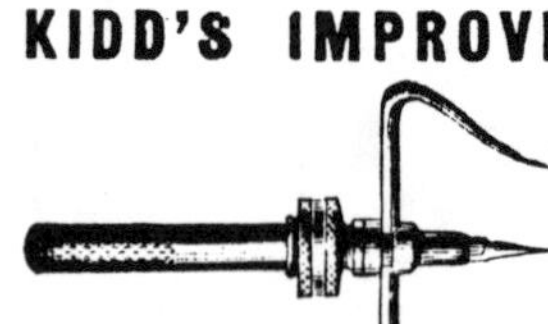

NO LIMIT TO CIRCLES DRAWN. LEAD PENCIL HOLDER WITH SAME.
SEND FOR CATALOGUE.

HARTFORD TOOL CO., HARTFORD, CONN.

American Machinist 1885

HAZELTON, Frederick D. Boston, MA.

Maker, c1876-1882, of "Hazelton's Improved Caliper Attachments," for which he received a patent on November 21, 1876 (No. 184524).

HAZELTON'S IMPROVED CALIPER ATTACHMENTS.

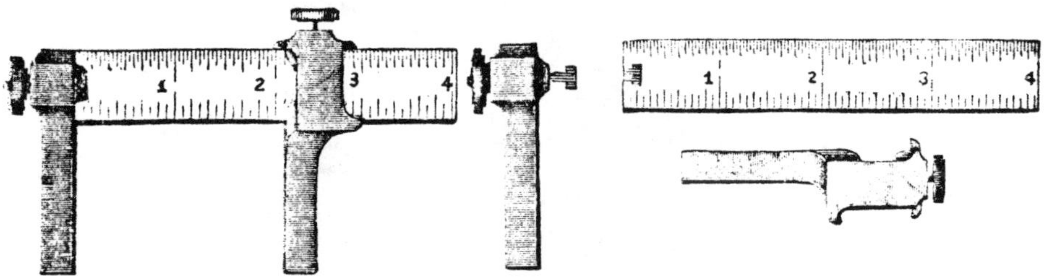

The attention of Machinists, and Mechanics generally, is called to our new Improved Caliper Attachments, which, in combination with the ordinary steel scale, makes a tool that, for general usefulness. accuracy and cheapness, cannot be excelled.

This combination may be readily adapted for various measuring purposes. The scale, with both jaws attached, makes a strong, neat, and correct Caliper Gauge; with the sliding jaw removed, a reliable Square; with end jaw removed, and sliding jaw attached, it may be conveniently used as a Depthing Gauge; with both jaws detached, we have a perfect Scale or Straight Edge. All these changes can be made instantly, and unlike most combination tools, this is complete and accurate in every charge. The jaws are hardened throughout. In the ordinary caliper square—where the scale and jaw are in one piece—if by constant use the jaws become worn, the divisions are no longer of use for accurate measurement; but in this tool, if the jaws are worn, by a little grinding on a straight surface they may be corrected in a few moments, and replaced as before.

The scales are divided in the most convenient manner, each edge being different. On one side we have 10ths and 64ths through the whole length; on the opposite side, 20ths and 100ths in like manner; and the sliding jaw is arranged as to read any one of the four divisions—an advantage found in no other tool of its class.

Inches.	4	6	12 in.
Price.	$4 00	4 50	6 00

1880

J.R. HERBERT CO. Brooklyn, NY

Maker of a speed indicator introduced in 1907.

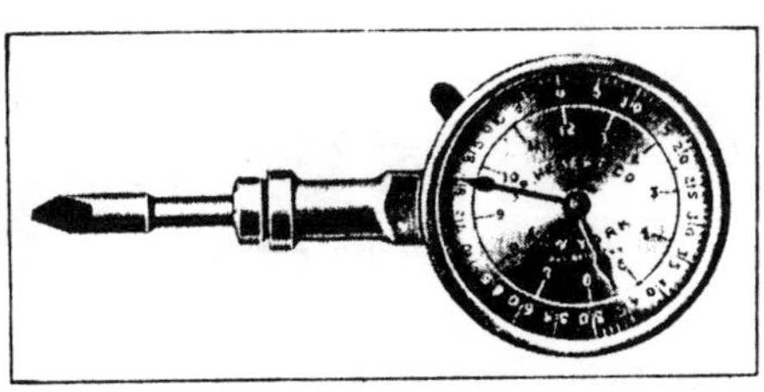

A SPEED INDICATOR

Machinery 1907

HERLTH MFG. CO., O.B. Hartford, CT

Maker of machinist's adjustable parallels in 1920. The tools appear very similar to those made later by the L.S. STARRETT CO., Athol, MA.

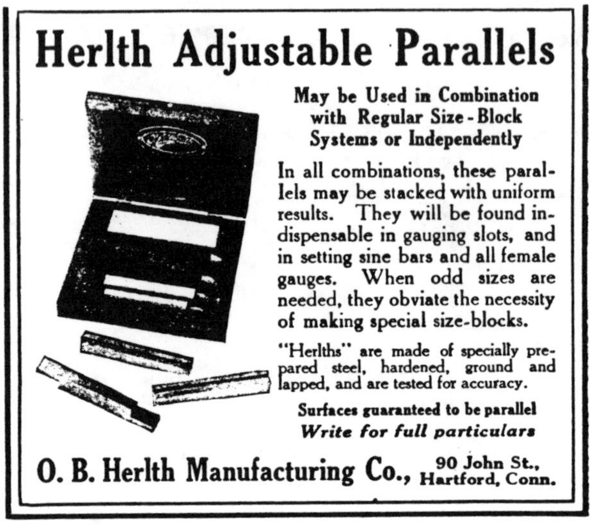

1920

HJORTH LATHE & TOOL CO. Boston, MA

Founded by Henry J. Hjorth in 1915 to make machinist's tools, lathes and upright drills. Hjorth had been granted patents for a surface gauge on May 31, 1904 (No. 761381) and a gauge for edged tools on February 29, 1916 (No. 1174009). The surface gauge was first made by the BEST TOOL CO., Boston, MA, in 1904 and later by the MECHANICS TOOL CO., Boston, MA. The HJORTH LATHE & TOOL CO. made the same surface gauge as well as a micrometer adjustment depth gauge, spring winders, and the Hjorth patent drill gauge.

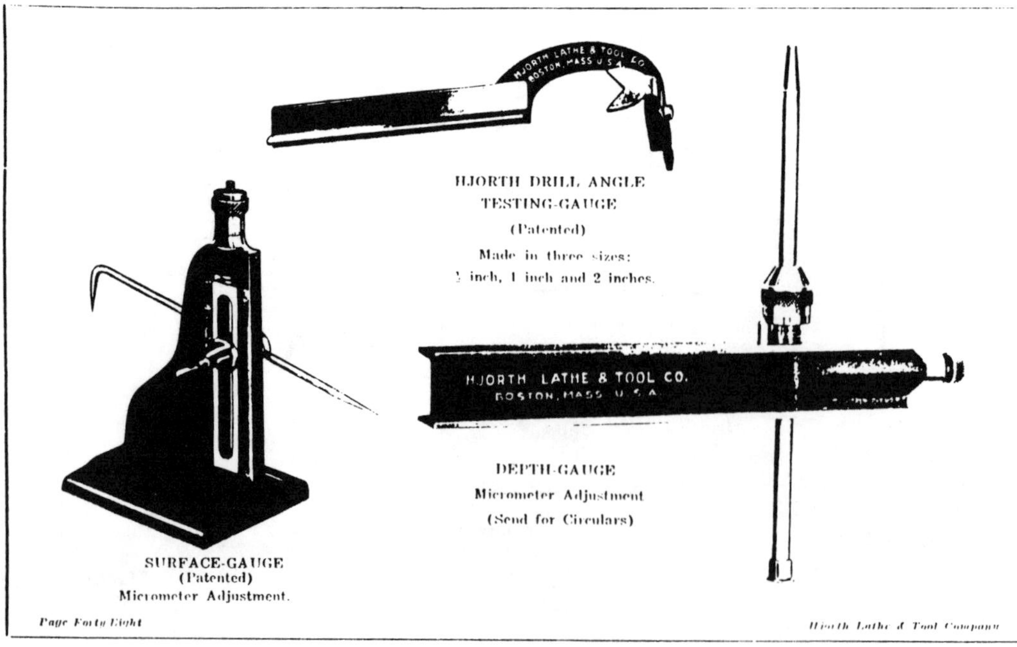

HOGGSON & PETTIS MFG. CO. New Haven, CT

Makers, in 1892, of a surface gauge patented January 26, 1892 (No. 467732), and February 2, 1892 (No. 468093), by James Carr. The gauge was made in three sizes; 6" and 9" stem, 12" and 18" stem, and 30" stem. One speciman examined is unmarked except for "PAT APL'D FOR" on the eccentric washer. Others are marked with the patent dates and the HOGGSON & PETTIS MFG. CO. name.

The company also made the Carr patent (May 27, 1890, No. 428890) rule clamp, the Carr patent scribing block (same patent as the surface gauge), and Carr's improved center square. See the JAMES CARR listing for illustrations. They also made turning tool holders under Carr's patent of July 18, 1899, which was assigned to HOGGSON & PETTIS MFG. CO. Hoggson & Pettis was listed as a maker of surface gauges and machinist's squares as late as 1905 and may have made them even later. *(continued next page)*

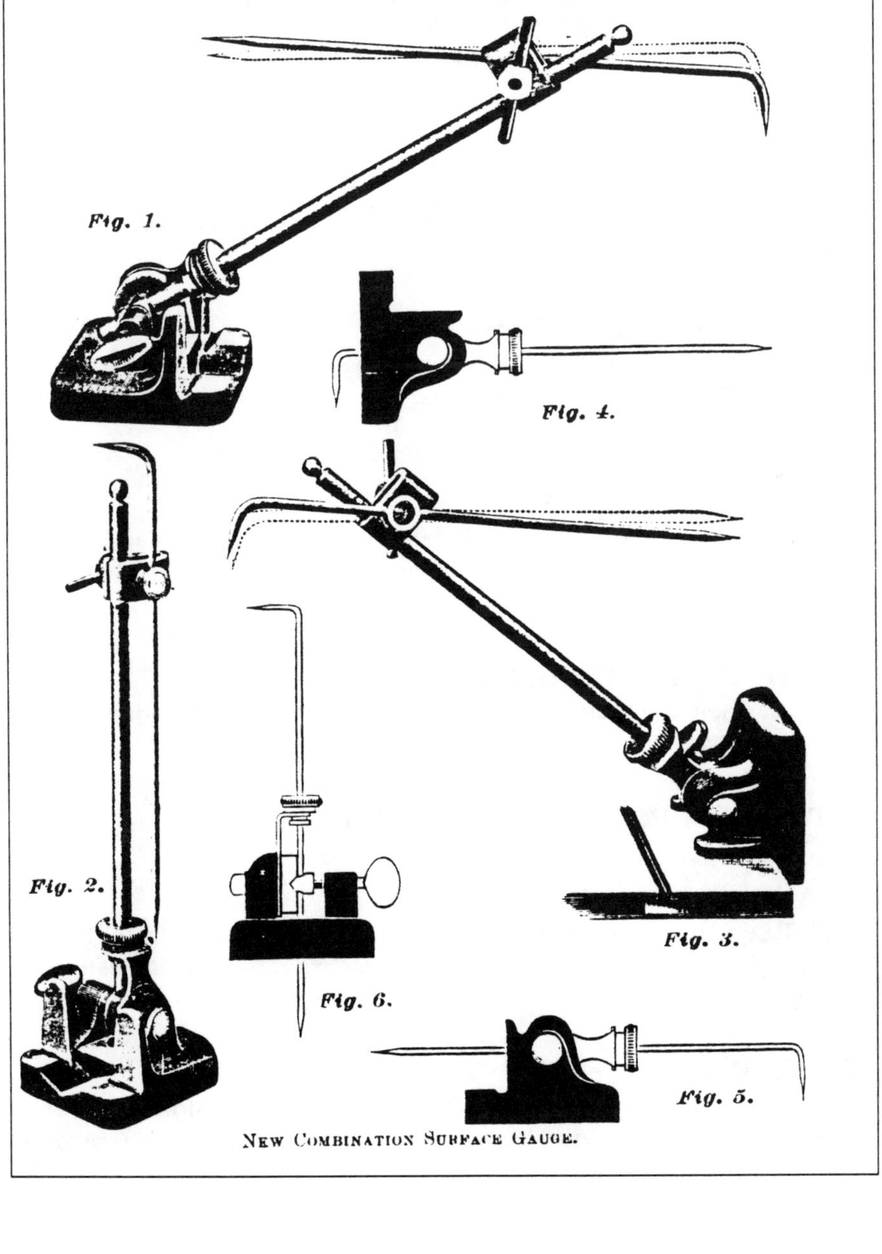

NEW COMBINATION SURFACE GAUGE.

American Machinist 1892

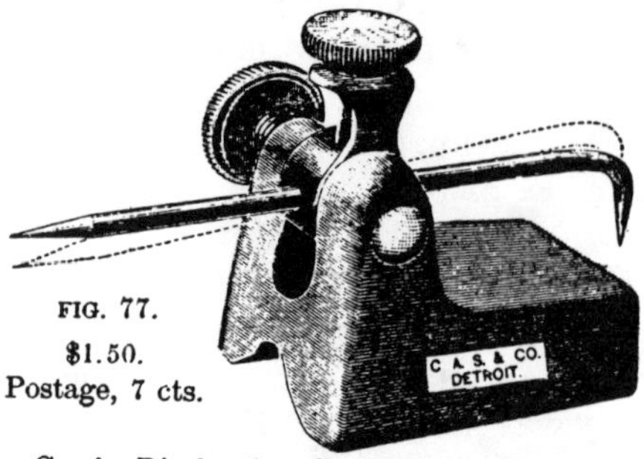

FIG. 77.
$1.50.
Postage, 7 cts.

Carr's Diminutive Scribing Block, or Surface Gauge (Fig. 77), is a new tool. The needle has an adjustment, by means of the top screw, and can be held rigid. The V slot in the base will be found very useful and desirable, in truing work from a boring bar, also for setting cutter in same. Forms a scratch and depth gauge.

HOLLADAY, GAVIN Lynn, MA

Started by Gavin Holladay and his brother George in 1854. They were the first makers of the center square patented July 6, 1852 (No. 9089) by Nathan Ames. Holladay sold the product to J.R. BROWN & SHARPE, Providence, RI, about 1856.

HORSTMANN CO., F.W. Irvington, NJ

Maker of indicating calipers patented by Frederick W. Horstmann on April 15, 1919 (No. 1300837) and May 26, 1925 (No. 1539356) and Henry J. Horstmann and Frederick W. Horstmann on November 28, 1922 (No. 1436656).

HOWELL TOOL CO. Orange, MA

Maker of machinist's try squares, c1900. Very little is known of the company. It may have had some relationship with GREBLE, TURNER & CO., Hamilton, OH, who claimed that its factory was located in Orange, MA.

HURLEY BROTHERS Hartford, CT

Introduced, in 1893, a finely adjustable trammel and a tool holder for boring or threading tools. *(see illustration next page)*

HYDE PATENT CALIPERS See C.P. FAY & CO.

IBSEN & CO. Milwaukee, WI

In April, 1917, H.B. Ibsen announced that he had severed his connection with WISMACH & CO., Milwaukee, WI, and had begun to manufacture reference gauges under the name of IBSEN & CO.

IDEAL TOOL CO. Rochester, NY

Maker of indicators and center testers from c1925 to c1965. The indicators were low priced and must have been very popular with machinists as they are quite plentiful. *(see illustration page 208)*

Trammel and Boring-Out Tool.

The accompanying cuts represent tools recently put upon the market by Hurley Brothers, 223 State street, Hartford, Conn. The trammel, Fig. 1, is designed for the use of pattern makers, draftsmen, &c. The tool is 14 inches in length, well finished and nickeled, and is referred to as having a beam of square brass tubing, which is very light, and points made from best tool steel and hardened; it also has a pencil attachment. The makers remark that after placing the attachment in the same radius as the steel point it can be changed from one radius to the other without changing the relation of one to the other. Attention is called to the fact that the points can be set approximately when brought down with fine adjustment at the end. At an additional cost an extension of 14 inches with a clamp is furnished when desired.

The boring-out or threading tool is made of round steel, turned at the point and

Fig. 1.—Trammel.

hardened. The tool holder has a V-slot milled in the top and bottom, one slot being small to hold a small tool. The tool is placed in the slot and the tool post screw is brought on the round steel to hold it fast in the slot. The point is made that the strap at the end acts as a center rest for the tool and that the round steel can be drawn backward or forward as the work may require. The tool is designed

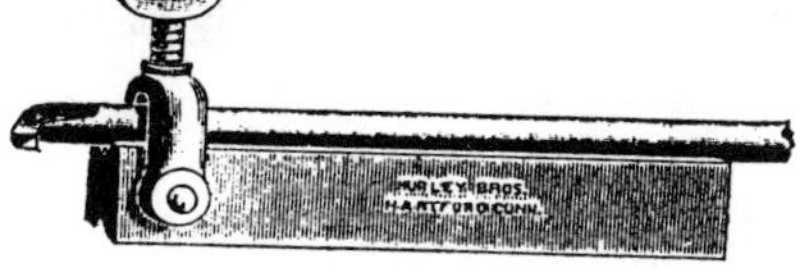

Fig. 2.—Boring-Out or Threading Tool.

to save forging and also to provide a stiff tool for boring or threading.

See entry for Hurley Bros., previous page

Iron Age, April 1893

Ideal Universal Test Indicator
Double Scale

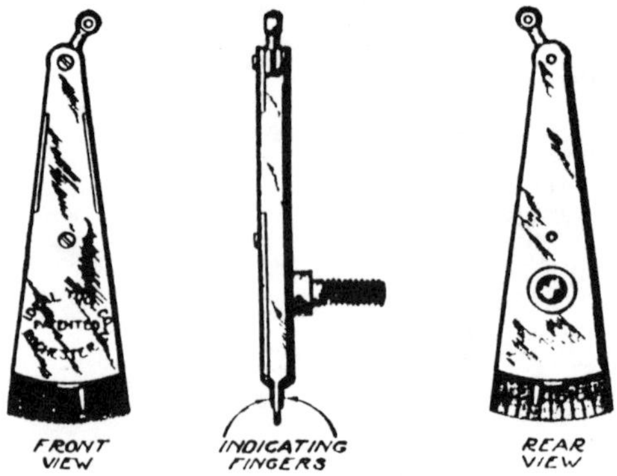

The advantages of an indicator with accurate readings from front or rear is obvious, especially when indicating holes, or where the indicator is fastened to a revolving spindle. Small, compact and sensitive. Indicates in any position.

All bearings bronze bushed and working parts of hardened tool steel. Mounted on hardened steel tool post holder. Contains compound levers. Scale graduated in thousandths, divisions 1/16th inch apart making reading easy. Contact point spherical in shape and can be turned to 90 degrees on either side of indicator by simply pushing it into desired position. Patented friction device protects contact point from accident.

Packed 1 in a carton. Weight 4 ounces.

Each.. $8.00

Ideal Universal Single Scale Test Indicators

Accurate and reliable, small, compact and sensitive. Indicates in any position. All bearings bronze bushed and working parts of hardened tool steel. Mounted on steel tool post holder. Contains compound levers. Scale graduated in thousandths, divisions 1/16th inch apart making reading easy. Contact point spherical in shape and can be turned to 90 degrees on either side of indicator. Friction device protects contact point from accident.

Each.. $6.25

Ideal Universal Center Testers

Used in connection with Ideal Test Indicator. Will true up a prick punch mark quicker and more accurately than a wiggler, either on work revolving or stationary.

Ideal Center Testers are designed to locate a prick punch mark exactly to center of a revolving spindle in a milling, drilling, boring or any other machine where work is stationary and spindle revolves. This tool has many uses.

Center Tester............ each $4.50

See entry for Ideal Tool Co., page 206

IVES MFG. CO. Bridgeport, CT

Maker of the Hayes patent surface and turning gauge c1904. The gauge was patented October 9, 1900 (No. 659368) by John E. Haberlin, William R. Haberlin, and Timothy F. Hayes. *(see illustration next page)*

IZARD-WARREN CO. Philadelphia, PA

Maker, in 1910, of the "Sterling Machinist's Precision Level." The levels were offered in six sizes, from 2 inches to 24 inches. "The bulbs are so ground as to indicate a variation of one-thousand inch in the length of the level." The body of the level was cast iron and the top was brass.

J.& H., Worcester MA

Maker of combination inside/outside calipers and dividers patented by C. Jilson on August 16, 1864 (No. 43854).

JAMES, C.B. Chicago, IL

Advertised for "Tool Agents" in south and west Pennsylvania, in 1889. The cut in the ad shows machinist's tools, but it is not clear if the company was a manufacturer or dealer. Probably the latter.

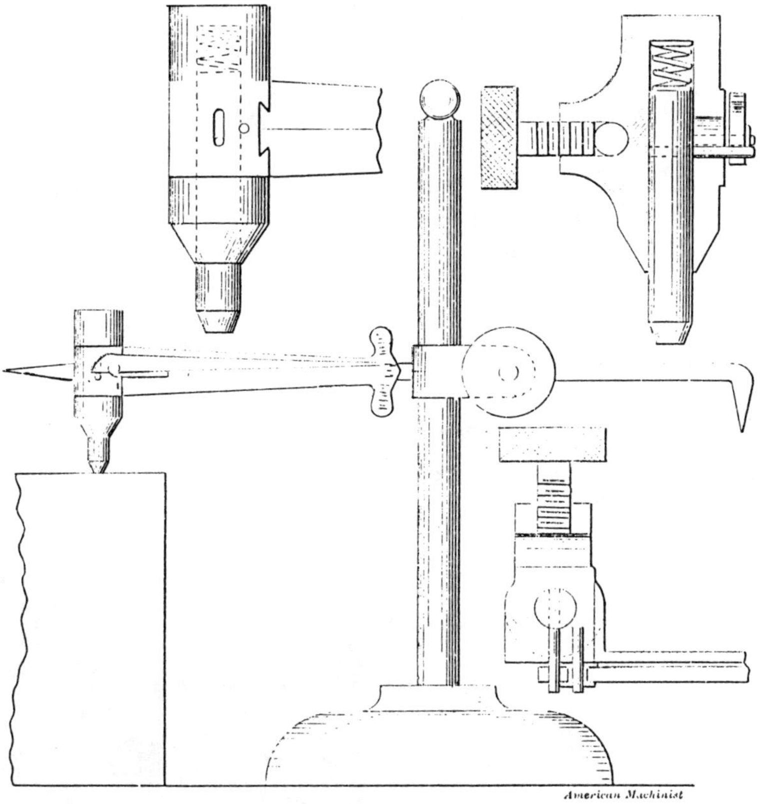

A TRUING GAGE.

1901

See entry for Ives Mfg. Co., previous page.

JAMES, THOMAS R. Chicago, IL

Maker of a micrometer attachment for caliper and surface gauges in 1901. He had patented the tool on May 22, 1900 (No. 649934).

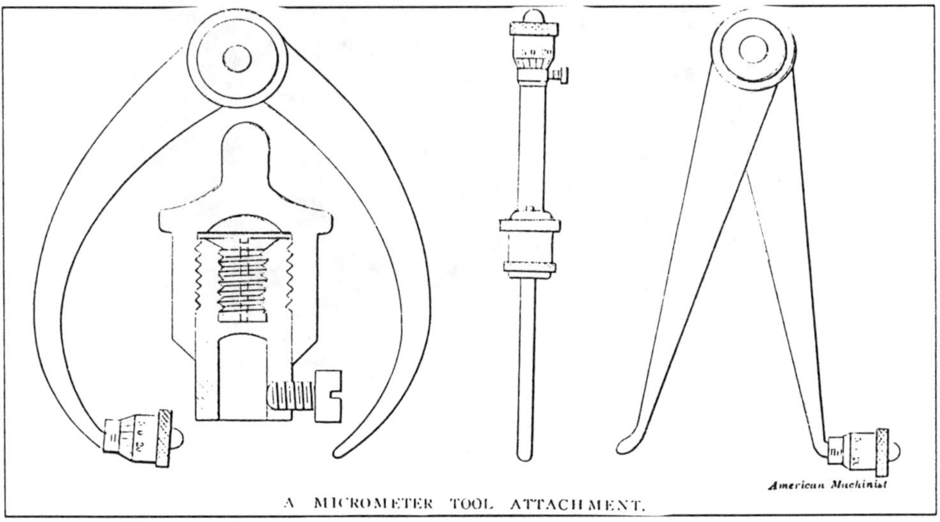

A MICROMETER TOOL ATTACHMENT.

American Machinist 1901

JAQUES, Fernando O. See CENTRAL TOOL CO. and WINSOR & JERAULD MFG. CO.

JENNINGS & CO., C.E. New York, NY

Established in 1876 by Charles E. Jennings. Jennings operated primarily as a hardware dealer but also appears to have been the manufacturer of some machinist's tools. Jennings introduced, in 1895, a line of machinist's hexagonal pocket levels in 2½", 3" and 3½" sizes and bench levels in 12", 18" and 24" sizes. He was also offering a line of machinist's tool chests in 1900. The firm continued as both a dealer and manufacturer until going out of business about 1924. *(continued next page)*

Special Machinists' or Jewelers' Tool Chest No. 40.

C. E. Jennings & Co., 101 Reade street, New York, are offering machinists' or jewelers' tool chests, as herewith illustrated. They are made of selected antique quartered oak or black walnut, well finished. Each chest is provided with a Yale pattern lock and two keys, nickel

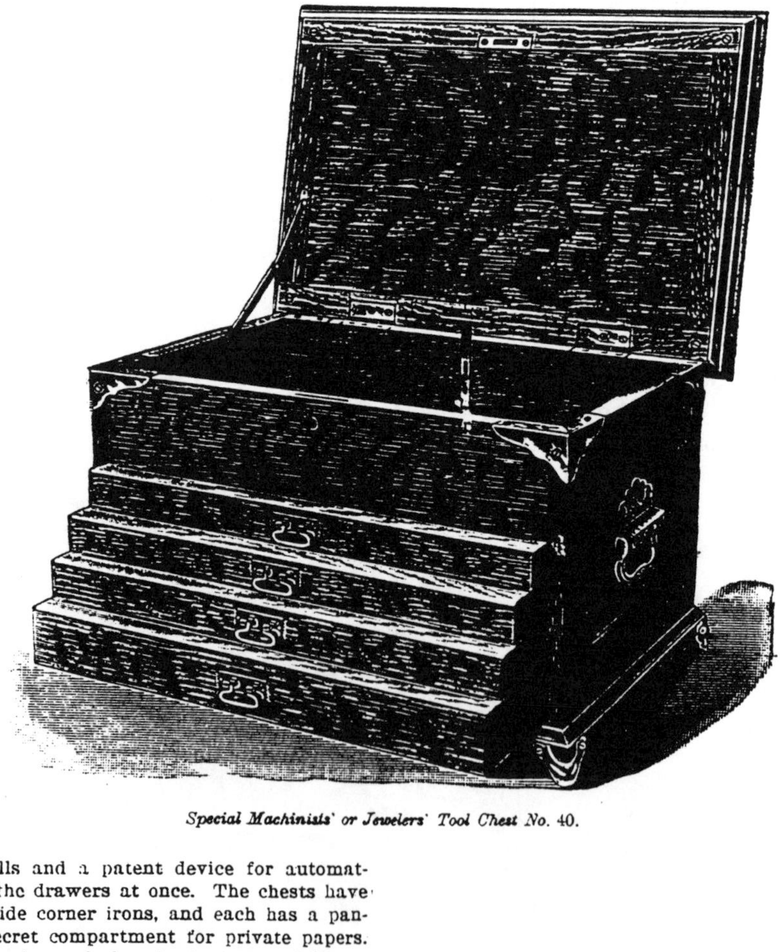

Special Machinists' or Jewelers' Tool Chest No. 40.

plated drawer pulls and a patent device for automatically locking all the drawers at once. The chests have nickel plated outside corner irons, and each has a paneled lid with a secret compartment for private papers. The outside dimensions of the chests are as follows: 18½ inches long, 12¼ inches wide and 11½ inches high.

MACHINISTS' LEVELS.—SEE PAGE 46.

American Machinist 1895

KELSEY, D.J. New Haven, CT

Operated by Duane J. Kelsey who patented a number of drafting tools. Kelsey introduced, in 1892, a right-angle protractor. The tool was made of German silver and was probably made for a draftsman's use. Many such tools, however, were used by machinists as well.

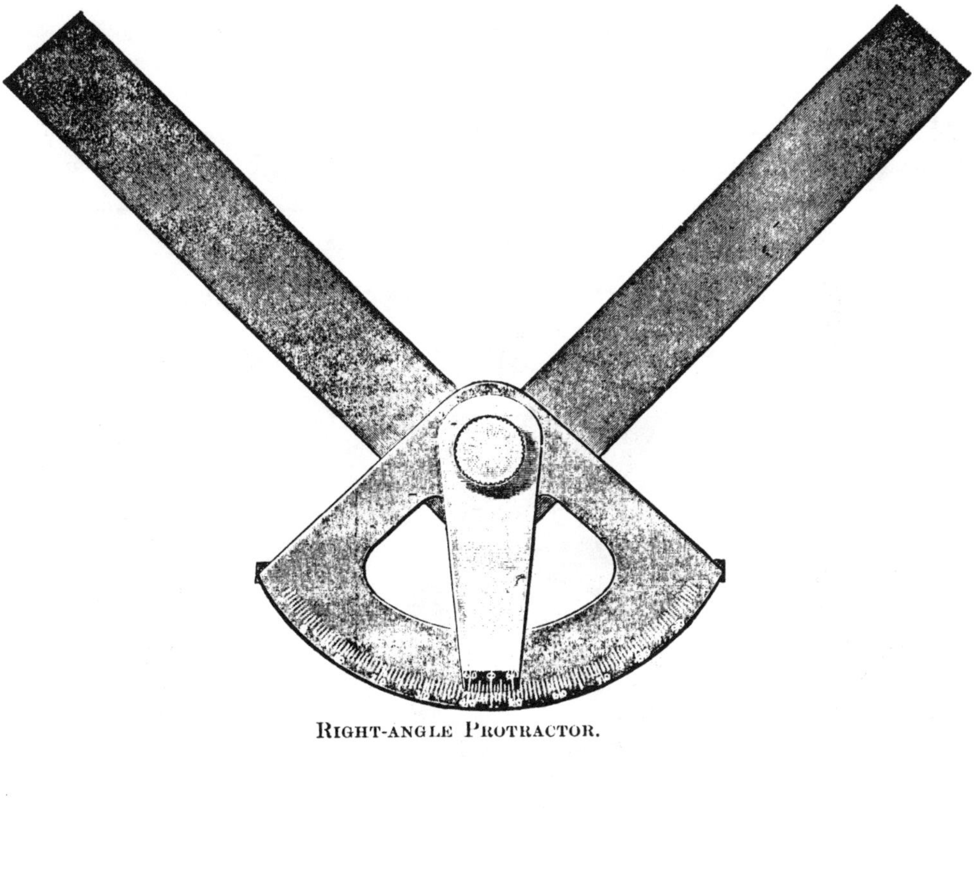

RIGHT-ANGLE PROTRACTOR.

American Machinist 1892

KENDRICK & DAVIS Lebanon, NH

Established by Frank B. Kendrick and William F. Davis in 1876 to make watchmaker's and machinist's tools. Maker of steel rules, c1903-1906. Its 1906 catalog offered toolmaker's rules with handles which appear to be made under the W.E. Severance patent of May 12, 1903 (No.)728137). See W.E. SEVERANCE for an illustration of the rules. The firm was incorporated in 1909 as the KENDRICK & DAVIS CO.

KENNELLY & CAIN (CRESCENT TOOL WORKS) Bridgeport, CT

Maker, in 1889, of Kennelly's Patent Protractor, patented October 2, 1883 (No. 285907), by Patrick Kennelly. The tool was offered as late as 1902. The same tool was later made by the L.S. STARRETT CO., Athol, MA.

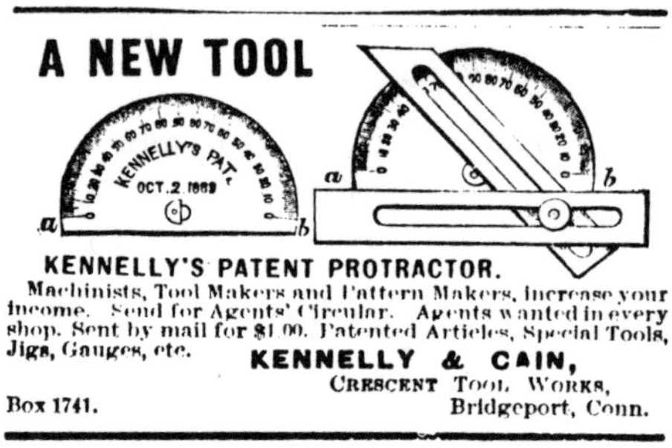

1889

KIMBALL & TALBOT Worcester, MA.

Maker of calipers and dividers patented January 27, 1863 (No. 37531) by David C. Talbot.

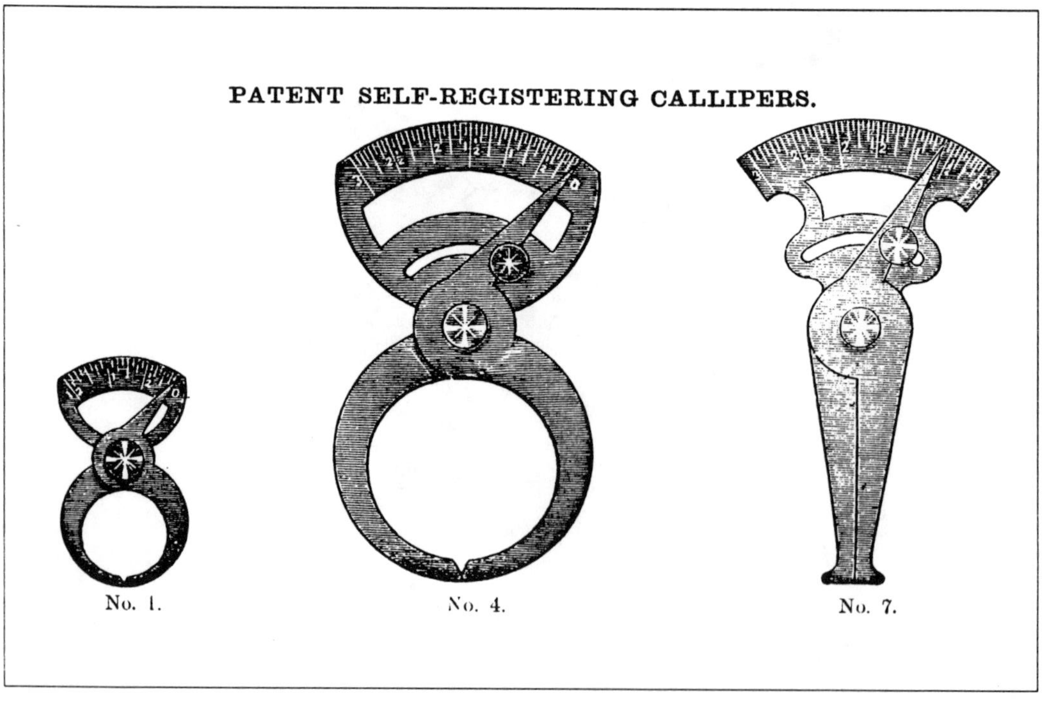

1870

KIRK, W.P. Cincinnati, OH

Maker, in 1914, of a "surface gauge of simple design which is particularly adapted for rapid manipulation". No examples have been observed.

KNAUEL TOOL WORKS Rock Island, IL

Maker, in 1920, of calipers and toolmaker's clamps in 1.25" and 2.25" sizes.

KNOEK, L.S. Hartford, CT

Maker of machinist's calipers which were copies of the Starrett "Yankee" design. Date of operation is unknown.

KOCH, H. Nyack, NY

Operated by Henry Koch who patented a test indicator on July 17, 1906 (No. 825911). The indicator appears to have been introduced in 1908. Koch's son joined him shortly after the start of the company and the name was changed to:

KOCH & SON, H. Nyack, NY

The company was later reorganized as:

KOCH & SONS, A.S. Lancaster, PA

Operated by Alfred Koch who was probably the son of Henry Koch. The firm continued to make the same H. KOCH test indicators as late as 1945. Toolmaker's pin vises were also offered in 1910.

Koch Universal Test Indicators

This indicator is simple and compact, has high magnifying power and soft, smooth action. It has two points of contact, a plunger and a bell crank, which gives it a broad, universal range. Each graduation represents .001 movement of plunger or a .002 movement of bell crank.

The safety feature is that the plunger operates away from instead of against the lever, preventing any of the delicate parts being broken by a sudden or excessive jolt of the plunger.

Price.......................................each **$5.00**

KOENIG, William A. Troy, NY

Maker of precision depth gauges furnished with 13" and 7" rules.

KOOPMAN, A.T. Chicago, IL

In 1888, Koopman introduced a new divider with adjustment for wear in the joint and with fine adjustment for setting the opening. In 1889, it introduced a guide gauge for use in lining locomotive guides and a "lightning" steel rule for machinists. The locomotive guide appears nearly identical to the one later offered by L.S. STARRETT.

In an 1889 ad the company notes that it "has all desirable tools of other manufacture" as well as its own dividers, guide gauges and steel rules. Apparently, it was also a dealer in machinist's tools. By 1902, the company was out of business and the passing of the Koopman calipers was lamented by Ferd. Wertheim in an article published in *American Machinist* magazine of December 4, 1902. See Appendix V.

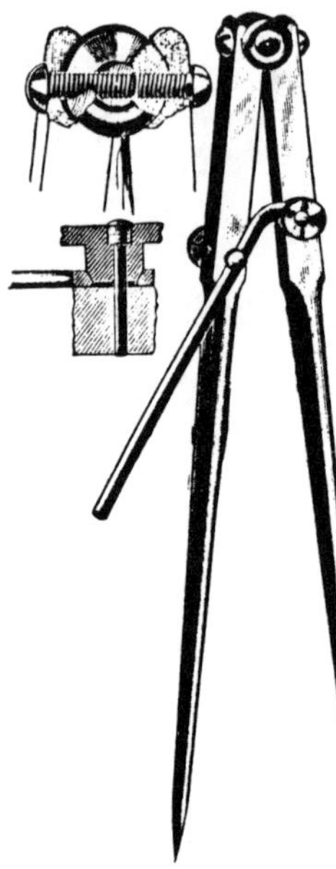

NEW DIVIDERS.

1888

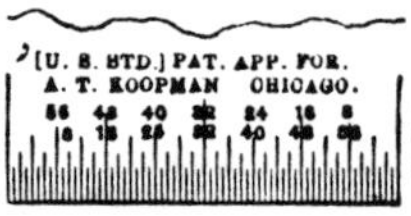

LIGHTNING SCALES FOR MACHINISTS.
See March 31 edition of this paper.
Also our New Dividers, Guide Gauge, &c., &c. Have all desirable tools of other manufacturers
Shop Agents Wanted Everywhere.
Send 10c. for Outfit.

1889

All American Machinist

KRAEUTER CO. Newark, NJ

Formed by August Kraeuter in 1879. The company made a wide variety of hand tools including calipers and dividers. All were of cheaper and lighter construction than most machinist's calipers but their low price, no doubt, appealed to many. A variety of Kraeuter tools are shown in the 1891, Machinist's Supply Co. catalog as reproduced below. It is interesting to note that a very similar selection of calipers is shown in a 1911 catalog. These calipers, however, were made by Wm. Johnson, also of Newark, NJ. *(see illustrations next two pages)*

LANE & HAKEWESSELL Hartford, CT

Maker, in 1890, of a machinist's protractor patented April 8, 1890 (No. 424977) by Reinhold Hakewessell and Fred K. Rand. Hakewessell is best known as the co-inventor, with Albert Henn, of the screw machines made by the NATIONAL-ACME CO. for many years.

LARRABEE & NORTH Chicago, IL

Maker of machinist's firm joint calipers, c1910.

No. 15. CENTRE CALIPERS.

	Each.
5 inch	$0 45
7 "	50
9 "	60
12 "	75

No. 16. KEY-HOLE CALIPERS.

Prices same as Centre Calipers above

No. 3. REGISTER CALIPERS, INSIDE AND OUTSIDE.

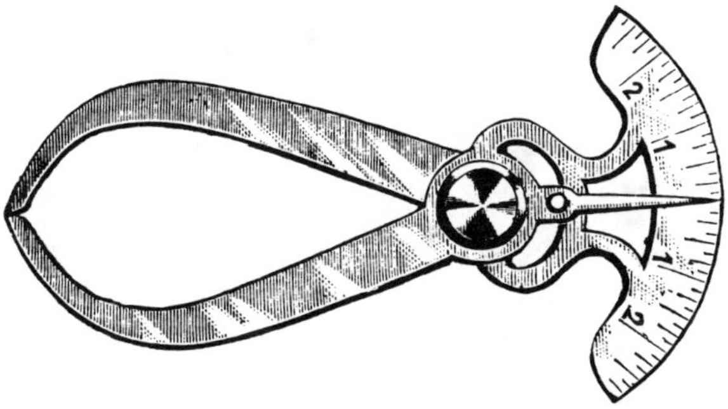

CAST STEEL, WITH SET SCREW, NICKEL PLATED.

	Each.
3 inch	$0 65
4 "	75
5 "	90
6 "	1 00

No. 4. REGISTER CALIPERS, OUTSIDE.

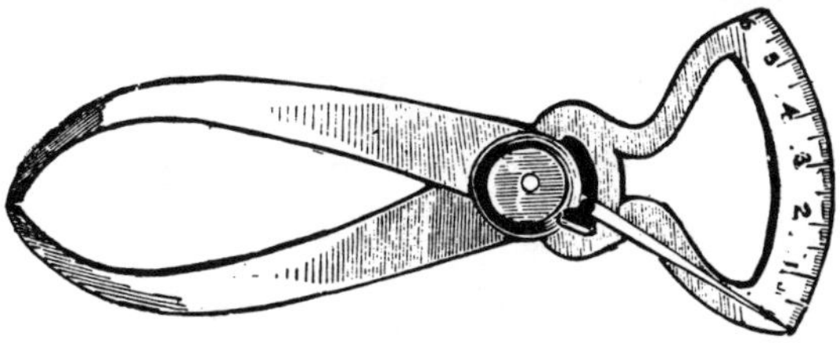

CAST STEEL, WITH SET SCREW, NICKEL PLATED.

	Each.
6 inch	$1 00
8 "	1 25

No. 5. REGISTER CALIPERS, INSIDE.

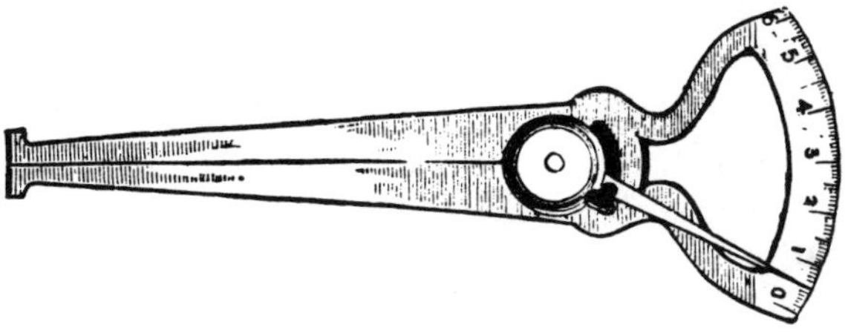

CAST STEEL, WITH SET SCREW, NICKEL PLATED.

	Each.
6 inch	$1 00
8 "	1 25

See entry for Kraeuter Co., previous page

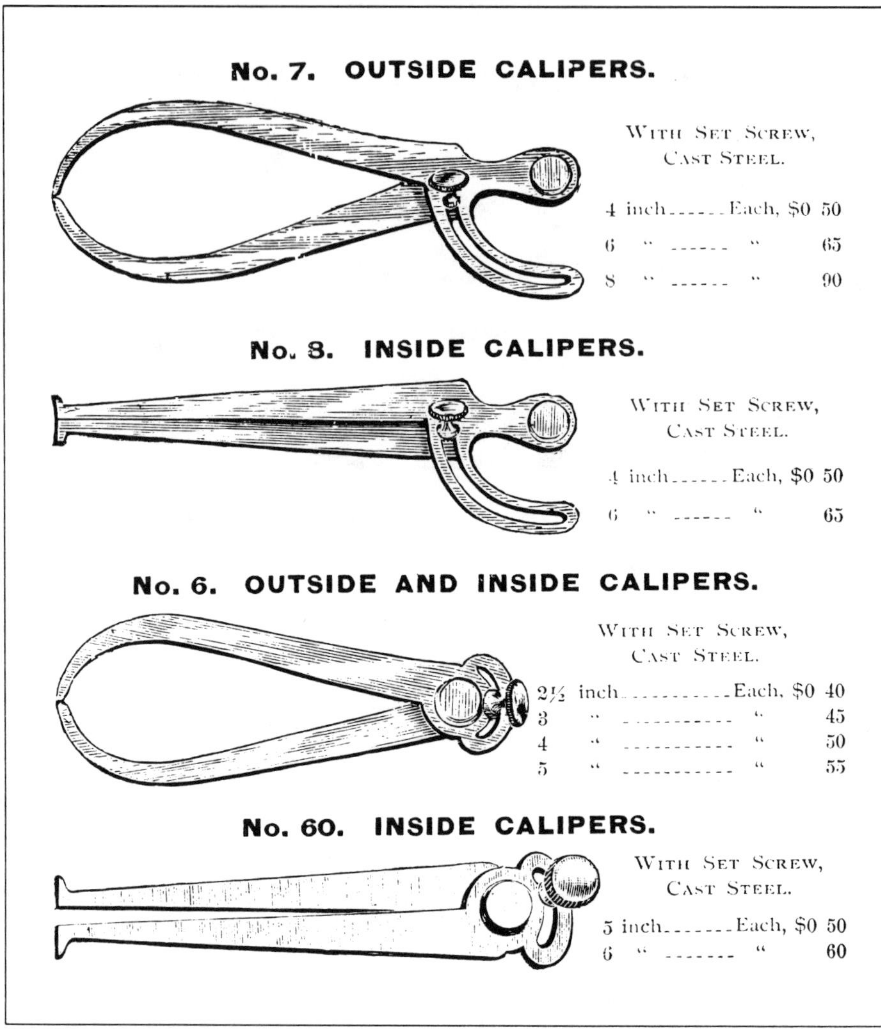
No. 7. OUTSIDE CALIPERS.

With Set Screw,
Cast Steel.

4 inch...... Each, $0 50
6 " ...... " 65
8 " ...... " 90

No. 8. INSIDE CALIPERS.

With Set Screw,
Cast Steel.

4 inch...... Each, $0 50
6 " ...... " 65

No. 6. OUTSIDE AND INSIDE CALIPERS.

With Set Screw,
Cast Steel.

2½ inch.......... Each, $0 40
3 " .... " 45
4 " .... " 50
5 " .... " 55

No. 60. INSIDE CALIPERS.

With Set Screw,
Cast Steel.

5 inch....... Each, $0 50
6 " .... " 60

See entry for Kraeuter Co., page 214

LAVIGNE MICROMETER CO. New Haven, CT

Operated by Joseph P. Lavigne who received the following machinist's tool patents:
Micrometer gauge, December 20, 1892 (No. 488280)
Micrometer gauge, December 20, 1892 (No.488281)
Micrometer gauge, November 21, 1893 (No. 508952)
Micrometer gauge, December 12, 1893 (No. 510688)
Micrometer gauge, February 20, 1894 (No. 515234)
Micrometer gauge, February 20, 1894 (No. 515235)
Micrometer surface gauge, February 20, 1894 (No. 515236)

The Henry McBride micrometer gauge patent of May 8, 1894 (No. 519541), was assigned to Lavigne. It is assumed that the firm was actually in production of at least some models of the micrometers, but none have yet been observed.

Joseph Lavigne was also the operator of the LAVIGNE MACHINE CO., New Haven, CT, maker of upright drills and other machine tools. The micrometer business may have been only a sideline. *(continued next page)*

In 1901, William M. Pratt, of the newly formed GOODELL-PRATT CO., bought the LAVIGNE MICROMETER CO., or at least the Lavigne patents, and began producing micrometers under the MASSACHUSETTS TOOL CO. name.

LINDHOLM & DAVIS New Haven, CT

Maker, in 1908, of a center test indictor patented by Thomas S. Dennis, Jr. and Arthur C. Lindholm on October 29, 1907 (No. 869483). The tool was designed to find the center of a jig button, prick punch mark, etc.

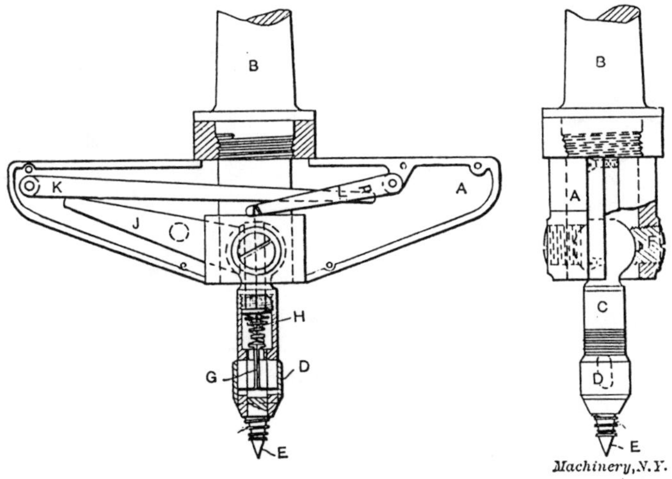

Fig. 4. Details of the Construction of the Instrument.

Machinery, November 1908

LINNAHAM, PETER J. Beverly, MA

Maker of indicating calipers for which he was granted a patent on May 23, 1911 (No. 993084). The sole example thus far examined is marked only, PAT. MAY 23, 1911.

LINTNER & SPORBORG Gloversville, NY

Maker of the "Paragon" ball-bearing speed indicators beginning in 1897. This is an odd looking device shaped like a small pistol, including a trigger. The device can be placed in position and running but the count does not start until the trigger is depressed, thus making the timing easier. The first model was covered by patents issued on November 14, 1893 (No. 508686) to W.F. Dial and June 5, 1894 (No. 520792) and October 16, 1894 (No. 527557) to W.T. Lintner. A later model was also covered by a patent issued May 29, 1900 (No. 650465) to Harvey Hubbell. *(see illustration next page)*

LOWE, H.A. Waltham, MA, later, Cleveland, OH

Inventor and maker of machinist's test indicators. Lowe's first indicator was introduced in 1902. The unit was made to mount directly in a lathe tool post and was furnished with an attachment for use as a center indicator. *(continued next page)*

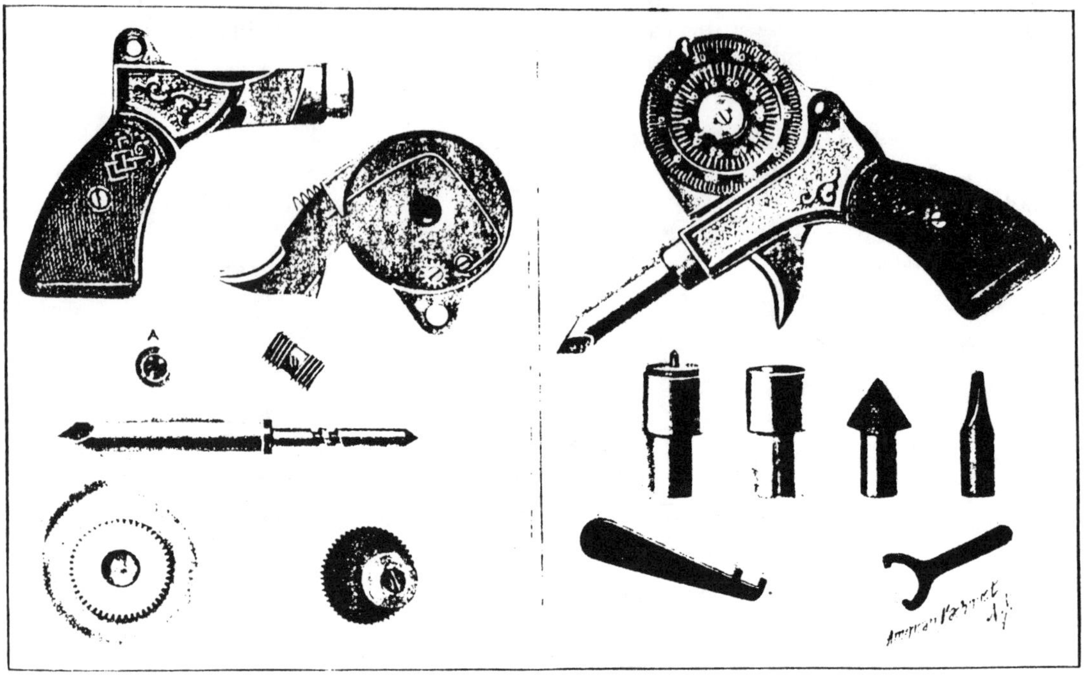

Fig. 2. PARAGON BALL BEARING SPEED INDICATOR. Fig. 1.

See entry for Lintner & Sporborg, previous page

About 1907, Lowe moved from Waltham, MA, to Cleveland, OH. He introduced the now famous "Last Word" dial indicator in 1914, the features of which were patented by Henry A. Lowe on June 29, 1915 (No. 1144367), September 4, 1923 (No. 146664), March 18, 1924 (No. 1486908), and May 25, 1926 (No. 1585778). The same indicators were later made by the L.S. STARRETT CO., Athol, MA, which bought the company, or the product line, sometime prior to 1938 when the "Last Word" indicators first appeared in the Starrett catalog. Lowe was listed as a "gauge maker" in a 1930 Cleveland, OH, directory.

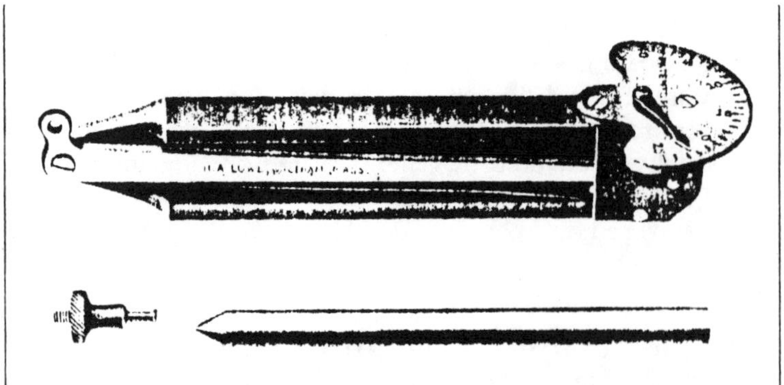

FIG. 1. NEW TEST INDICATOR.

Machinery 1902

LUFKIN RULE CO. Cleveland, OH, later Saginaw, MI

Founded in 1869 by Edward T. Lufkin who was a maker of board and log rules, including those covered by his patent of April 7, 1874 (No. 149321). In 1884 the firm was acquired by Fred Buck (1858-1938). Buck incorporated as the LUFKIN RULE CO. on January 1, 1885, and on November 1, 1892, moved the company to Saginaw, MI.

Tools offered in the 1901 catalog were: steel and linen measuring tapes, folding steel rules, circumference rules, steel and hickory board rules, log rules, log calipers, and pattern rules for sheet metal workers. Except for steel rules and shrink rules listed in the 1905 and 1908 catalogs, no machinist's tools appear to have been made until the early 1920's. Once started, Lufkin offered a complete line of machinist's tools, tool chests, and other supplies until discontinuing the line about 1967. Nearly all Lufkin machinist's tools are almost identical to the same model made by the L.S. STARRETT CO., or the BROWN & SHARPE MFG. CO.

M.T. CO., Greenfield, MA. See MASSACHUSETTS TOOL CO.

MACHEK & CO., A.J. Milwaukee, WI

Introduced, in 1905, tubular micrometer frames based on the Frank M. Davis patent of August 15, 1905 (No. 797409). The frames were "made of a single piece of sheet metal formed into a tube of rectangular cross-section. On the ends are fastened two sleeves fitted with locking devices for fastening in place any standard make of detached micrometer head on one side and the adjustable anvil on the other. Four frames comprise the set for measuring from 0 to 12 inches." Davis had been making the micrometers under his own name for a short time when Machek bought him out in 1905. By 1913, Machek was the owner of the WELLES CALIPER CO., Milwaukee, WI, and was offering the tubular micrometer frames along with the line of WELLES calipers. See F.M. DAVIS and DAVIS & THOMPSON CO.

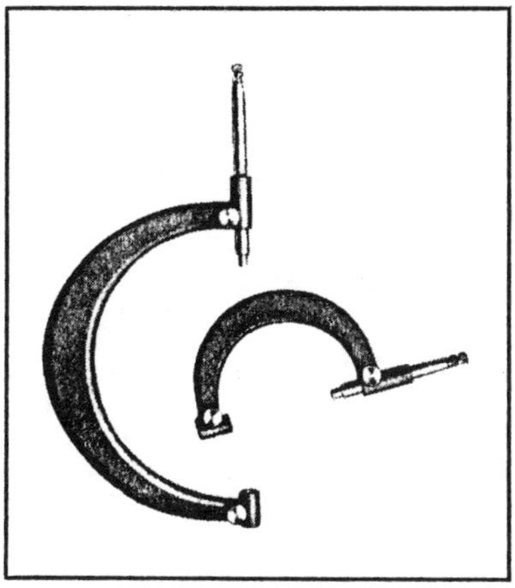

1905

FIG. I. TUBULAR MICROMETER FRAME.

MALMQUIST, CARL W. New Haven, CT

Inventor and maker of a depth micrometer, patented October 21, 1890 (No. 438851).

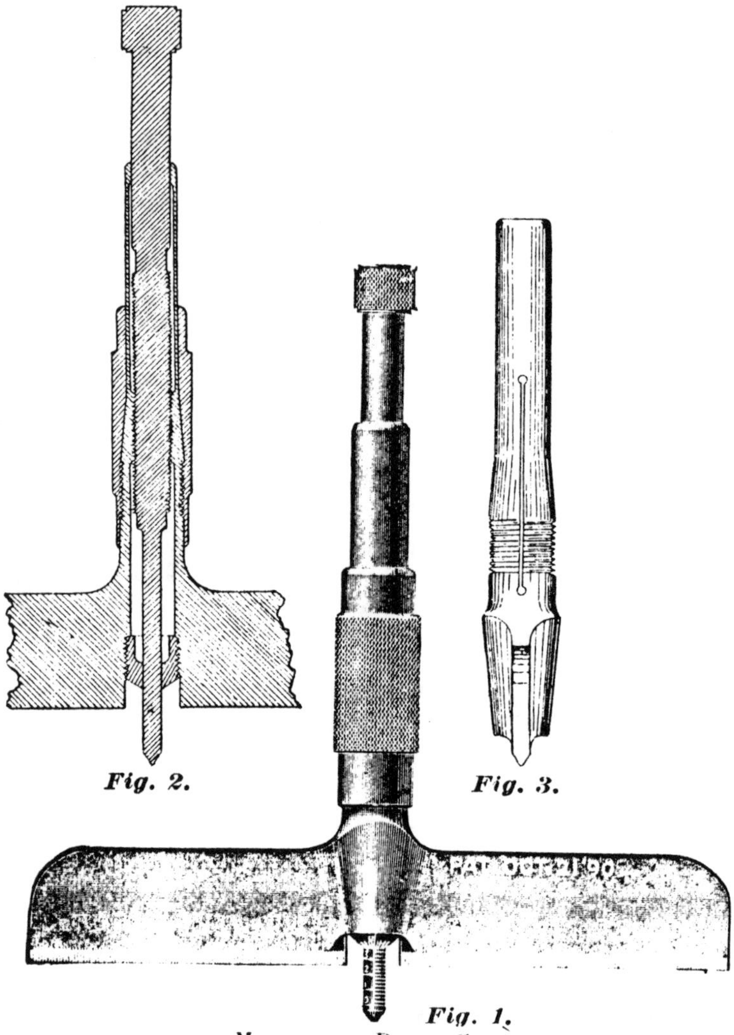

MICROMETER DEPTH GAUGE.

American Machinist 1890

MARBACH, FRANK G. Medina, OH

Inventor and maker of a combination inside/outside micrometer patented January 13, 1903 (No. 718266), and introduced in 1904, and a set of radius gauges (he called them concave and convex gauges) introduced in 1905. The radius gauges were made in sets ranging from $\frac{1}{8}$" to 1". He was granted a second micrometer patent on November 28, 1905 (No. 805480). A ratchet tap wrench was introduced in 1906. By 1910, he was operating as the F.G. MARBACH CO., "makers of emery wheel dressers, concave and convex gauges, machinist's metal tool boxes and ratchet wrenches". *(continued next page)*

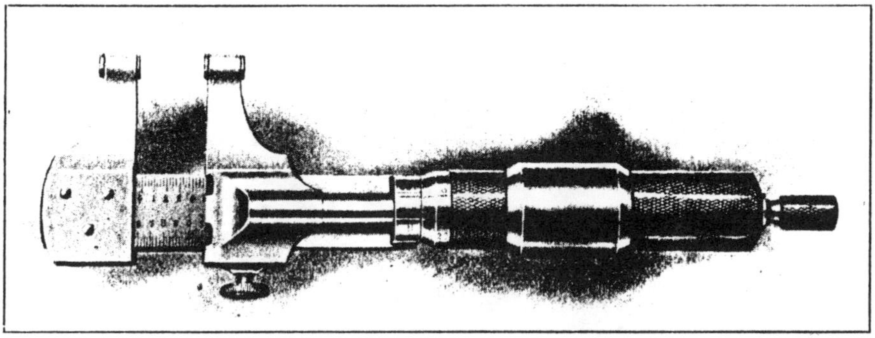

1904

AN OUTSIDE AND INSIDE MICROMETER.

Concave and Convex Gauge.

A tool of obvious utility is being put on the market by Patterson, Gottfried & Hunter, 146-150 Centre street,

Concave and Convex Gauge.

New York. It consists of eight tempered steel gauges measuring from ⅛ to 1 inch wide, held together by a central rivet, as shown in the accompanying cut. One end of each gauge is accurately ground to measure a concave circle, the diameter of which is stamped upon it, while the other end measures the convex arc of the same size circle. The makers state that this tool serves as a guide for turning work accurately in a lathe and enables a workman to make absolute duplicates. They refer to it as essential to pattern makers and suggest that other uses will occur to machinists, as, for example, grinding round nose tools.

Iron Age, April 12, 1906

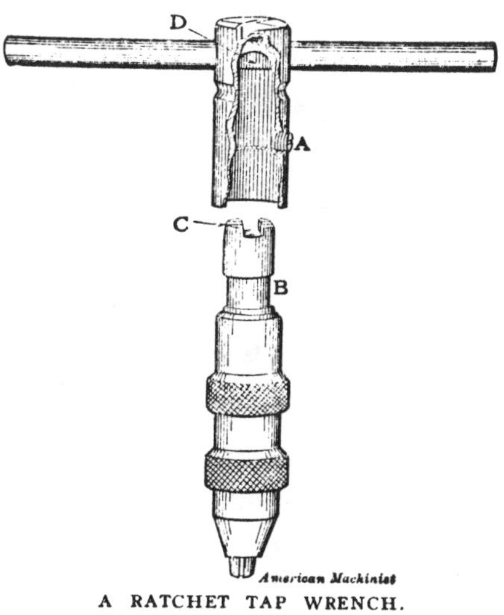

A RATCHET TAP WRENCH.

1906

MASON REGULATOR CO. Boston, MA

Offered for sale, in 1893, a depth gauge which "we have found very useful in our own shop."

1893

MASSACHUSETTS TOOL CO. Greenfield, MA

A captive company formed by GOODELL-PRATT CO., Greenfield, MA, in 1900 to make a wide variety of machinist's tools. It was capitalized at $25,000 and had about 5,000 square feet of manufacturing space. The early products were a line of machinist's rules and adjustable notch center gauges acquired by the purchase of COFFIN & LEIGHTON CO., Syracuse, NY, in October, 1901, and micrometers acquired by the purchase of the LAVIGNE MICROMETER CO., New Haven, CT, about the same time. It introduced, in 1901, a combination depth and outside micrometer based on the Joseph P. Lavigne patent of February 20, 1894 (No. 515234).

Machinist's tools listed in the MASSACHUSETTS TOOL CO. section of the 1911 GOODELL-PRATT catalog were: A wide variety of micrometers based on the Lavigne patents; a six-inch beam micrometer based on a December 20, 1892 (No. 488280) Lavigne patent; depth micrometers; inside micrometers based on the Henry McBride patent of February 29, 1890 (No. 519541); keyseat rule blocks; rule clamps; a wide variety of steel rules and straight edges; depth gauges; combination squares; protractors; solid and sliding blade squares; center squares; center gauges including the adjustable notch type; screw pitch gauges; scribers; clamps; V-blocks; levels based on the Richardson's patents; a fairly complete line of calipers and dividers similar to the Starrett Fay and Yankee style; surface gauges; and trammels.

Sometime prior to 1920, the GOODELL-PRATT CO. closed down the company as a separate operation but continued to make the same tools under the MASSACHUSETTS TOOL CO. name and later under the GOODELL-PRATT name. Dealer catalogs of the 1920's show that auto mechanics, rather than machinists, were the primary market during that time. *(continued next two pages)*

NEW MICROMETERS.

Instruments of Precision. Standards of Accuracy.

HALF THOUSANDTHS MEASUREMENTS TO ONE INCH.

No. 3

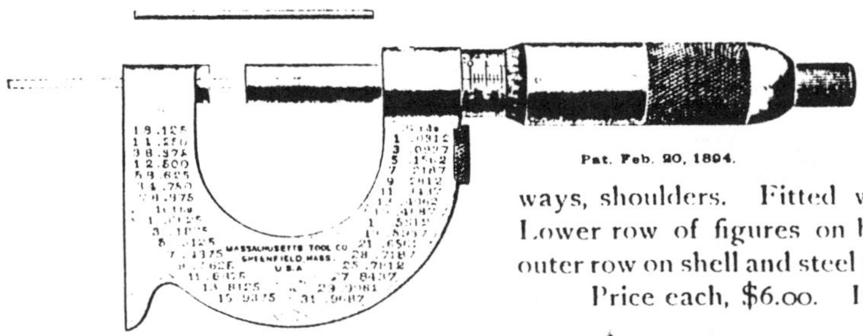

Double Graduation, Round end Shell. Depth Plug Attachment for measurement of holes, slots, keyways, shoulders. Fitted with patent locking device. Lower row of figures on barrel in combination with outer row on shell and steel plug used as a depth gauge.
Price each, $6.00. In leather case, $6.50.

No. 2

Round end bevel edge frame Special knurling and round end shell. Fitted with patent locking device.

Price each - - $5.50.
Price in leather case, $6.00.

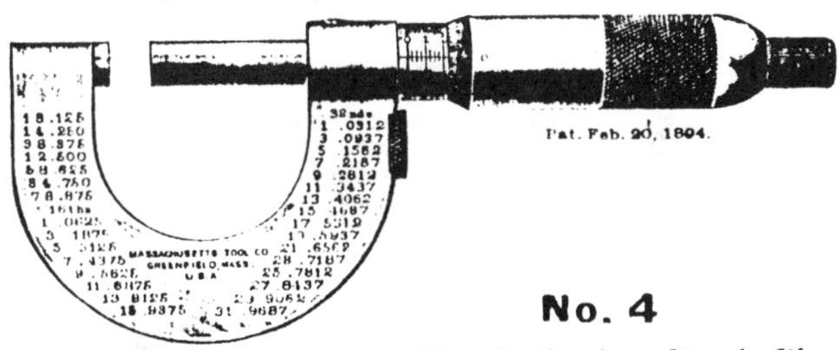

No. 4

Double Graduation, Depth Plug Attachment for measurement of holes slots, keyways, shoulders. Fitted with patent locking device. Lower row of figures on barrel in combination with outer row on shell and steel plug used as a depth gauge.

Price each - - $6.00.
Price in leather case, $6.50.

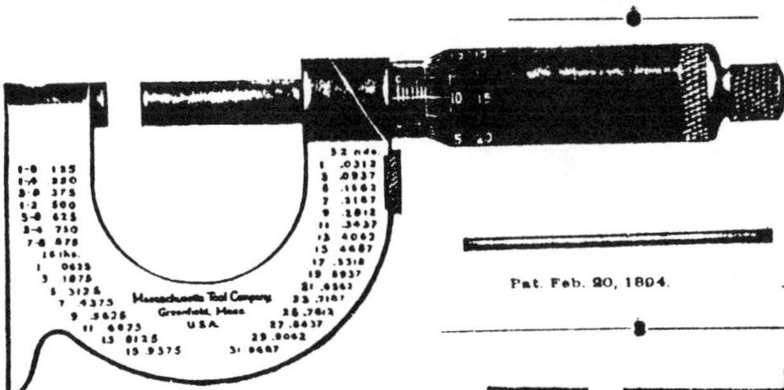

No. 1

Round end bevel edge frame Shell end slightly beveled with regular knurling. Fitted with patent locking device.

Price each - - $5.50.
Price in leather case $6.00.

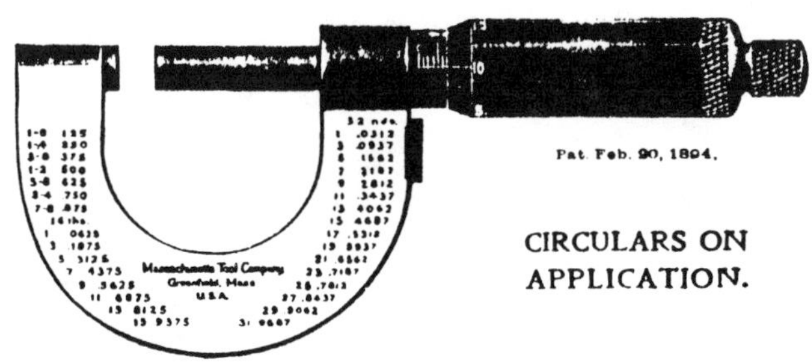

CIRCULARS ON APPLICATION.

Special discount of 15 per cent. including payment of postage by us on all orders received before August 1st, 1901.

MASSACHUSETTS TOOL COMPANY,

GREENFIELD, MASS., U. S. A.

Combination Square and Bevel Protractor.

Massachusetts Tool Company, Greenfield, Mass., is introducing the combination square and bevel protractor shown in the accompanying illustration, made in 9, 12, 18 and 24 inch sizes. The device can be used, as shown, as a combination square or as a bevel protractor. The blades are of crucible steel with a spring temper and are accurately graduated. They are fastened to the heads by round cornered bolts, which slide in a round clamping

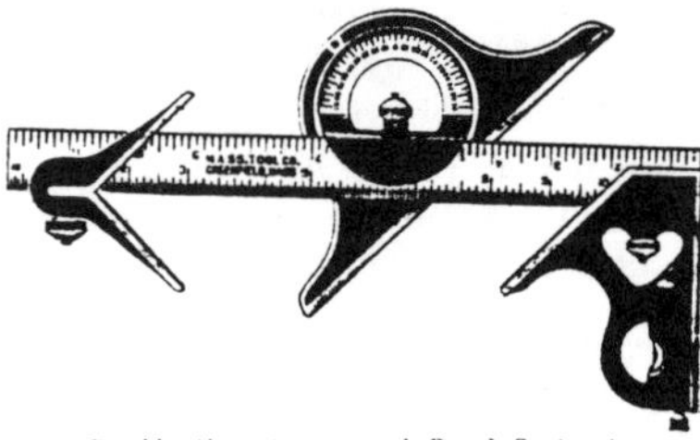

Combination Square and Bevel Protractor.

groove in the blade. The heads are of steel, designed so as to give the most rigid working surfaces without cumbersome weight, and accurately made with ground angles. They are finished in black enamel in recessed parts and polished on all wearing surfaces. The blade is held in a revolving turret of the bevel protractor by a round ended bolt. The turret is accurately fitted, and engine graduated to 90 degrees either side of zero, care being taken to insure its being at right angle to the face of the head. It carries a level, set and fastened to the side of the turret.

Surface Gauge No. 115.

The surface gauge shown herewith has a solid iron base and stands squarely on the work. The spindle has a fine adjustment by means of a screw with a knurled

Surface Gauge No. 115.

head nut, and after setting can be locked firmly by tightening the screw shown in the cut. The base is finished in black enamel, with wearing surfaces polished. The scriber is made of tool steel, carefully tempered. The gauge is 9 inches high. This device is put on the market by Massachusetts Tool Company, Greenfield, Mass.

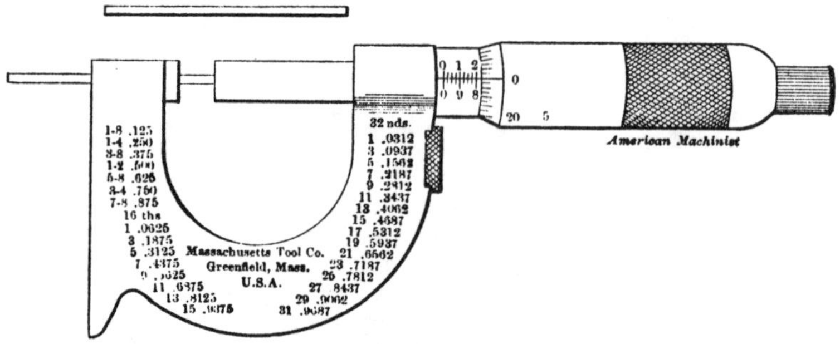

AN IMPROVED MICROMETER.

MAUSER, ALFONS Hartford, CT

Introduced, in 1892, a surface gauge for which he received a patent on June 4, 1892 (No. 476806). Although he indicated Hartford, CT as his address, he was living in Germany when he applied for the patent.

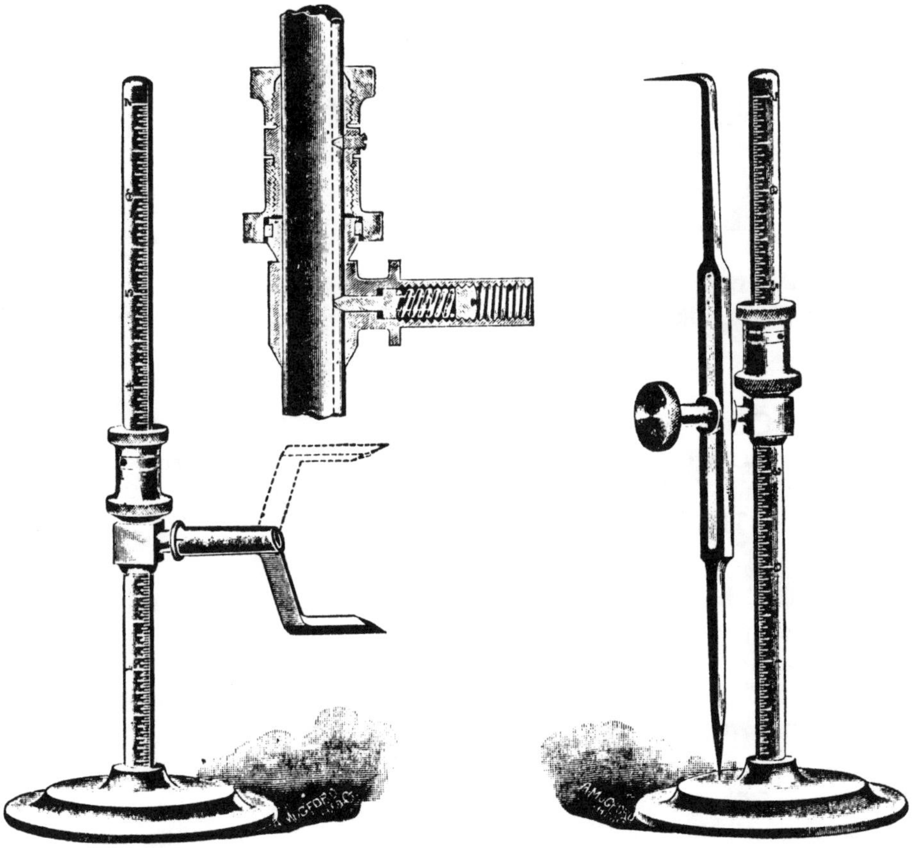

Fig. 1. A NEW SURFACE GAUGE. *Fig. 2.*

American Machinist 1892

MCLURE & LEVESQUE Hartford, CT

Maker, in 1903, of sets of hardened, ground and lapped size blocks. The blocks were made of $^{13}/_{16}$" hexagonal steel and ranged in thickness from $^{1}/_{16}$" to 1".

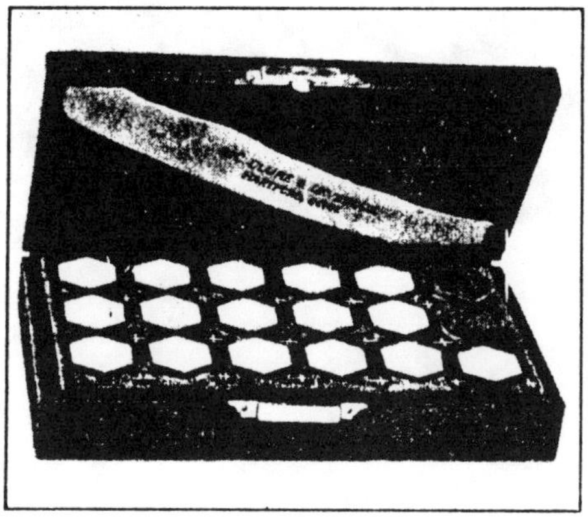

SIZE BLOCKS.

McDONNELL ODOMETER CO. Chicago, IL

Maker of "Lightning" speed indicators in 1885. The same unit was listed as late as 1913.

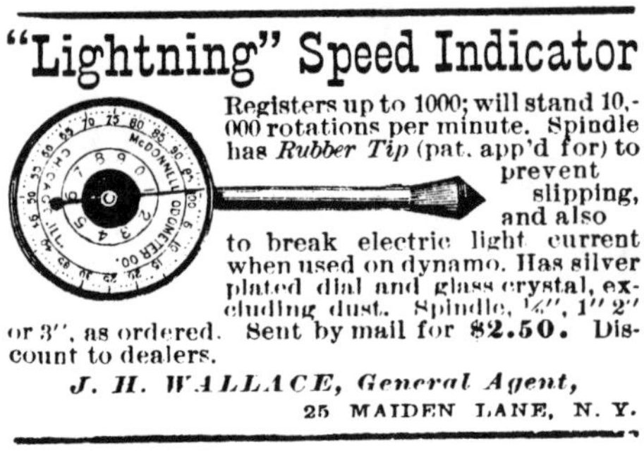

American Machinist 1885

McKNIGHT, G.L. Worcester, MA

Inventor and maker of finely adjustable calipers, patented April 30, 1867 (No. 64240) and May 28, 1867 (No. 65253).

MECHANICS TOOL CO. Boston, MA

Maker, c1910, of a surface gauge, patented May 31, 1904 (No. 761381) by Henry J. Hjorth. The gauge had been made by the BEST TOOL CO., Boston, MA, beginning in 1904, and was later made by the HJORTH LATHE & TOOL CO., Boston, about 1915.

MEIGS-POWELL CO. Milwaukee, WI

Makers of Welles Patent calipers, spring calipers, and adjustable firm joint calipers. The spring calipers are a copy of the Starrett "Yankee" caliper and the adjustable firm joint calipers appear to be copies of the Starrett 1901 patent calipers. The firm sold out to the PERDIEU TOOL MFG. CO., Milwaukee, WI, when it was formed in 1922.

MEISEL PRESS & MFG. CO. Boston, MA

Introduced, in 1904, sets of size blocks made of hardened steel and ground to size. Three different sets were offered; from ¼" to 1", ¼" to 2", and ¼" to 3" in increments of 1/16".

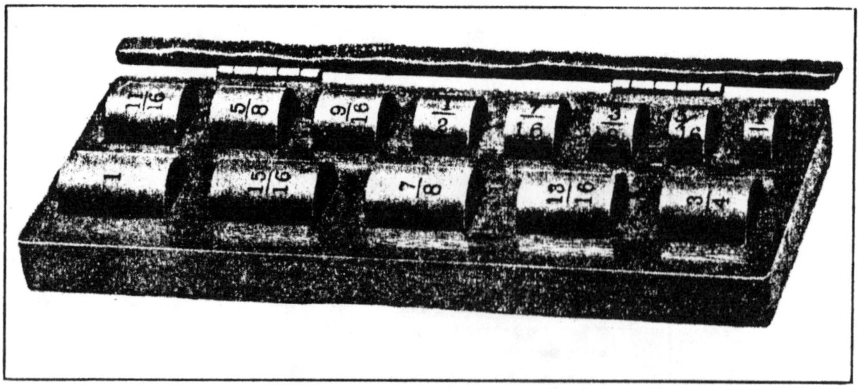

A SET OF SIZE BLOCKS. *1904*

MELDRUM-GABRIELSON CORP. Syracuse, NY

Makers of "Syracuse" adjustable limit snap gauges in 1921. Alexander Meldrum had been granted a patent for a setting gauge on July 25, 1916 (No. 1192089), and had assigned it to the L.S. STARRETT CO., Athol, MA. Meldrum and Carl Gabrielson assigned their snap gauge patent of September 30, 1924 (No. 1510189) to the company. Carl Gabrielson also received a patent for a snap gauge on October 9, 1923 (No. 1469953).

MELICK CLINOMETER CO. St. Louis, MO

Maker, in 1890, of a "clinometer for machinist's use in striking levels or angles".

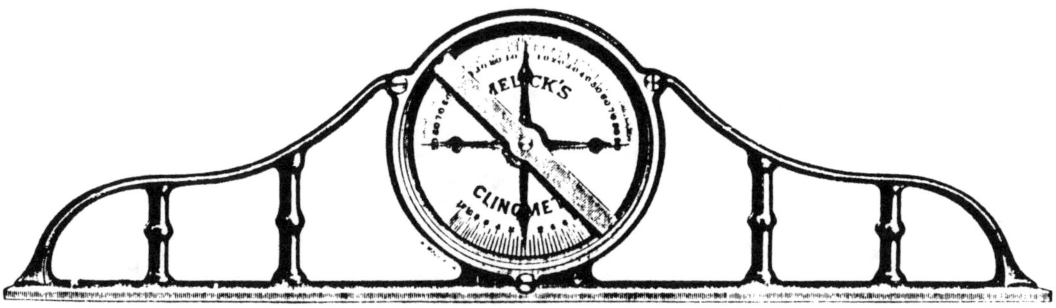

THE MELICK CLINOMETER.

American Machinist 1890

MERIDAN MACHINE PRODUCTS CORP. Detroit, MI

The A. Olevin Micrometer patent of July 22, 1919 (No. 1310431), was assigned to this company. It is not known if the micrometer was actually produced.

MICROMETER CALIPER CO. St. Louis, MO

The Charles E. Coe micrometer patent of May 1, 1900 (No. 648390) was assigned to this company. It is not known if the micrometer was actually produced.

MILLERS FALLS CO. Millers Falls, MA

Founded by Henry L. Pratt when he reorganized the firm of GUNN & AMIDON in 1868. Major products were bit-braces, spiral screwdrivers and push drills. In 1884, the firm was making a surface gauge patented by Charles W. Smith on June 13, 1882 (No. 391488). This was to be its only machinist's tool of the period. In 1931, the MILLERS FALLS CO. absorbed the GOODELL-PRATT CO. which had been an important maker of machinist's tools. The line was continued and, as late as 1938, GOODELL-PRATT machinist's tools were listed separately in Millers Falls catalogs and were still marked GOODELL-PRATT. Beginning in 1949, micrometers and other machinist's tools were marked with the MILLERS FALLS CO. name only. *(illustration next page)*

MODEL TOOL & GAUGE CO. Bridgeport, CT

Introduced, in 1916, the "Simplex" sine bar "which can be used in conjunction with a table of natural sines and cosines for the accurate determination of angles. It measures 5 inches from center to center of the disks and forms a useful addition to the toolmaker's kit."

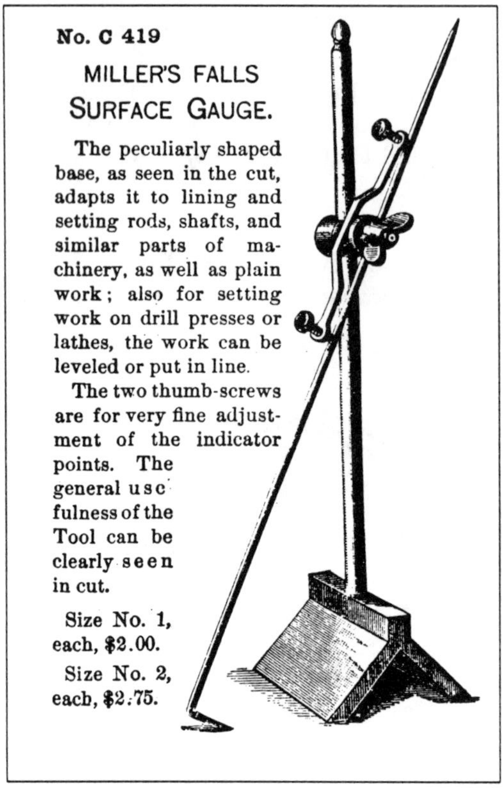

See entry for Miller's Falls, previous page

MODERN CALIPER & TOOL WORKS Chicago, IL

Combination inside/outside calipers with a helical center spring, marked E.& T. MFG. CO. have been found in boxes marked with this company's name. Nothing else is presently known.

MOLDENKE, RICHARD G. New York, NY

Moldenke introduced, in 1889, a combination tool which appears to be a combination of a surface gauge, V block, and square. No examples have been located and it is not certain that the tool was actually made. *(see illustration next page)*

MONTGOMERY & CO. New York, NY

Founded by James M. Montgomery who had been a partner in Peter A. Frasse & Co., New York, NY, until leaving in 1876. Montgomery was a dealer in a wide variety of tools and was one of the major dealers in machinist's tools. Montgomery offered a small (3.5" post) surface gauge in 1890 as its own make. In 1898, it advertised the Blodgett combination twist drill, thread pitch, centre and tap drill gauge. At that time it claimed to be "Tool Makers" so it may have made some of its own products. *(see illustration next page)*

See entry for R. Moldenke, previous page

New Combination Tool.

This tool is made entirely of steel, and is so designed that it may be put to a great variety of uses. The base-block has in three of its faces a square or V-groove, while in the upper side is a slot in which is fitted the blade, which is held in position by a thumb-screw. The blade is so shaped as to form a square, either inside or outside, as also a hexagon or nut gauge. The upright inclined edge is half way between the two steel pins shown below, and is at right angles to a line joining them, thus forming a center square. This upright blade carries, by means of the slot and thumb-screw, a slotted angle blade and a scribing needle, the needle being employed as a surface gauge or scriber, in which latter case the square groove in the base-block being placed on the shaft or arbor, a circle on any surface at a right or other angle concentric with the shaft can be made. With equal facility it answers as a depth gauge. The slotted angle blade referred to is used in connection with an angle or degree index which the reverse side of the upright blade carries. For this use the angle blade is released from the thumb-screw, and the post shown below is slipped into the hole in the end of this plate. The graduation on the reverse side of the upright blade indicates correctly angles respectively of 80, 45, 60, 90, 120 and 125 or polygon degrees. The manufacturer of this tool, Richard G. G.

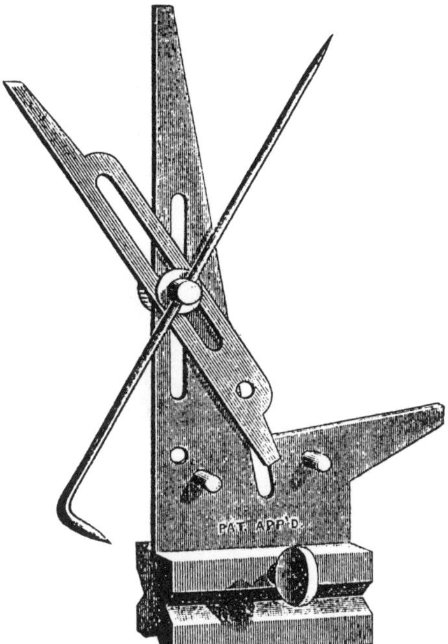

Square, Bevel and Surface Gauge.

Moldenke, of 1368 Broadway, New York, names the following uses to which the tool may be put: Straight edge, inside square, outside square, inside centering square, outside centering square, hexagon gauge, T-square, surface scratch-gauge, circular scratch-gauge, depth-gauge, adjustable dividers, key-seat rule, V-block, universal bevel-gauge and protractor for special angles.

MINIATURE SURFACE GAUGES.

This little instrument has all the features of a regular surface gauge. It stands 3½ inches high, but it is of somewhat heavier proportions than the larger ones, to give it extra stiffness. It is used for all purposes for which scratch blocks and gauges, costing twice as much are adapted, has double the range of work and is a handsome tool.

It is especially adapted for tool makers and others in laying out dies, jigs, fixtures and all small jobs. By passing straight end of scriber through a hole in the base, it becomes an accurate depth gauge, and by reversing scriber, it forms a hand scratch gauge and a good one. The tool is finely finished all over and with the exception of cast iron base, is all steel. Its many uses and low price should give it a place in every mechanic's kit in the country.

Each.................$1.00 | Postage................10c.

See entry for Montgomery, previous page

MORGAN MACHINE CO. Cleveland, OH

Maker of bevel protractors in 1889.

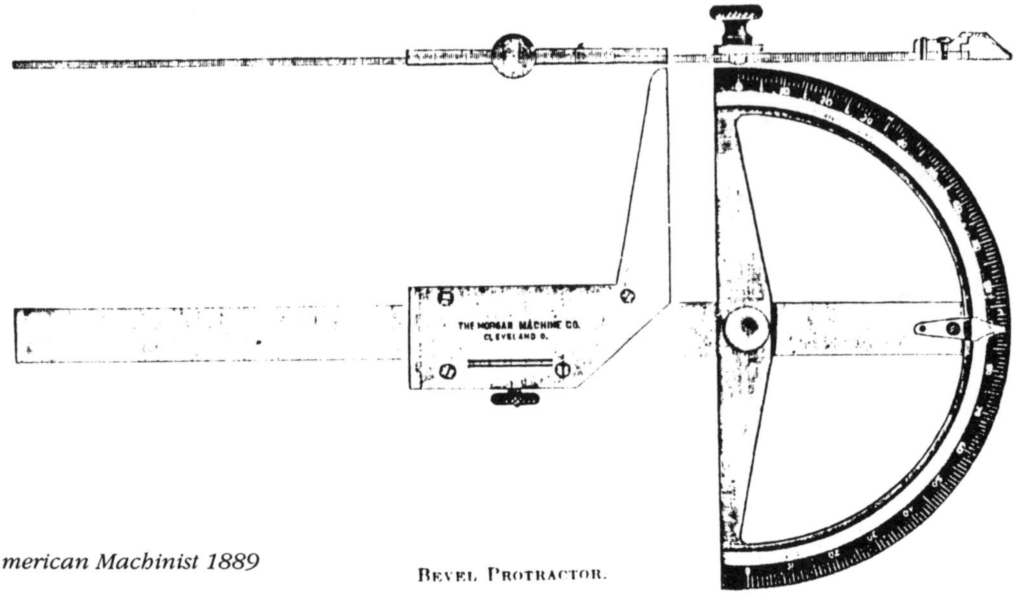

American Machinist 1889

BEVEL PROTRACTOR.

MORSE TWIST DRILL & MACHINE CO. New Bedford, MA

Founded by Stephan A. Morse in 1864 with a capital of $30,000. Morse is credited with the invention of the modern twist drill. Its primary products were drills and other cutting tools, but like most cutting tool makers it offered various gauges for grinding and sizing cutting tools. In 1905, it was listed as a maker of ring, plug, screw-cutting and drill gauges. By 1929 the company was capitalized at $2,000,000 and employed about 1000. Products are usually marked M.T.D.& M. Co. *(see illustration next page)*

MUNCH, ARTHUR St. Paul, MN

Inventor and maker of adjustable calipers introduced in 1895. The calipers were patented December 2, 1890 (No. 441685), when Munch was living in Milwaukee, WI.

A NEW CALIPER.

1895

Morse Gauges for Grinding Drills

For Use With Straight Shank Drills

Style No. 1

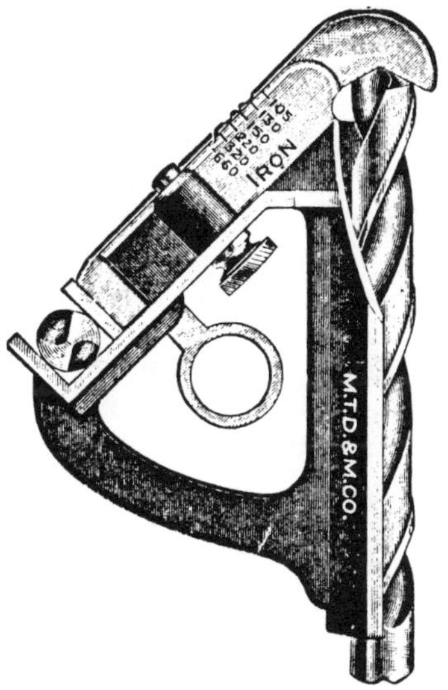

For Drills ¾ Inch and Smaller

Style No. 2

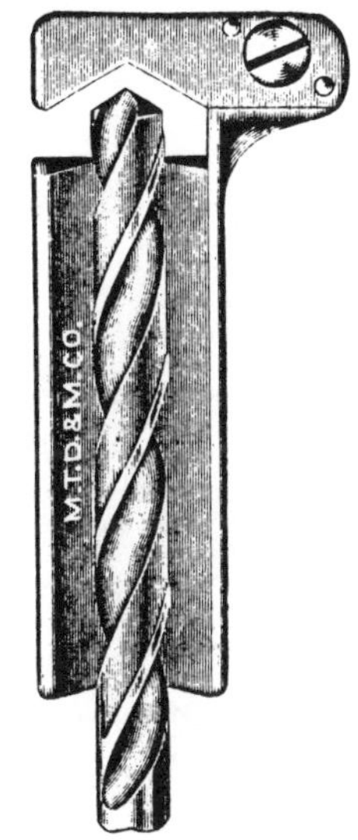

For Drills ¾ In. and Smaller

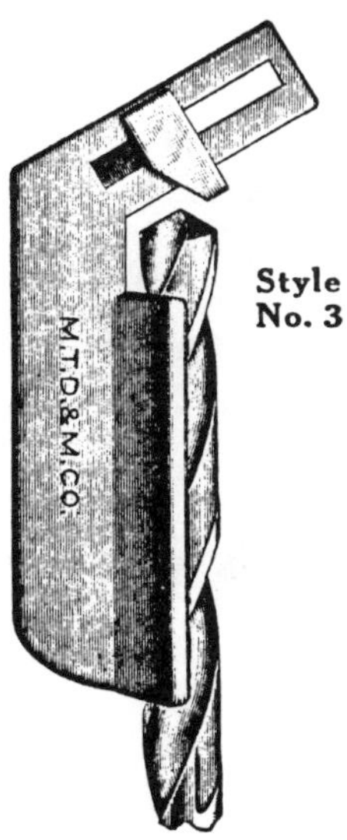

Style No. 3

For Drills ¾ Inch and Smaller

Style No. 4

Made in 2 Sizes

For ¼ Inch and Smaller
For ¾ Inch and Smaller

Prices on application.

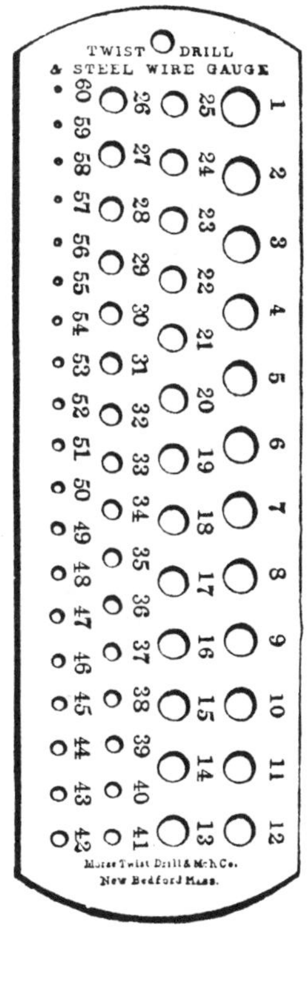

See entry for Morse Twist Drill, previous page

NASH ENGINEERING CO. Brooklyn, NY

Maker, in 1910, of a micrometer plug gauge "intended for measuring small holes or slots with great precision." The tool was graduated to read to .0001" and had a range of 1⁄16 inch.

NEW STANDARD CORP. York, PA

Makers of the "Kwick Chek" hole size gauge.

NICHOLSON, W.T. Providence, RI.

Patentee, on May 1, 1860 (No. 28104), of a cast iron level made by STANLEY RULE & LEVEL CO., New Britain, CT. Nicholson later founded the NICHOLSON FILE CO. where he received a number of patents for files and file handles much used by machinists.

NORRIS, SAMUEL Springfield, MA

Maker of calipers and dividers patented July 21, 1863 (No. 39281), by C. A. Fairfield. This same design, in modified form, was later used by J. STEVENS & CO., Chicopee Falls, MA.

NORTON EMERY WHEEL CO. Worcester, MA

A large maker of grinders and grinding wheels, formed in 1885. The firm was also the manufacturer, in 1896 and 1897, of the Bath indicator, patented by John Bath on May 22, 1894 (No. 520201), and October 2, 1894 (No. 526960). See JOHN BATH for illustrations of the tool.

O&S MFG. CO. Plainville, CT

Makers of a cast iron bench micrometer with open screw threads.

OHIO-NATIONAL MFG. CO. Chillicothe, OH

Maker, in 1911, of tool maker's and machinist's tool cases.

TOOL CASES

DRAFTSMEN

You probably have more instruments than can be placed in the small case which contained your first set, and yet you certainly desire to keep them all together. We have a tool case in which you can keep all your instruments, pencils, note book, etc., under lock and key, where they will not be injured or meddled with.

TOOL MAKERS

You will find our tool maker's case the very thing you need. It is strongly constructed and nicely finished, has felt-lined drawers, brass handles and a strap handle which makes it easily carried.

MACHINISTS

Don't have a tool box made to order and pay two prices for it. Our tool case for machinists will surely please you.

Write for descriptive matter and prices.

The Ohio-National Mfg. Company

CHILLICOTHE, OHIO, U. S. A.

ORDWAY, NEHEMIAH Hartford, CT

Introduced a toolmaker's gauge in 1926. The gauge was designed for gauging drill points, but could also be used as a square or as a taper gauge in filing the clearance in dies.

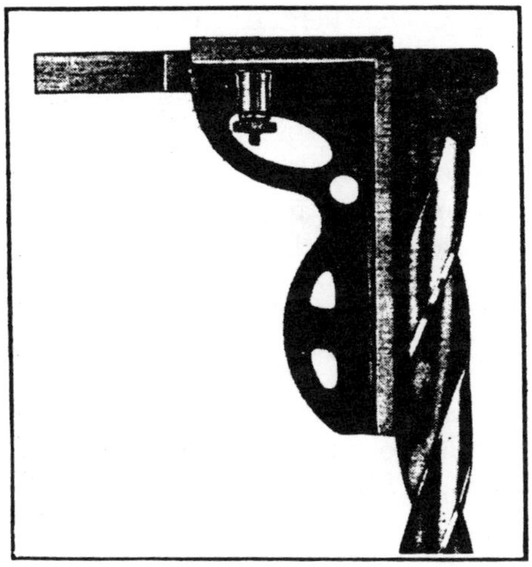

American Machinist 1926

Ordway Toolmakers' Gage

OSBERG & JOHANSON Dorchester, MA

Makers, in 1919, of an indicator "suitable for truing up center marks on work to be bored, for testing lathe centers, aligning headstocks and tailstocks of lathes, etc".

OSLUND TOOL & DIE CO. Hartford, CT

Maker of machinist's test indicators in 1936.

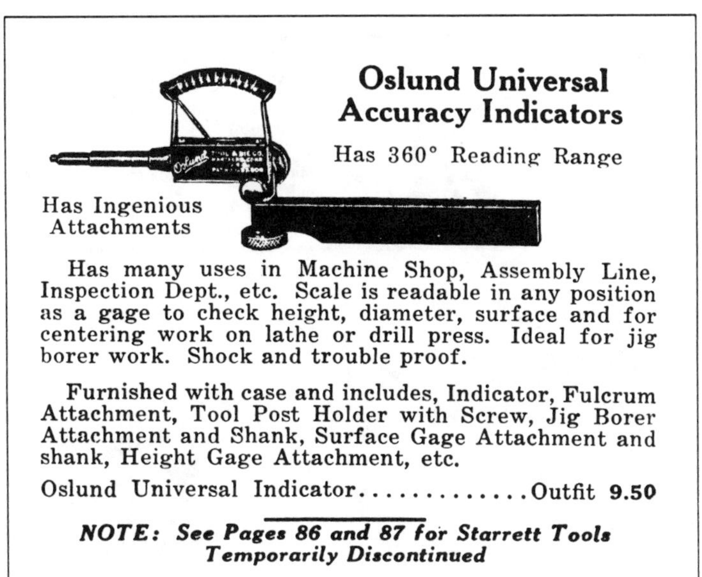

1942

PECK, STOW & WILCOX CO. Southington, CT

Formed in 1870 by the merger of the PECK, SMITH MFG. CO., S. STOW MFG. CO., and the ROYS & WILCOX CO. The company was primarily a maker of tinner's tools, but it also made several types of calipers which were used by machinists, including a transfer caliper patented October 11, 1887 (No. 371489) by James H. Culver. Perhaps the most interesting machinist's tool is a "SELF REGISTERING CALIPER" patented July 1, 1873, (No. 140554) by Henry Smith. Smith was almost certainly an employee as his patent was witnessed by Stephen Walkley, the company secretary. The tool is calibrated on one side in "ENGLISH WIRE GAUGE" numbers and, on the other, in fractions of an inch from 7⁄16" to 1⁄64", and decimal from 1⁄100" to 5⁄100". The only example now known is very well made and appears to be a very useful tool. The company was listed as a maker of machinist's squares and calipers in 1905. Many of its products are marked only P.S.& W. CO.

PERDIEU TOOL MFG. CO. Milwaukee, WI

Formed in 1922 to take over the business and factory of the MEIGS-POWELL CO., maker of Welles Patent calipers. The company continued to make the Welles calipers as well as a line of calipers and dividers which were copies of the Starret "Yankee" line. Rugiley A. Perdieu was president.

PERKINS & SON, INC., B.F. Holyoke, MA

The Henry Bartlett thickness gauge patent of August 24, 1920 (No. 1350769) was assigned to this company.

PHILLIPS, ARTHUR B. Whitman, MA

Maker, in 1887, of a planer and surface gauge patented November 16, 1886 (No. 352584) by Benjamin E.D. Stafford and Fenner E. Whipple. By 1904, the partnership of STAFFORD & WHIPPLE had been formed to make the same tool.

1887, both illustrations

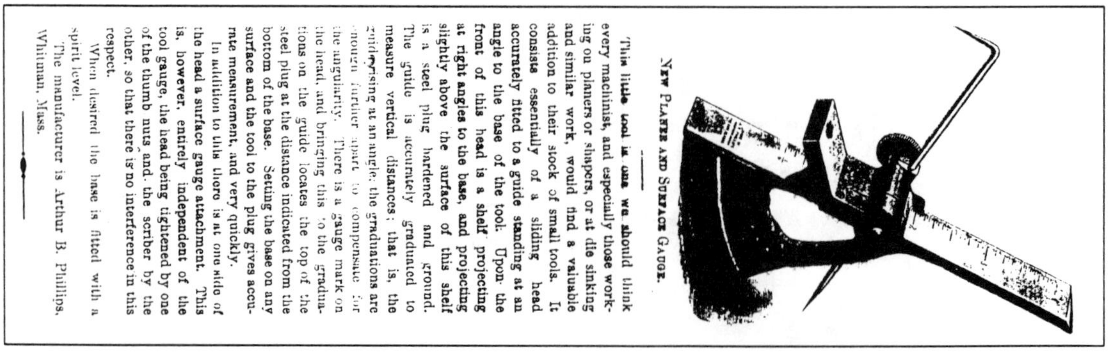

NEW PLANER AND SURFACE GAUGE.

This little tool is one we should think every machinist, and especially those working on planers or shapers, or at die sinking and similar work, would find a valuable addition to their stock of small tools. It consists essentially of a sliding head accurately fitted to a guide standing at an angle to the base of the tool. Upon the front of this head is a shelf projecting at right angles to the base, and projecting slightly above the surface of this shelf is a steel plug hardened and ground. The guide is accurately graduated to measure vertical distances; that is, the guide rising at an angle, the graduations are enough further apart to compensate for the angularity. There is a gauge mark on the head, and bringing this to the graduations on the guide locates the top of the steel plug at the distance indicated from the bottom of the base. Setting the base on any surface and the tool to the plug gives accurate measurement, and very quickly.

In addition to this there is at one side of the head a surface gauge attachment. This is, however, entirely independent of the tool gauge, the head being tightened by one of the thumb nuts and the scriber by the other, so that there is no interference in this respect.

When desired the base is fitted with a spirit level.

The manufacturer is Arthur B. Phillips, Whitman, Mass.

PHOENIX ROLL WORKS Pittsburg, PA

Maker of a combination inside/outside calipers, patented April 10, 1888 (No. 380836) by O.D. Warfield. Warfield was closely associated with J. STEVENS & CO. of Chicopee Falls, MA, but apparently also had some relationship with the Phoenix Roll Works.

PITKIN MACH'Y CO., A.B. Providence, RI

Advertised, in 1892, the Standard revolution counter which was identical, even to the ad copy, to a revolution counter advertised in 1885 by ULSTER MACHINE CO. This may have been a company name change or they may have bought the product from USLTER MACHINE CO. The revolution counter was patented September 26, 1882 (No. 264900), by William G. Richards.

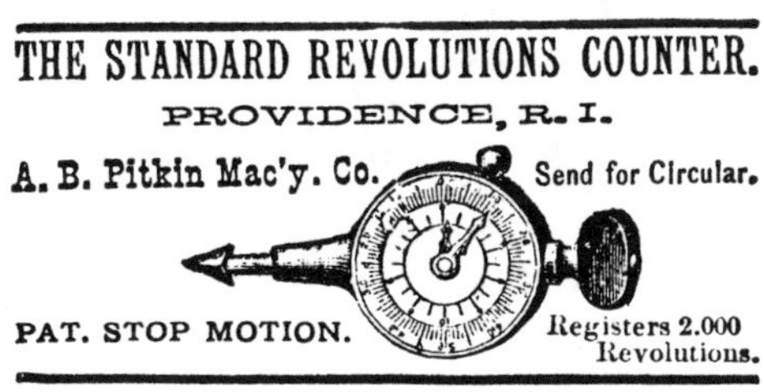

1892

POND MACHINE CO., L.W. Worcester, MA.

Founded by Lucius W. Pond (1826-1889) in 1854, the company was a maker of large machine tools, primarily planers. Pond also made revolvers during the Civil War, and beginning about 1867, began making "Pond's Improved Caliper" under the Edward Wright patent of September 24, 1867 (No. 69292). Wright may have been an employee of the L.W. Pond Machine Co. before starting his own company. See WRIGHT MACHINE CO.

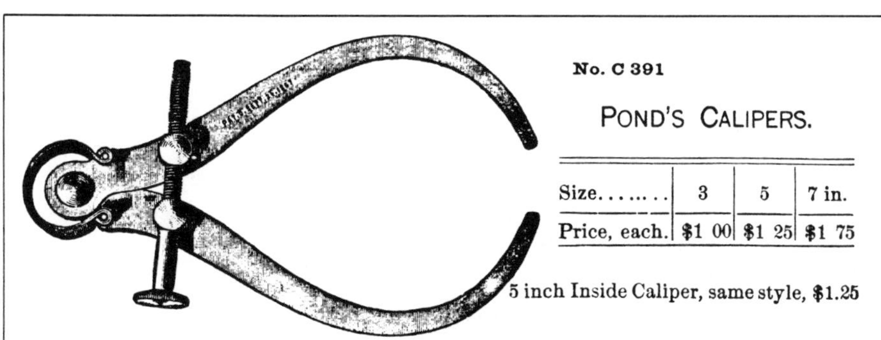

1884

PRATT & WHITNEY CO. Hartford, CT

Founded in 1860 by Francis A. Pratt (1827-1902) and Amos Whitney (1832-1920) to make machine tools. The partners incorporated the firm in 1869. Machine tools were always the primary product but it offered, at various times, a variety of machinist's tools, mostly measuring devices. It advertised, in 1880, ratchet drills, screw plates, and a bevel protractor patented by Ambrose Swazey on July 13, 1875 (No. 165519). In 1889 it was offering adjustable caliper gauges.

By 1892, it was making very accurate end measure blocks very similar to today's gauge blocks. Fred J. Miller, in a speech delivered in 1892, described the blocks as follows: "—a small bar of steel one inch and a quarter long and when I say it is one inch and a quarter long I do not mean that is about that long. I mean that the finest and most perfect appliances known to mechanics would fail to show that it varied from that length as much as 1/50000 part of an inch, if at all. It is called a standard end measure piece."

By 1886, it was offering "HARTFORD" calipers and gauges that were the same as those that had been offered by the HARTFORD TOOL CO., which had been founded and operated by a group of Pratt & Whitney employees. The American Railway Master Mechanics Assn. decimal gauge design patent of June 30, 1896 (No. 25719), was licensed to the Pratt & Whitney Co. Limit gauges, standard gauges, and screw thread gauges were offered in 1897. Screw pitch gauges, and sets of "knife-edge" straight edges were available in 1902. The company was listed as a maker of calipers, gauges, straight edges and micrometers in 1905.

P&W began making Hoke blocks in 1919 and offered a variety of taper gauges, plug and ring gauges, snap gauges, levels, and micrometers in 1929. The 1930 catalog listed precision levels, end measures, star gauges, spline gauges, Woodruff key gauges, taper gauges, Hoke blocks, "Mikechecks", toolmaker's flats, cylindrical gauges, thread gauges, snap gauges, ring gauges, and railroad gauges. *(see illustrations next two pages)*

PULVER MACHINIST TOOL CO. Chicago, IL

Listed in a 1930 directory as a manufacturer of machinist's tools. No products have been observed.

PUTNAM, J. & S.W. Fitchburg, MA

Founded by brothers John Putnam (1810-1888) and Salmon W. Putnam (1815-1872) in 1836 to make machine tools, steam engines, and steel rules. It appears to have been the first commercial maker of steel rules for machinists. In 1854, the firm was reorganized as:

PUTNAM MACHINE CO. Fitchburg, MA

It was the maker of a rule for laying out gear teeth, patented May 8, 1877 (No. 190447) by John Putnam. Also listed as a maker of iron machinist's levels and straight edges in 1905. The company went out of business about 1912.

QUEEN & CO. Philadelphia, PA

Maker of a portable tachometer, "adapted for high speed engines, electric dynamos, etc."

PORTABLE TACHOMETER.

Adapted for High Speed Engines, Electric Dynamos, Etc.

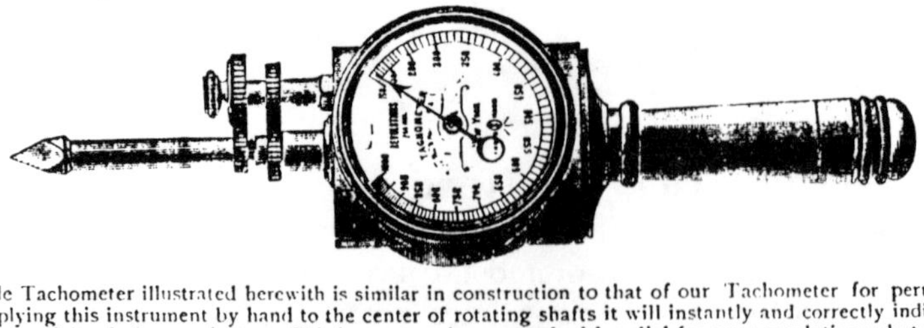

The Portable Tachometer illustrated herewith is similar in construction to that of our Tachometer for permanent attachment. By applying this instrument by hand to the center of rotating shafts it will instantly and correctly indicate the number of revolutions of the shaft per minute. This instrument is arranged with a dial for 1000 revolutions; but with it are also furnished two sets of geared wheels, in proportion of 1 to 2, by which the speed can either be reduced to 500 or increased to 2000 revolutions. These wheels are easily attached to an arbor with locknut provided for that purpose. Instruments for speeds higher than 2000 revolutions or lower than 500 revolutions are made to order.

On unscrewing the end of handle will be found a set screw, by means of which any lost motion may be taken up when it becomes necessary. The dimensions of the Portable Tachometer are: Diameter of dial, 2½ inches; depth of case, 3 inches; entire length, including handle and detachable point, 11 inches; total weight, including morocco case, 5 lbs.

The instrument is of excellent workmanship, plated and finished. It is contained in an elegant morocco case, for easy transportation. We confidently recommend it for its adaptability in obtaining accurate indications of the variations in speed of fast-running machinery.

Each $55.00

1902

THE PRATT & WHITNEY CO.'S

ADJUSTABLE CALIPER GAUGE

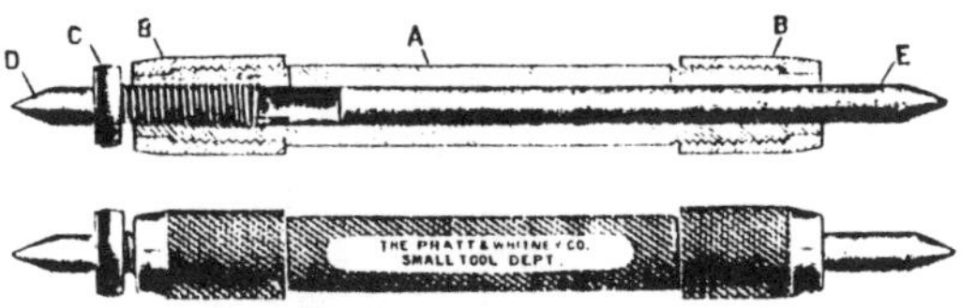

THIS little Tool will be found a convenient and economical substitute for the wire, usually made by repeated filing and "drawing," to get the exact size. Two chucks are arranged, one on each end of a hollow tube. The hole in one end of tube is smooth, so that by loosening the chuck, the wire may be quickly slid out or in. In the other end a fine thread is tapped, and a knurled wire is thread to fit. In use, in taking a size, the smooth wire is moved out to give nearly the diameter required, and fastened; then the fine adjustment is made by the threaded wire. Ends of wire are hardened. Any large diameter may be taken by changing the plain wire for a longer one. This tool will be found convenient in taking distances between uprights, &c., being positive. Holder, full nickel plate. Diameter of holder, ½ inch Diameter of steel wire, $\frac{5}{32}$ inch.

WEIGHTS.—No. 1 Gauge, without wires, ⅝ oz.; No. 2 Gauge, without wires, 1 oz; wires, 2 inch, ¼ oz.; 3⅜-inch, ⅜ oz; 5 inch, ½ oz.; 9-inch, ⅝ oz.; 13 inch, 1¼ oz.; 20 inch, 1⅞ oz.

1902

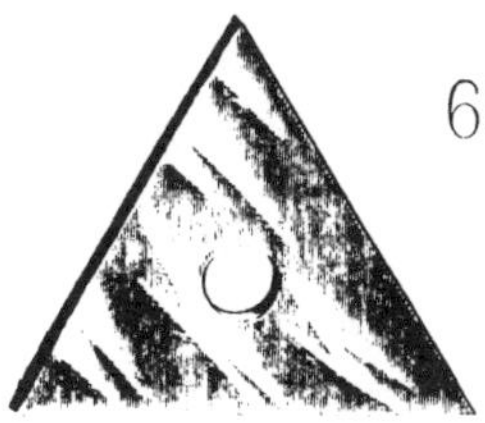

THE PRATT & WHITNEY CO.'S

60-DEGREE HARDENED AND GROUND STEEL TRIANGLES

THESE Triangles are offered as gauges which are practically perfect, possessing, at the same time, durability and wearing qualities, and furnishing in themselves means for easily and accurately testing the angle represented.

May 12, 1898. **AMERICAN MACHINIST**

STANDARD GAUGES.

If your BOLT IRON is not uniform in diameter when received in one lot or at different times and causes annoyance in adjusting your heading or threading dies and expense, when too large, for more weight than is required, you can protect yourself by buying and using, to test iron, as delivered, BAR IRON LIMIT GAUGES made by the undersigned. If your PIPE and FITTINGS WON'T CONNECT WITHOUT RETHREADING, you can use PIPE THREAD GAUGES, BRIGG'S STANDARD, to test them on receipt.

The manufacturers have the gauges and will use them rather than have their goods rejected because not made to the gauges when so ordered.

Illustrated in Gauge Catalogue.

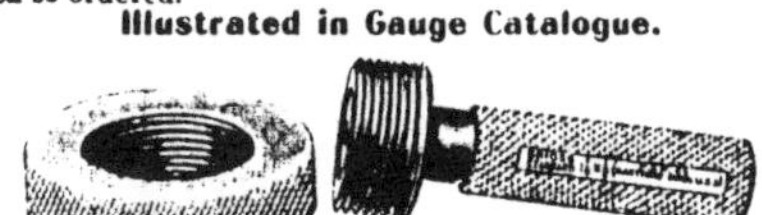

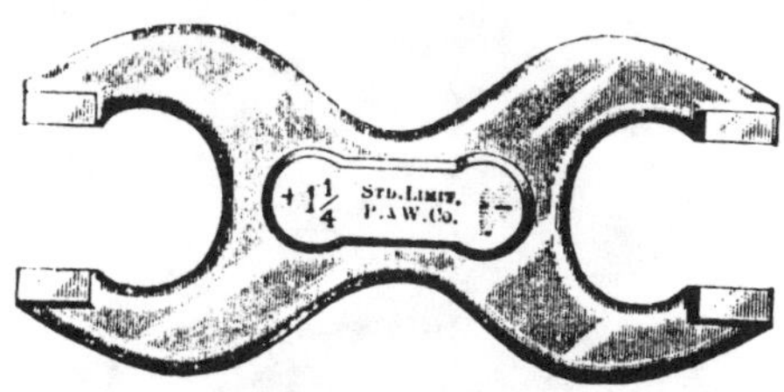

THE PRATT & WHITNEY CO.,
Hartford, Conn., U. S. A.

New York: 123 Liberty St. Boston: 144 Pearl St.
Chicago: 42 South Clinton St.

1902

SCREW PITCH GAUGES

Pratt & Whitney

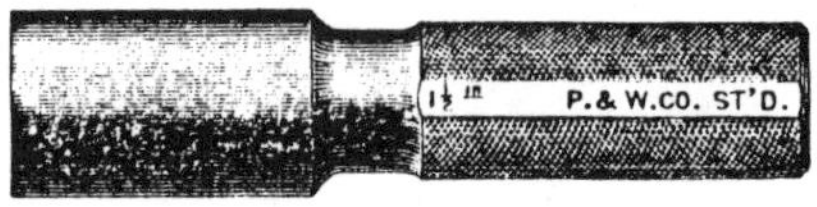

Fig. 2.

STANDARD "KNIFE-EDGE" STRAIGHT-EDGES

Made to the A. R. A. Standard with all slots accurately ground after hardening.

List No. R-120. Price each.................................$7 50

RANDALL, FRANK E. Waltham, MA

Founded in 1888 by Frank E. Randall when he left the Howard Watch Co., Waltham MA. He began by making dial gauges and his early products were covered by the J. Logan patent of August 21, 1883 (No. 283627). By 1927 Randall was making dial indicators under the trade name of RANDALL & STICKNEY.

REED & PRINCE MFG. CO. Worcester, MA

Incorporated in 1902, the officers were: Edgar Reed, President; William L. Ames, Treasurer; E. Howard Reed, Secretary; and Thomas Prince and Chester T. Reed, directors. The company began making micrometers in 1910, introducing an improved model with fixed anvil in 1915. Templet gauges were offered in 1914. On June 15, 1916, E. Howard Reed, son of Edgar Reed took over the micrometer business and formed a new company: the

REED SMALL TOOL WORKS Worcester, MA.

The new company continued to make a line of micrometers in 1" to 6" sizes. After 1921 the company made micrometers under contract for other companies such as Cenco, Fleming, and K.R. Wilson Co. An inside micrometer with an unusual height-gauge attachment and a set of inside micrometers was offered in 1922. The Robert G. De La Mater micrometer patent of October 25, 1927 (No. 1646393), was assigned to the company as was the Edgar Howard Reed micrometer patent of July 30, 1935 (No. 2009519). In 1946, the company merged with Rolled Thread Die Co. to form Reed Rolled Thread Die Co. The micrometer line was probably discontinued at this time and sold to the TUBULAR MICROMETER CO. Reed boxes are found with TUBULAR MICROMETER CO. labels pasted over the Reed label.

REICH MFG. CO., J.R. Dayton, OH

Operated by Julius Reich, inventor and maker of an indicator, patented August 31, 1937 (No. 2091376). Reich also received a patent for a surface test indicator on November 21, 1922 (No. 1436111).

Reich 3-Way Surface Test Indicators

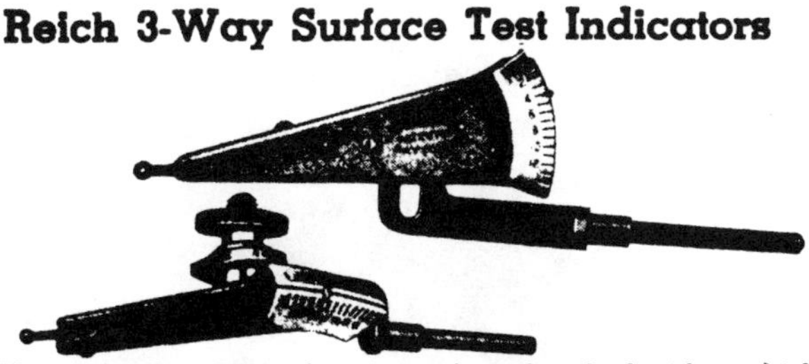

Tapered indicating surface is graduated on both sides, which enables readings to be taken from either side, or from the end. Indicator is small, compact, and of sturdy construction. The feeler or contact-point is .110 inches in diameter and may be initially set at any position within 90° from each side of its neutral position. It is provided with conical bearings set in adjustable seats which are mounted in the housing.

Outfit furnished consists of three parts—the indicator itself, an adjustable mounting arm, and a universal holder. These facilitate its use with practically every tool, gage, or machine.

Reich 3-Way Surface Test Indicator .... Each **5.00**

REIFFEL & THORN New York, NY

Operated by Charles Reiffel and Nickolaus Thorn who were the inventors and makers of machinist's calipers patented November 21, 1848 (No. 5929). Their patent is the earliest patent known for a machinist tool.

REISNER MFG. CO. Hagerstown, MD

Made dial indicators patented by William H. Reisner on December 21, 1897 (No. 595865), November 1, 1904 (No. 773983), and August 30, 1910 (No. 968884). The indicators were introduced to the machine trades in 1905. This was an early design of a practical dial indicator. In fact, the 1897 patent refers to the device as a "Lens Measure" and it was apparently designed as an optician's tool. The later patents refer to the device as an "Indicating Surface-gauge". In 1920, the L.S. STARRETT CO. bought out Reisner. The same indictors were made by L.S. STARRETT CO. still marked with the above patent dates. A later indicator patent issued to William H. Reisner on June 13, 1922 (No. 1419306) was assigned to the L.S. STARRETT CO.

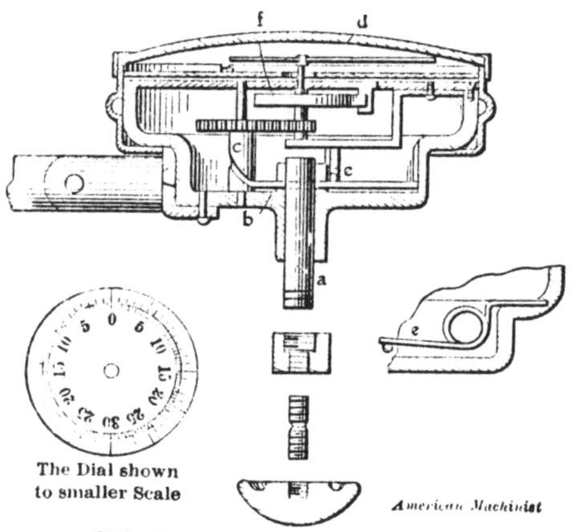

FIG. 2. DIAL TEST INDICATOR.

1905

REMACLE, LEOPOLD New York, NY

Inventor and maker of a micrometer caliper patented November 20, 1883(No. 288736). The micrometers were sold through FRASSE & CO.

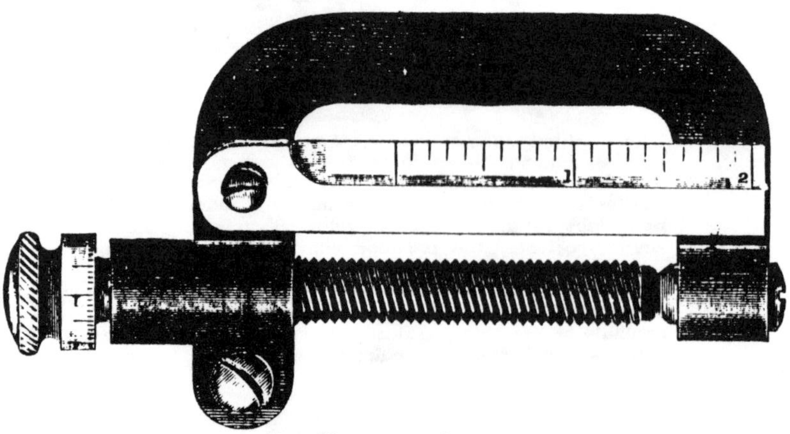

NEW MICROMETER CALIPER.

American Machinist 1883

REMINGTON TOOL & MACHINE CO. Boston, MA

Its 1909 catalog offered drill grinding gauges and micrometer surface gauges. No examples have been observed.

RHODE ISLAND TOOL CO. Providence, RI

Formed in 1883 as a reorganization of the PROVIDENCE TOOL CO. William B. Dart (1845-1897) was treasurer and general manager until his death in 1897. Dart's son, William C. Dart (1869-1946), served as president from 1908 until his retirement in 1944. The micrometer gauge patent of December 27, 1887 (No. 375688), issued to George Whitehead, was assigned to the Rhode Island Tool Co.

RICHARDSON, NATHANIEL Athol, MA

Richardson operated as a general machine shop from about 1835 until his death in 1883. In his later years, however, the shop was run by his sons, George H. Richardson and C. Fred Richardson. In 1878, Laroy S. Starrett rented a small part of the shop and began making his newly invented combination square. Richardson severed ties with Starrett in 1880 when the ATHOL MACHINE CO. threatened legal action against Starrett and anyone else engaged in the making of the combination square. Nathaniel died in 1883 and his son, C. Fred Richardson bought out his brother and began operating as:

RICHARDSON, C.F. Athol, MA

The spring caliper patent of March 22, 1881 (No. 239072), issued to Frederick Thomas, was assigned to C.F. Richardson. Richardson introduced the new caliper in 1886 as a "sliding-nut" caliper. Soon thereafter, Richardson sold the patent to J. STEVENS ARMS & TOOL CO., Chicopee Falls, MA, which offered the sliding-nut feature as an option on its line of calipers. These calipers, marked with the Richardson name are very scarce. Richardson made transits, levels and other small tools, advertising iron bench and pocket levels as early as 1887. His 1890 catalog offered: iron levels, leveling and squaring instruments, carpenter's and architect's transits and "Boss" scratch gauges. He introduced 4",6",8" adjustable bench levels in 1892, and later offered adjustable grade levels made under his patent of October 27, 1896 (No. 570056). About 1895, Richardson's son became a partner and the company became:

RICHARDSON & SON, C.F. Athol, MA

The company's products remained the same, but they began selling the levels through other tool makers located in Athol. Richardson levels were included in the following catalogs: ATHOL MACHINE CO., Athol, MA, 1893; L.S. STARRETT, Athol, MA, 1895; STANDARD TOOL CO., Athol, MA, 1897; STANDARD TOOL CO., Athol, MA, 1901, (small levels and transits only). The company reorganized January 1, 1903, as the RICHARDSON-OLIVER CO. when it merged with the Oliver & Whitney Co. of Athol, MA. The line of iron levels was sold to the GOODELL-PRATT CO., Greenfield, MA, in 1904. In 1905, the L.S. STARRETT CO. announced that it had purchased the remaining leveling instrument and transit business of the RICHARDSON-OLIVER CO. and would make them under the STARRETT name. In 1907, Richardson sold his shop, his house and his father's house to the L.S. STARRETT CO. Starrett donated the land to the YMCA for the erection of a new Y building. *(continued next two pages)*

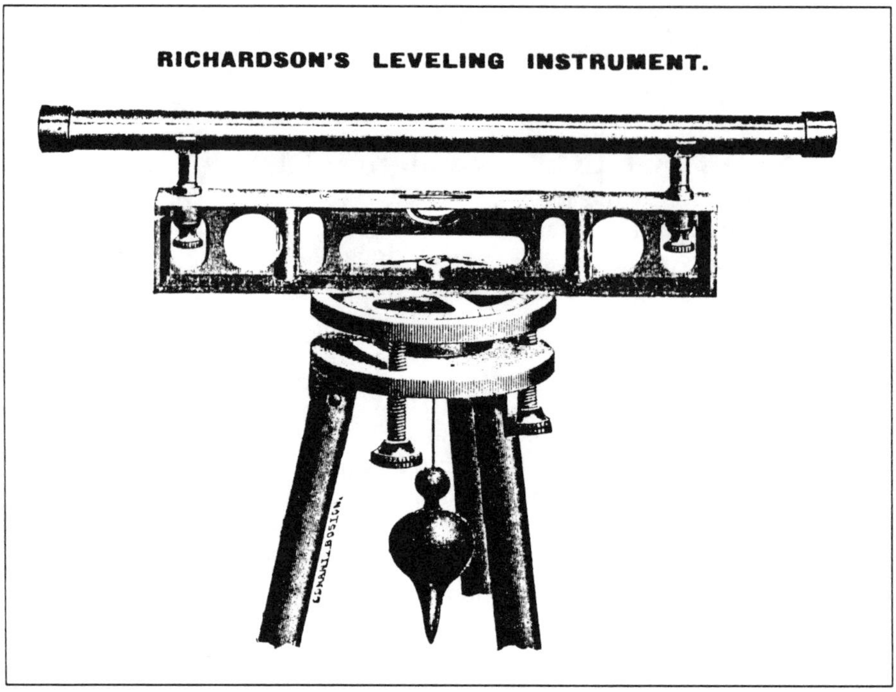

1888

C. F. RICHARDSON'S MACHINE SHOP.

NICKEL-PLATED POCKET LEVEL.

No. 11. 2½ inch............Price, $7 20 per dozen.............Price, 70c each
No. 12. 3½ " .... " 9 00 " " .... " 90c "

1891

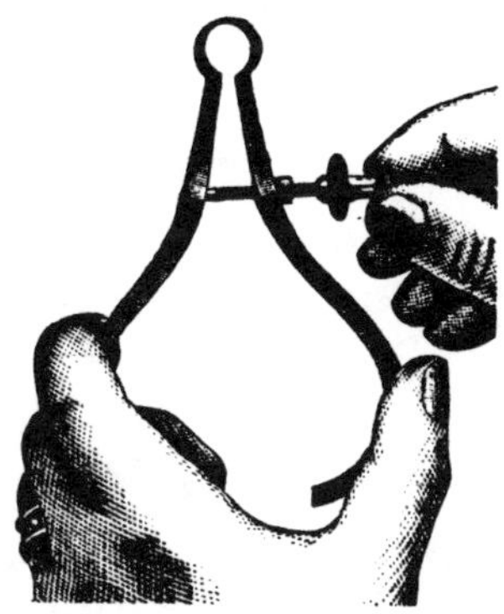

Sliding-nut Caliper.

In this caliper the inner end of the nut is beveled on the outside to fit a corresponding inside bevel on the washer. The nut is cut in two parts, horizontally, from the inner end up to the small knurled end at the outer end. This makes a spring of the nut, the normal condition of the nut being to remain out of contact with the thread of the screw. When the caliper is held as shown in the engraving, the nut can be slid along to any desired position on the screw, then by releasing the hold on the caliper the bow-spring forces the washer on the coned end of the nut, binding the latter on the threads, when the adjustment can be completed as with the ordinary solid nut. Since but little motion of the nut, when locked to the screw, is required, a screw of fine pitch is used, which facilitates close adjustment. The manufacturer of this caliper is C. F. Richardson, Athol, Mass.

Nickel Plated Pocket Level.

No.  11.

2½-In. 60 cents. 3½-In. 75 cents.
By Mail on Receipt of Price.

1887

IRON LEVELS,

Manufactured by
C. F. RICHARDSON, Athol, Mass.

1887

NEW ADJUSTABLE BENCH LEVEL.

WITH GROUND AND GRADUATED VIAL.

4-6 & 8 Inch.  Three Sizes.

MANUFACTURED BY
C. F. RICHARDSON,
ATHOL, MASS.

1892

All illustrations from American Machinist

RIEHLE BROS. TESTING MACHINE CO. Philadelphia, PA

Makers, beginning in 1892, of a special micrometer caliper patented by John D. Sloane on July 7, 1891 (No. 455644), and March 27, 1894 (No. 517073). The micrometer was designed for boiler inspectors to measure the thickness of boiler sheets. The hook was designed to go through a small hole drilled in the sheet for that purpose. The micrometer line was sold to the L.S. STARRETT CO., Athol, MA, who continued to offer the tools for many years.

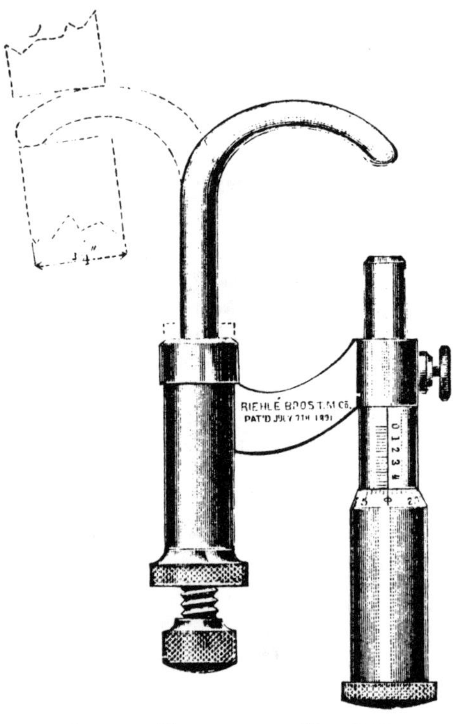

SPECIAL MICROMETER CALIPER.

American Machinist 1892

RIEKER INSTRUMENT CO., Philadelphia, PA

The company began making a line of precision levels in 1920. Examples observed are of very high quality.

RIGGS & BABIN Waterbury, CT

Operated by Gardiner G. Riggs and Arsene Babin who were joint patentees of a test indicator on October 25, 1904 (No. 773425). The partners introduced the indicator in February, 1906. By November of the same year, the partnership was dissolved and Babin was offering a slightly different indicator under his name only. *(continued next page)*

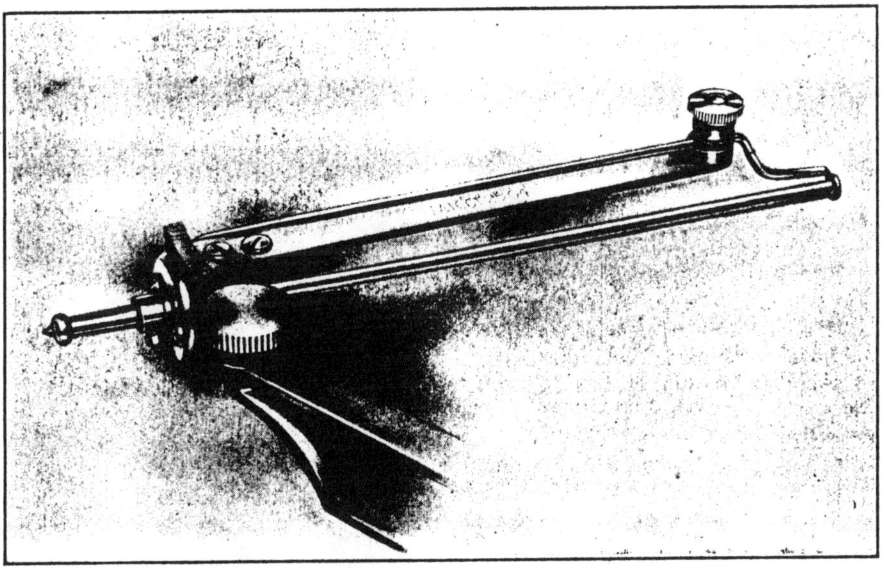

FIG. 1. TEST INDICATOR.

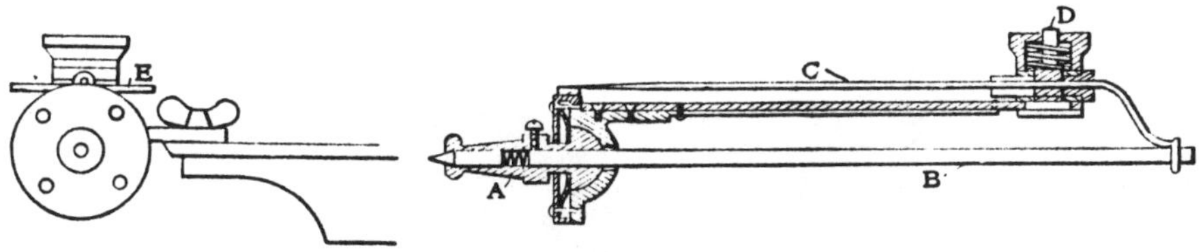

FIG. 2. TEST INDICATOR.

ROACH & RIDLON Westfield, MA

Introduced, in 1904, a universal test indicator with a swivel mount to allow use in a lathe or as a height gauge when used with a drill press. The unit was marked "Pat. Applied For". *(see illustration next page)*

ROBINSON CO., C.E. Orange, MA

Maker, in 1913, of a test indicator "equipped with a reversible dial which is graduated on both sides, one side giving readings in inches and the other in millimeters".

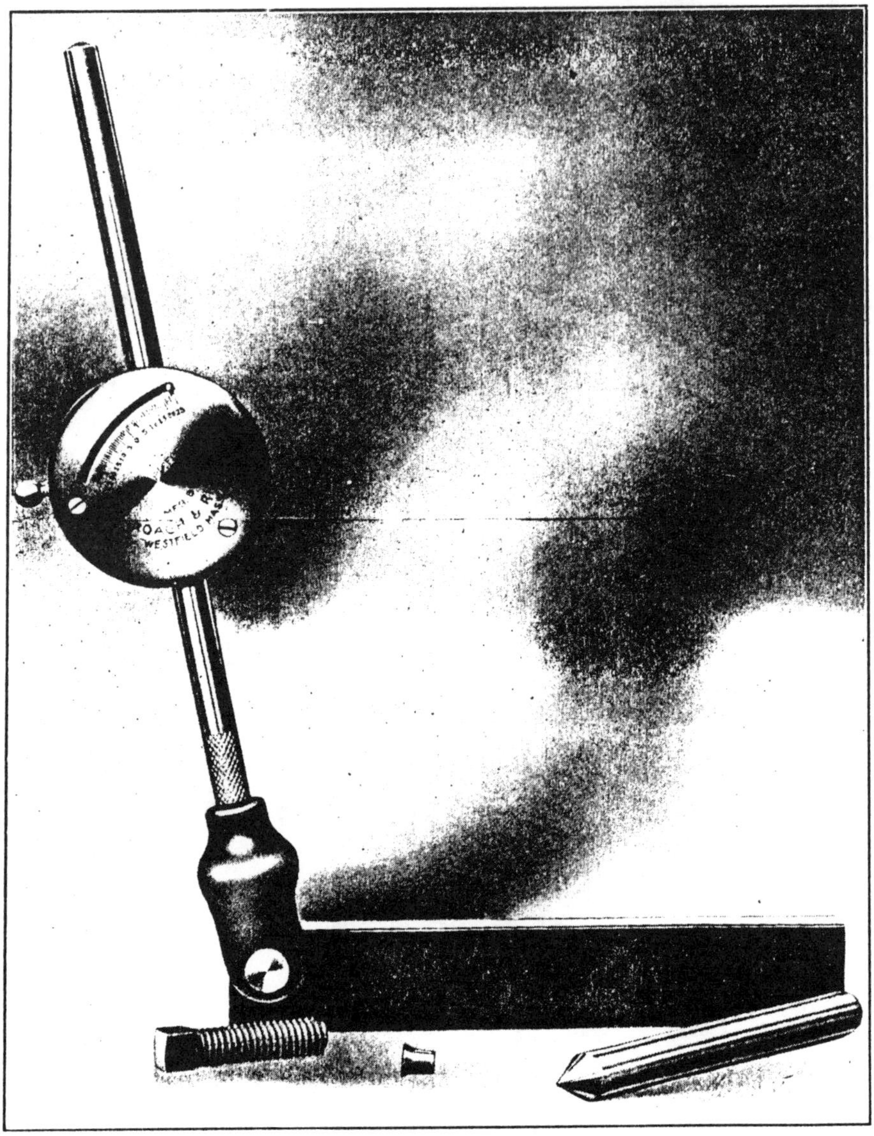

THE R. & R. UNIVERSAL TEST INDICATOR AND HIGHT GAGE.

Machinery 1904

See entry for Roach & Ridlon, previous page

ROBINSON CO., M.W. New York, NY and Rockfall, CT

Founded by Marcus W. Robinson (1820-1892) in 1869 to operate as the sole selling agent for SMITH & WESSON. Beginning in the mid 1870's it became a manufacturer and dealer in small tools and hardware. In 1893, the M.W. Robinson Co. bought the DAVIS LEVEL & TOOL CO. of Springfield, MA, and continued production of the levels at their Rockfall, CT, factory. Davis style levels made by Robinson are marked M.W.R. CO. *(continued next page)*

Robinson Pin Vises.

The M. W. Robinson Company, 79 Chambers street, New York, has just put on the market two pin vises, as

Fig. 1.—Robinson Pin Vise, No. 1.

here shown, similar in principle but proportioned to meet varying requirements. Fig. 1 illustrates the No. 1 vise, which is 5¼ in. long over all, with chuck ⅞ in. outer diameter, and pin capacity from 0 to 11-64 in. Fig. 2 reproduces the smaller size, No. 2, for more delicate purposes, which is 4½ in. long, with chuck ⅝ in. in diameter and a pin capacity of 0 to 5-64 in. Both have hollow handles, are finely polished, nickeled and knurled, and

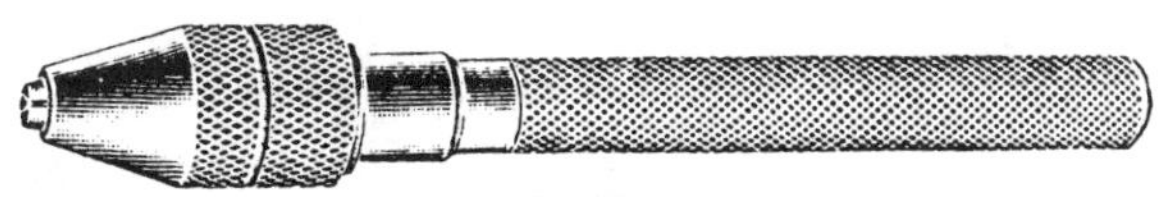

Fig. 2.—Robinson Pin Vise for Delicate Work, No. 2.

while suitable for innumerable purposes in working metals are especially designed for the use of machinists, jewelers and kindred mechanics. Both the material and workmanship are of a fine order for a discriminating trade, the chuck jaws being made of high grade tool steel, slotted transversely near their base, so that they are easily and positively carried for opening or closing by means of three longitudinal spring steel members operated by revolving the tubular handle, there being no spiral springs or other working parts to easily get out of order. Each tool is put up in a double pasteboard box attractive in appearance and labeled on top and end.

Iron Age
January 16, 1908

ROGERS BOAT, GAUGE and DRILL WORKS, JOHN M. Gloucester City, NJ

An important maker of fixed caliper gauges beginning about 1897. It also made the Richard's patent measuring machines and gauge standards. Its catalog #7, issued in 1904, offered: fixed caliper gauges, internal and external cylindrical gauges and limit gauges, measuring machines, gauge standards, and reference disks. In 1906 the name was changed to:

ROGERS WORKS, INC., JOHN M. Gloucester City, NJ

The company continued the line of fixed caliper, ring and plug gauges along with sets of standard reference disks. They also continued to make "Richards" precision measuring instruments. The Frank H. Daniels' patent of February 8, 1921 (No. 1367537), for an adjustable limit gauge was assigned to the company. Reported to have been a maker of the STAFFORD & WHIPPLE planer and surface gauge, probably after STAFFORD & WHIPPLE went out of business. *(continued next two pages)*

Measuring Instruments .....of Precision.

OVER 85,000 OF OUR GAUGES IN USE.

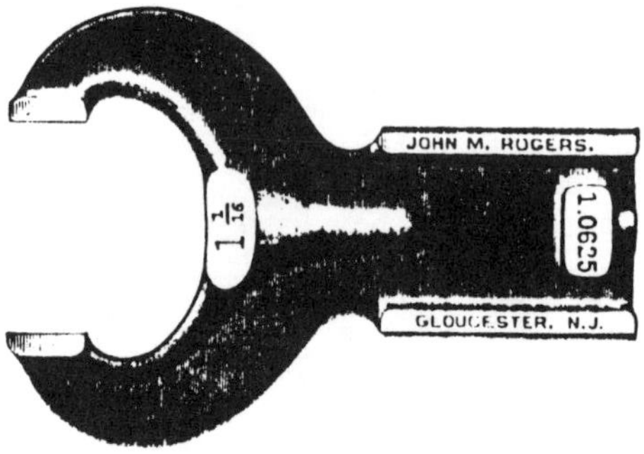

Fixed Caliper Gauges.

A set of Caliper Gauges, with a corrective standard for keeping them in proper adjustment, constitutes a set of working implements sufficient to meet every requirement for maintaining uniform sizes.

Carefully hardened to prevent wear and accurately ground to size.

Made in sizes from 1-4 inch to 6 inches.

Crescent Pattern.

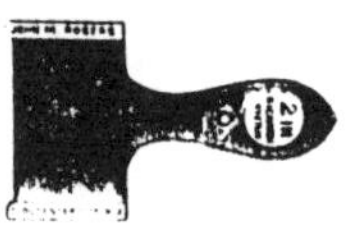
Flat Bar Pattern.

Made in sizes from ¼ inch to 12 inches.

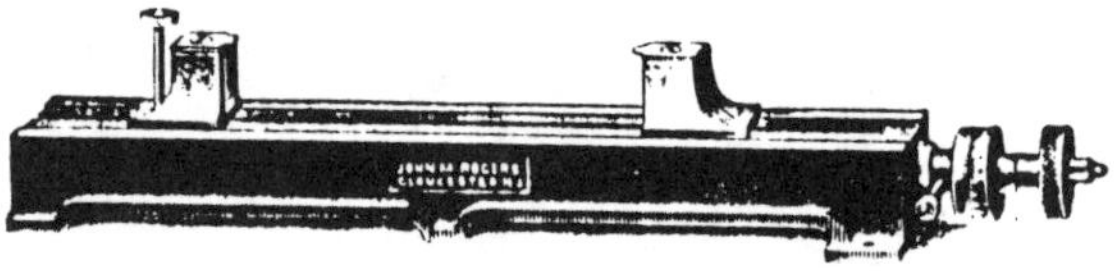

Standard Measuring Machine

For use in the tool-room in preparing templates and originating standards. Will measure differences of 1-10,000 of an inch.

Limit Gauges

Made with any desired limit of variation.

Corrective Gauge Standards.

These discs are employed for testing and correcting fixed gauges, for setting calipers and as reference to prove dimensions within their range. Each disc separate and ground independently to size.

SEND FOR 1900 CATALOGUE SHOWING FULL LINE.

These tools may be at the Paris Exposition in charge of A. Janssens.

THE JOHN M. ROGERS, BOAT, GAUGE AND DRILL WORKS,

Gloucester City, New Jersey, U. S. A.

FOREIGN AGENTS.—Chas. Churchill & Co., London, E. C., England. C. W. Burton, Griffiths & Co., London, E. C., England. Selig, Sonnenthal & Co., London, E. C., England. De Fries & Co., Dusseldorf and Berlin, Germany. A. Janssens, Paris, France.

Machinery, June 1900

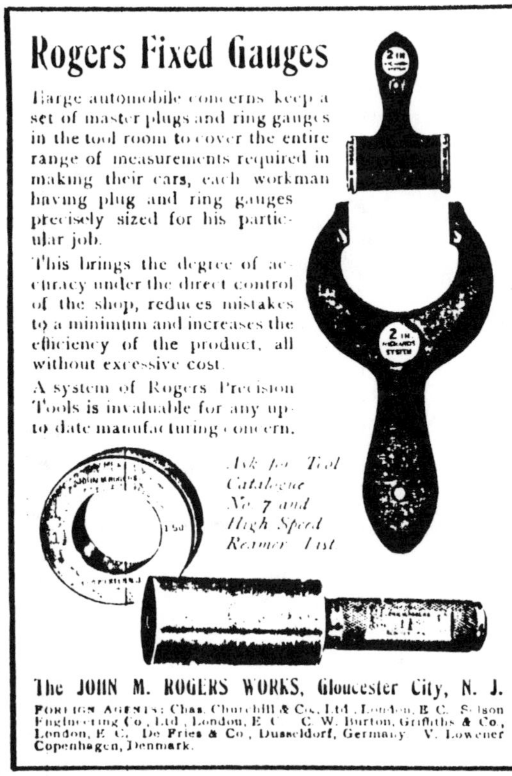

1909-1910

RUSSOM, C.H. Brooklyn, NY

Operated by Charles H. Russom who was the maker, in 1890, of a micrometer gauge which he patented February 5, 1889 (No. 397442). *(see illustration next page)*

S & B CO. Boston, MA

Makers of rules and combination squares, probably operating c1910-1920. Examples are fairly common and are of medium quality.

ST. JOHNSBURY TOOL CO. St. Johnsbury, VT

Founded by Isaiah J. Robinson about 1873. Makers of a variety of bevels and try squares, including some machinist's squares. The company went out of business about 1885.

SAWYER, SYLVANUS Fitchburg, MA

On October 12, 1865, Sawyer placed the following ad in the FITCHBURG REVEILLE:

> SYLVANUS SAWYER successor to C.H. KELTON PRACTICAL MACHINIST Newtons Lane, Fitchburg, Mass Manufacturer of Woodworth's celebrated wood planer of all sizes, chair machinery, shafting, mill work, etc.

Sylvanus Sawyer received patents for Calipers & Dividers on April 9, 1867 (No. 63656) and on July 7, 1868 (No. 79603). It is not known if these calipers were actually produced.

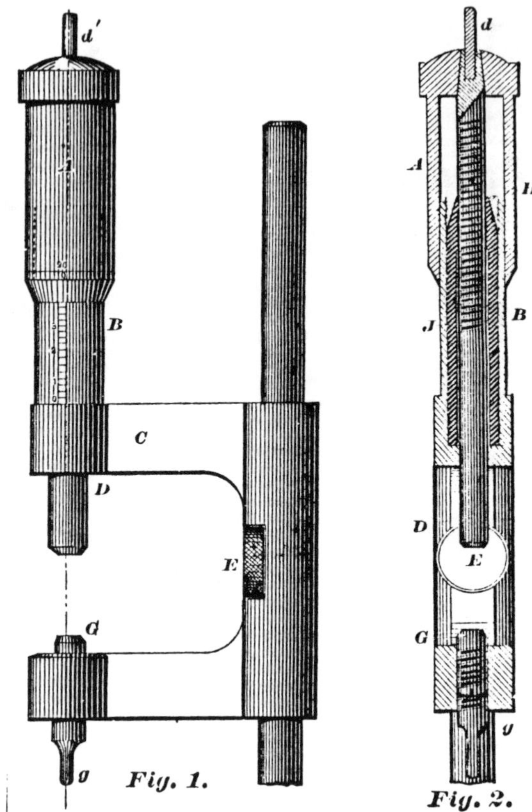

Fig. 1. Fig. 2.

Russom's Micrometer Caliper.

American Machinist 1890

See entry for Russom, previous page

SCALBOM MFG. CO., Chicago, IL

Makers, in 1921, of a diemaker's square patented March 9, 1920 (No. 1333223) by Oscar L. Scalbom.

SCALBOM DIE SQUARE AND BEVEL GAGE

The die square and bevel gage shown in the accompanying illustration is essentially a diemaker's tool, but it will also be found useful by toolmakers and machinists. It is made by the Scalbom Mfg. Co., 11 S. Desplaines St., Chicago, Ill. The tool consists of a body, a thin blade which can be swiveled to any desired angle relative to the axis of the body, and an arrangement for clamping the blade after it has been set. This construction renders the tool suitable for laying out angular lines on a surface at right angles to the one against which the body is held. The tool also will be found convenient for ascertaining the taper of a hole. When the blade is inserted in a hole, the recess in the body permits a view of the blade edge for its entire length. Blades of different thicknesses can be furnished. All parts are made of steel and are hardened and ground.

Die Square and Bevel Gage made by the Scalbom Mfg. Co.

SCHAEFER, FRED Minneapolis, MN

Inventor and maker of a combination spirit level and key seat rule introduced in 1894.

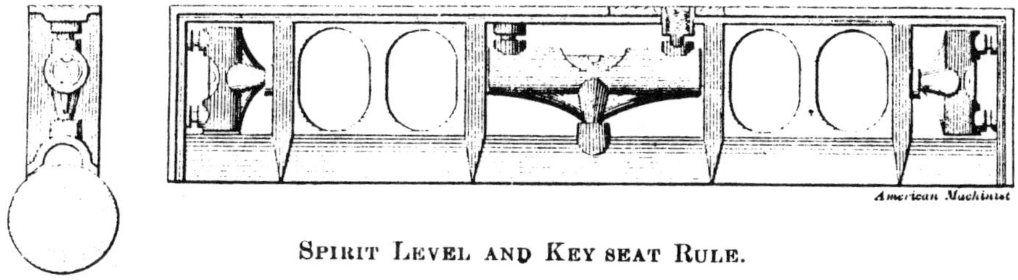

1884

SPIRIT LEVEL AND KEY SEAT RULE.

SCHELLENBACH-HUNT TOOL CO. Cincinnati, OH

Introduced, in 1908, a Universal Micrometer and Surface Gauge. The tool was patented June 1, 1909 (No. 923446), by Charles O. Schellenbach.

FIG. 1. UNIVERSAL MICROMETER SURFACE GAGE

FIG. 2. IN USE VERTICALLY

Machinery 1908

SCHOLLHORN & CO. WILLIAM, New Haven, CT

Founded by William Schollhorn (1834-1890) in 1863 to operate as a machine builder and, later, a maker of "Excelsior" calipers and dividers under the William Schollhorn and Frank P. Pfleghar patent of January 9, 1866 (No. 51967). The firm was incorporated in September, 1891, as WILLIAM SCHOLLHORN & CO., INC. It continued to make the calipers and dividers as late as 1892, but from 1891 on their major product was the Bernard patent parallel jaw pliers. *(continued next page)*

No. C 402

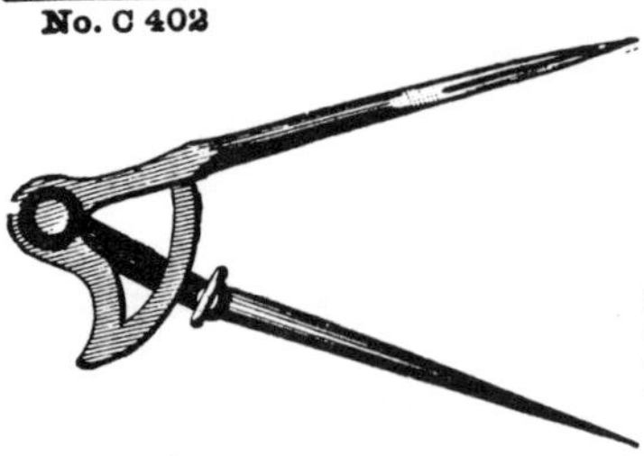

EXCELSIOR DIVIDERS.

PRICE PER DOZEN.

5	6	7	8	9	10	12 inch.
$7 00	$7 50	$9 00	$10 25	$12 50	$13 75	$16 50
PRICE EACH.						
$0 60	$0 65	$0 75	$0 85	$1 05	$1 15	$1 40

The segment fastening is superior to any other style, and the points, which are most essential on a divider, are of *tempered, genuine Stubs' Steel Wire inserted in each leg,* as shown in the cut, by the mechanical device of being drilled, reamed and driven in tight, and then ground to a sharp point.

No. C 403

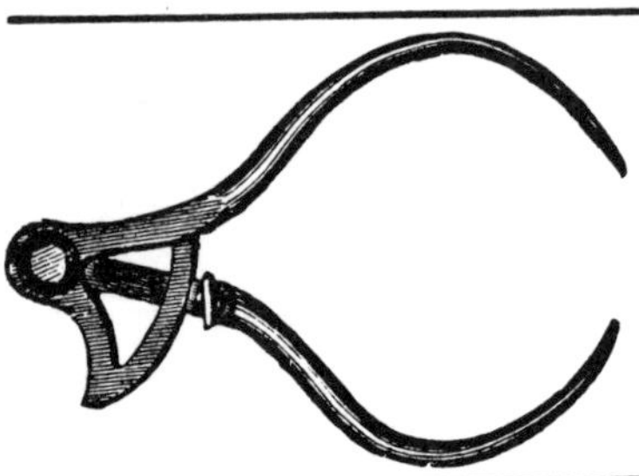

EXCELSIOR CALIPERS.

	6	8	10
Price per dozen.......	$8 25	$11 00	$13 75
Price each	0 70	0 90	1 15

1884

SCHRADE, GEORGE, New York, NY

Maker, in 1898, of the "Time Saver" drill and tap drill gauge, copyright 1898. By 1902, the identical tool was being made by the L.S. STARRETT CO. of Athol, MA.

Time Saver.

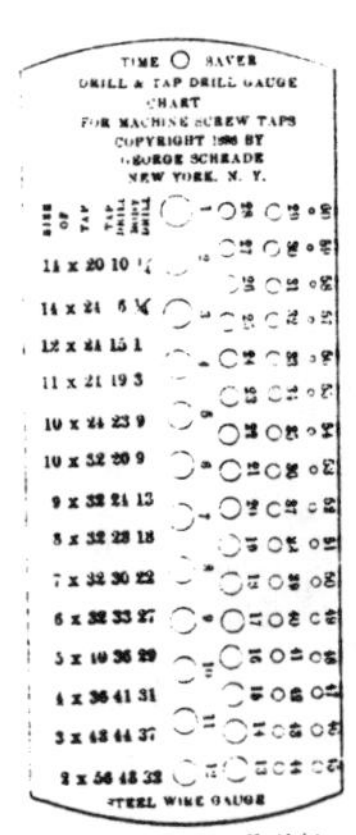

Half Size,
Price $1.50.

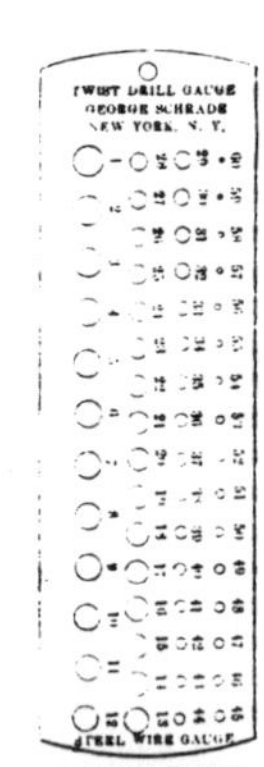

Half Size,
Price, $1.00.

Drill and Tap Drill Gauge chart fills a long felt want; does away with guessing when a drill is needed for a tap or body of a screw. The tap drill sizes usually found in catalogues are for a full thread, which necessitates working the tap through very slowly, and very often is the cause of broken taps. In all kinds of manufacturing where screws are put in to stay the thread need not be full. The sizes on chart are chosen after ten years' experience with taps and work where one tap is used to complete the threading, instead of two (taper and plug). There is no danger of breaking taps in annealed tool steel where the sizes of drills are chosen according to the chart. The gauges are made on special machinery, by experienced workmen, from the most suitable material, hardened and polished. Special charts and gauges made to order. Chart, $1.50; Drill Gauge, $1.00. Mailed on receipt of price. Sent on trial to responsible parties. Discounts to the trade.

Geo. Schrade,

1337 Webster Avenue, New York, N. Y.

SCHULTZ, AUGUST L. Cleveland, OH

Inventor and maker, in 1900, of a combination tool which appears to be a combined surface gauge and combination square. Schultz received his patent on September 4, 1900 (No. 657136).

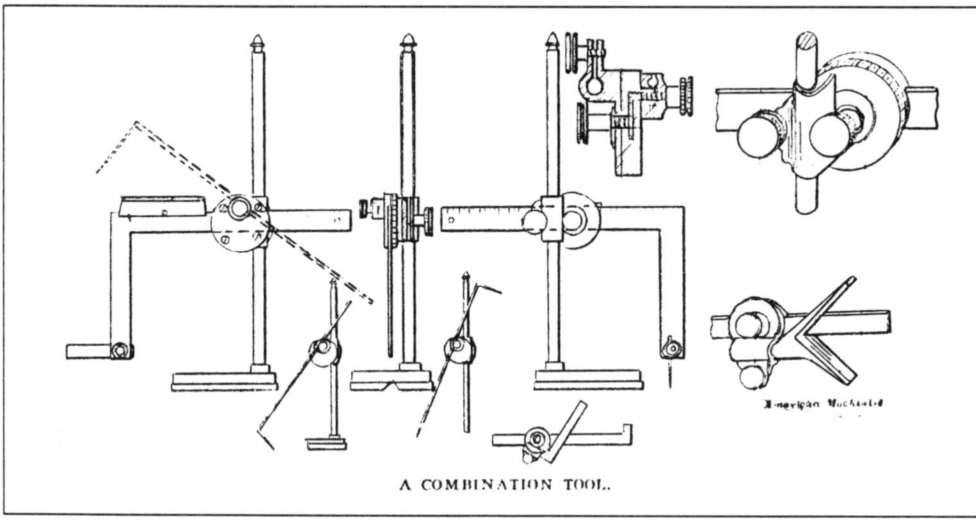

A COMBINATION TOOL.

1900

SEAVER, FREDERIC A. Hartford, CT

Inventor of a Key Centering Tool patented July 23, 1889 (No. 407617). The tool may never have been made as Seaver was advertising only to sell the patent rights in 1890.

1890

SEVERANCE, W.E. Revere, MA

Operated by William E. Severance, inventor and maker of toolmaker's rules which he patented May 12, 1903 (No. 728137). The rules were introduced in 1904 and were available in lengths of 1/4", 3/8", 1/2", 3/4" and 1". By 1906, the KENDRICK & DAVIS CO., was offering the Severance rules in its catalog. It is not known if it had taken over production or just bought the rules from Severance. *(continued next page)*

Steel Rules with Handle.

W. E. Severance, Revere, Mass., is offering the handled rules shown herewith. They have tempered blades, which are held by friction at any angle. The knurled handle makes them very convenient, as they can be held steadily in the hand, and by placing at an angle may be used in many places where it is impossible to get with the end graduations of an ordinary rule. The rules are handy in setting up screw machines, in milling machine work, in lathe work where there are grooves and short shoulders to be turned, and many times the tail stock center is in the way in measuring small work. They are especially valuable in tool and die making, it is explained; in fact, in all kinds of machine work, large and small, there are many places where the rules can be used to advantage. The back corners of the scale blade are rounded, which

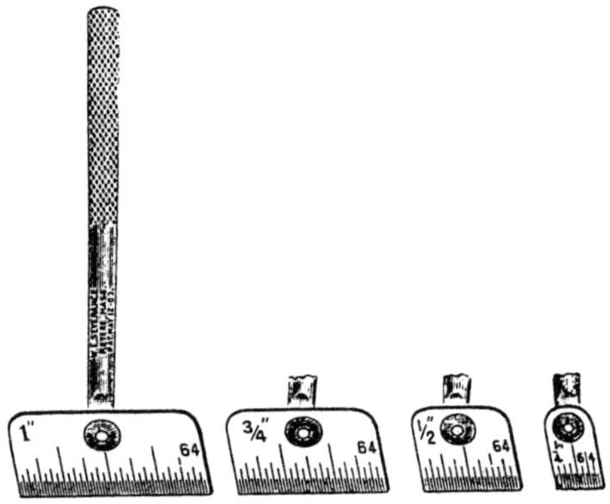

Steel Rules with Handle.

permits it to be moved with the thumb while holding in the hand. In addition to the sizes shown in the illustrations the rules are made in 3/8 inch, and are graduated with 100, 64, 50, 32, 20 and 16 lines to the inch.

Iron Age, June 23, 1904

SEWARD, ERNEST R. Hartford, CT

Inventor of a thread micrometer introduced in 1908. No patent has been found.

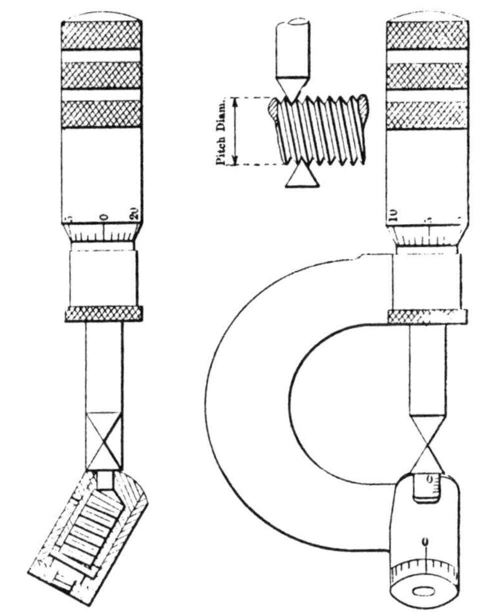

FIG. 2. THREAD MICROMETER

1908

SEYMOUR & CO. William, New York, NY

Reported as makers of machinist's try squares. Its period of operation is unknown.

SHAUGHNESSY TOOL CO. New York, NY

Maker of a surface gauge patented April 16, 1916 (No. 1179142) by Bernard F. Shaughnessy. The tool is of odd construction, with the base, which can be fit onto a rule, made of a heavy non-ferrous alloy. *(see illustration next page)*

SHULTS PROTRACTOR & SPECIALTY CO., Chicago, IL

Makers of a combination rule and protractor patented April 8, 1919 (No. 1299978), by Silas C. MacDowney, Chicago, IL.

SIMPLEX TOOL CO., Woonsocket, RI

Introduced, in 1916, a precision square "accurately ground to a fine edge on two sides and flat on opposite sides. The fine edges in contact with the work enable the detection of an error less that .0001" in 3 inches." This description also fits squares listed in the TAFT-PIERCE CO. 1929 catalog. It appears likely that Simplex was bought out by TAFT-PIERCE.

SLOCUM, AVRAM & SLOCUM LABORATORIES, INC., New York, NY

Maker, in 1917, of a toolmaker's square with indicator. The tool was covered by the A. Schachat patent of June 17, 1919 (No. 1306722), which was assigned to this company. The company also made limit snap gauges, plug gauges, sine bars, and thread measuring machines.

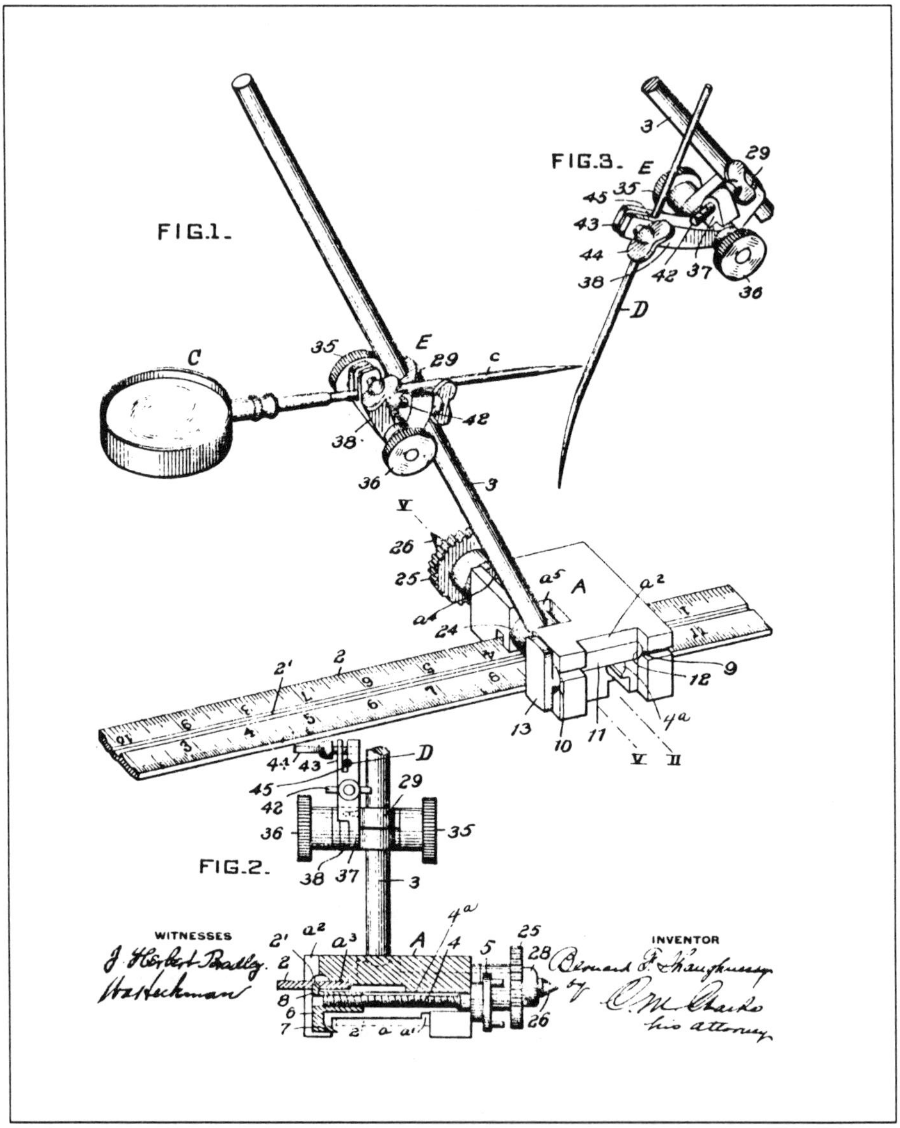

See entry for Shaughnessy Tool Co., previous page

Patent drawing 1916

J.T. SLOCOMB & CO. Providence, RI

Founded in 1891 by John Tibbets Slocomb to make a center drill he had invented. This drill is still the most common type in use today. Slocomb was born in Canada in 1864, coming to the U.S. as an infant. At age 16 he began his machinist's apprenticeship in Bangor, ME. He later worked for the Brainard Milling Machine Co. at Hyde Park, MA, H.B. Smith Machine Co. in Smithville, NJ, Holyoke Machine Co., Holyoke, MA, Brown & Sharpe Mfg. Co., Providence, RI, from 1885 to 1890, and the Rhode Island Locomotive Works in Providence. He claims to have seen his first micrometer at the Brainard Milling Machine Co. in 1885.

Besides the center drills, his first products were a combination depth and inside micrometer introduced in 1892 and center indicators produced in 1894. The combination depth and inside micrometer was offered in an improved version in 1893. He began making outside micrometers in 1894. The first micrometers were made in the 3 to 8 inch sizes but were a failure and he began making the one and two inch sizes. His first micrometers, shown in *The American Machinist* magazine in 1894, were the first to be made with an I-beam frame. In 1895, Slocomb offered a further improved version of his inside micrometer with a tubular brass extension piece. Slocomb recieved micrometer patents on May 12, 1896 (No. 559820) and April 16, 1897 (No. 26884). The former patent was for a means of adjusting for wear and the latter was a design patent for the I-beam style frame.

Slocomb introduced, in 1903, a micrometer on which the decimal equivalents marked on the body "instead of being sunk in are raised above the surface." He claimed that the pressure required to do this added to the stiffness of the frame. The 1909 catalog offered: inside, outside and micrometer calipers, screw thread micrometers, tube micrometers, quick-acting micrometers, standand reference discs and end measures. The 1912 issue offered: micrometers to 18", internal micrometers, standard reference discs and end measures, screw thread micrometers, and special micrometers.

Slocomb sold the company to J.H. Drury, sales manager of the Union Twist Drill Co., in 1914. Drury incorporated the firm as the J.T. SLOCOMB CO. and kept Slocomb on as general manager. On May 15, 1917 (No. 1226396), Slocomb received a patent for a tube micrometer. This appears to be the first of the type and is still made today. The company introduced the "Davenport-Slocomb" direct reading micrometer in 1920, making it under the William S. Davenport patent of December 26, 1916 (No. 1210051). A duplex micrometer for use as a snap gauge was introduced in 1921. *(continued next four pages)*

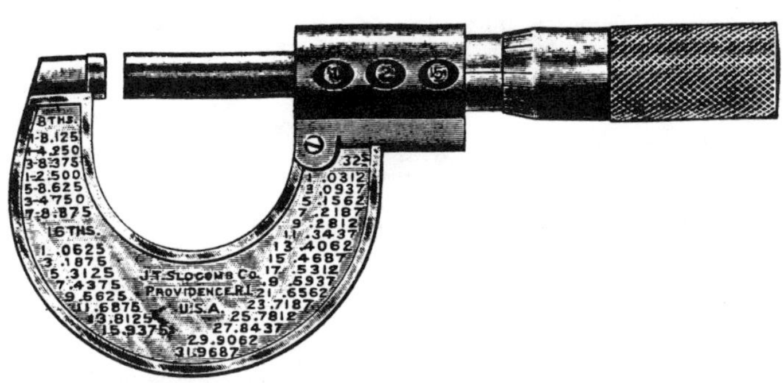

DAVENPORT-SLOCOMB DIRECT READING MICROMETER

1925

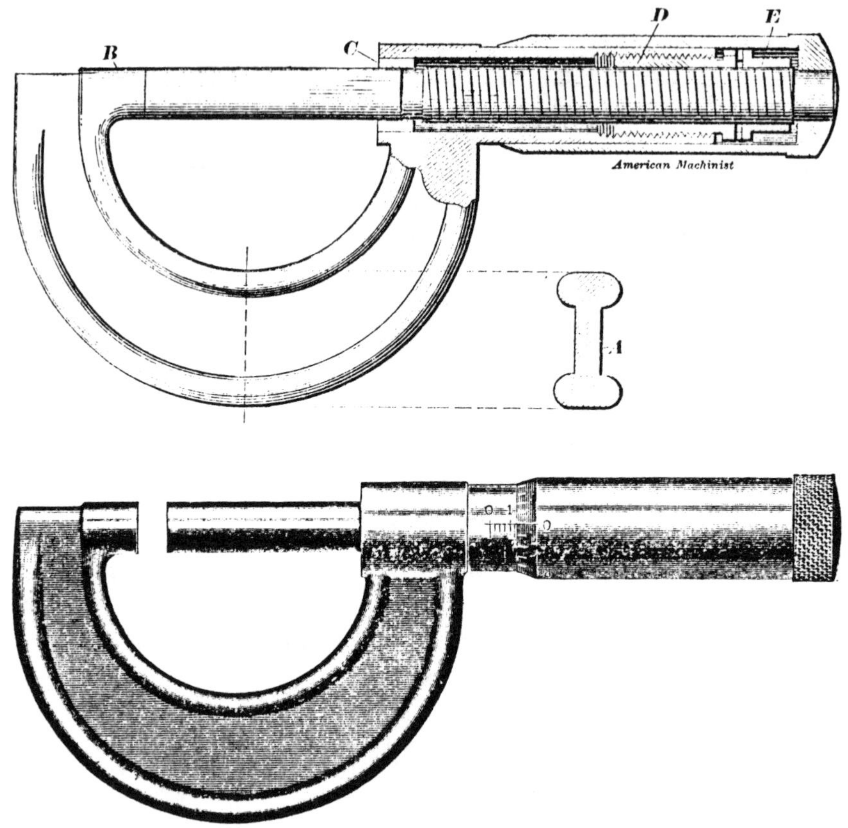

1894

A New Micrometer Caliper.

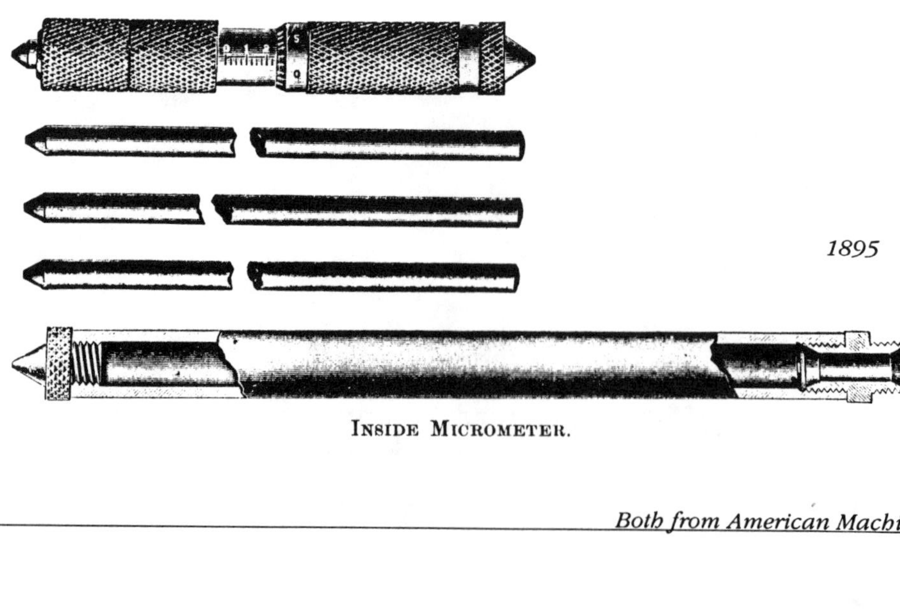

1895

Inside Micrometer.

Both from American Machinist

Direct Reading and Quick Acting Micrometer

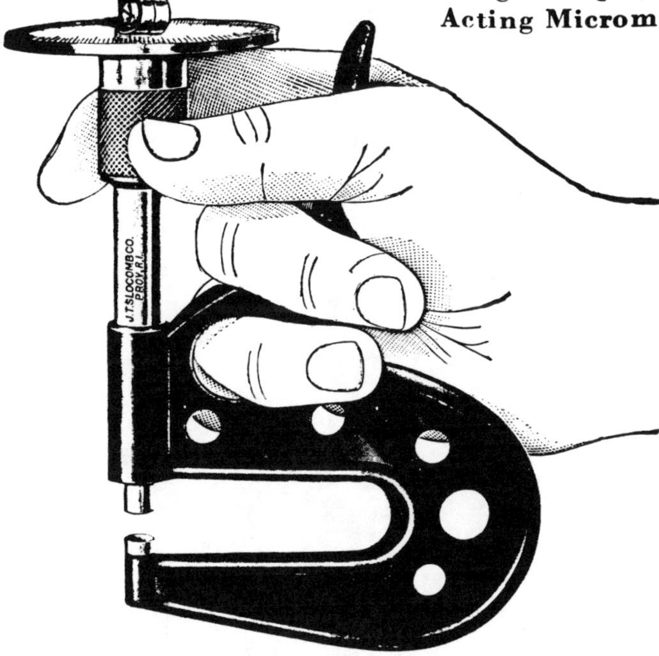

1925

QUANTITY PRODUCTION MICROMETER

Pat. June 18, 1918

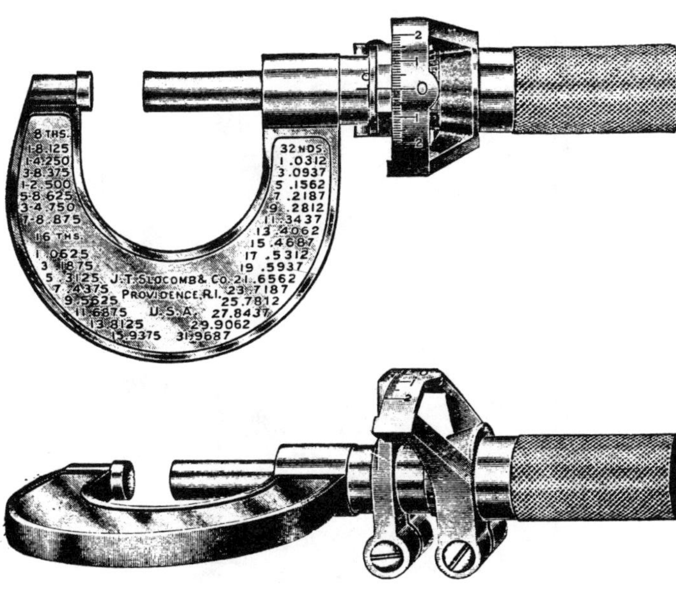

1925

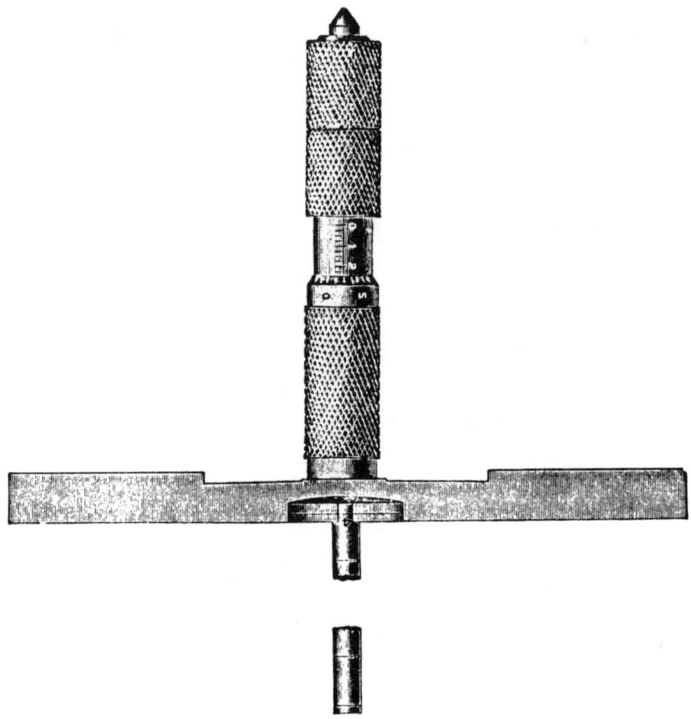

INSIDE MICROMETER AND DEPTH GAUGE.

1893

American Machinist

Machinery
1900

New Six-Inch Bench Micrometer

1925

American Machinist 1891

American Machinist 1894

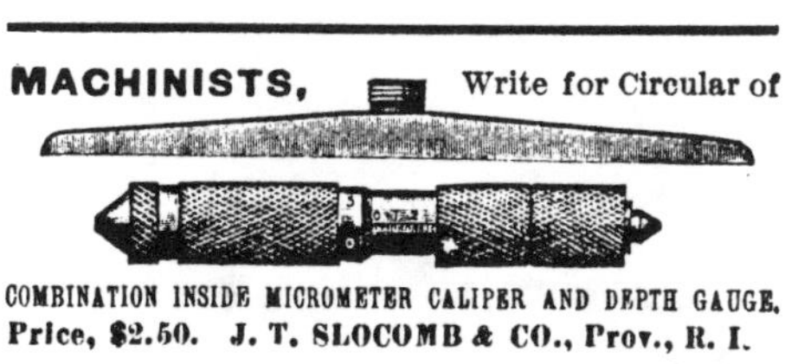

American Machinist 1892

American Machinist 1893

SMITH, CHARLES W., Hyde Park, MA

Inventor and maker of a surface gauge, patented June 13, 1882 (No. 259596). The tool was made by the MILLERS FALLS CO. in 1884.

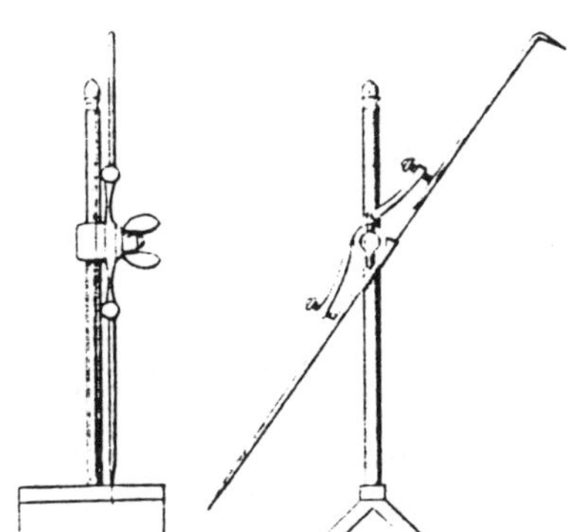

NEW SURFACE GAUGE.

American Machinist 1882

SMITH, EDGAR Boston, MA

Inventor of a surface gauge patented October 23, 1888 (No. 391488). The only known specimen is unmarked but is obviously factory made. A description and illustration of the gauge appears in an 1896 French catalog. In the illustration, the gauge is clearly marked C & F, BOSTON, which would be CHANDLER & FARQUHAR, Boston, MA. Smith was probably the maker and sold the gauges through CHANDLER & FARQUHAR.

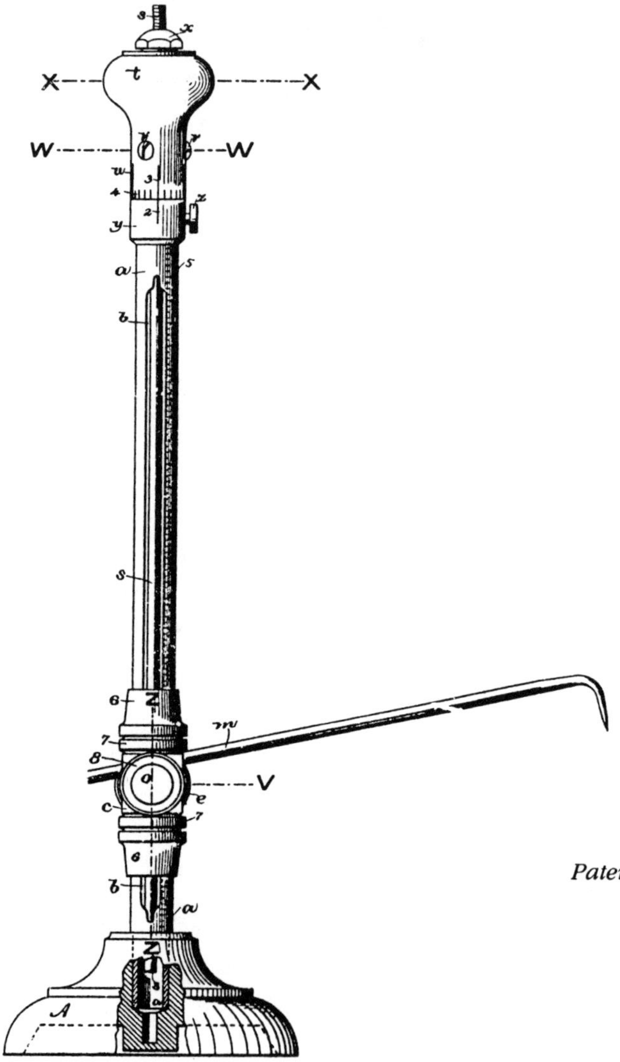

Patent drawing 1888

SMITH, EDWARD C. Brooklyn, NY

Inventor and manufacturer of machinist's calipers patented April 4, 1882 (No. 256062).

SMITH, E.G. Columbia, PA, later, Tampa, FL

Founded by Ernst G. Smith about 1890. Smith's first products were "beam calipers, plain or vernier reading to $\frac{1}{128}$". The calipers were made in 5 sizes from 4" to 12" and in the metric system. These calipers, when examined, look very European in style and the numbers appear to have been marked with single stamps while the Smith name and address are roll marked. Smith may have, in fact, imported his calipers from Europe, probably Germany. A one-inch micrometer, offered in 1900, is certainly of European make, probably French.

Smith introduced two models of a surface gauge in 1891. One model had fine adjustment for height and one did not. In 1892, he introduced a micrometer caliper made in one-half and one inch sizes. The design had an odd feature in that the anvil was not secured to the frame and would "slip if the measuring screw be in any way forced too hard." The idea was to avoid damage but it must have been a very inconvenient feature for the user.

In 1895, Smith introduced his Spherometer, a device for measuring spherical surfaces. It was made in two styles, one with a 1mm pitch screw and 100 divisions on the disk to read $\frac{1}{100}$mm, and the other with a $\frac{1}{2}$mm pitch screw and 500 divisions on the disk to read $\frac{1}{1000}$mm.

In a 1903 circular, Smith offered some slightly shopworn calipers and a few hundred "Which Way" pocket levels "only slightly rusted" at sale prices. He also offered $\frac{1}{2}$" square parallel strips at two cents an inch and "a combination micrometer which any mechanic can construct himself, either singly or in quantities". In 1905, Smith offered his "Columbia" vernier caliper fitted with a pair of adjustable points at one end of the jaws. The points were inserted in holes drilled at an angle with the jaw faces which were split and fitted with clamping screws. In 1909, Smith offered a friction roller attachment for his beam calipers. This was much the same device as is still in use today.

Introduced, in 1910, a Universal Angle Gauge consisting of "two straightedges that are connected by the curved piece shown. As the three adjustable connecting joints permit the straightedges to be placed at any angle with each other, this gauge can be employed for a variety of purposes". Catalogs issued in 1899 and 1911 offered calipers, pocket levels, "Columbia" spherometers, and steel rules. Smith moved to Tampa, FL in 1915. His only known product made after the move is a pocket level. *(continued next three pages)*

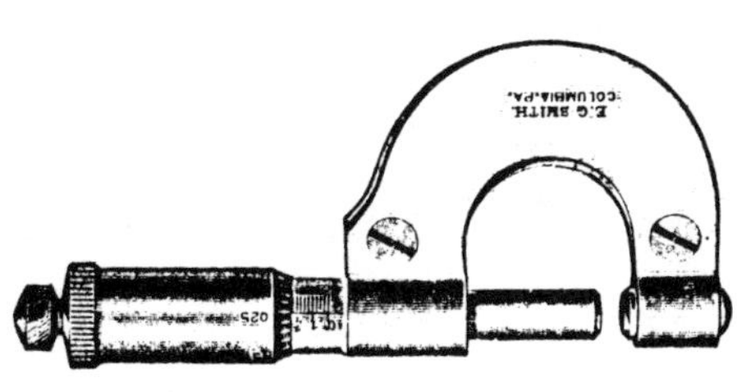

1892

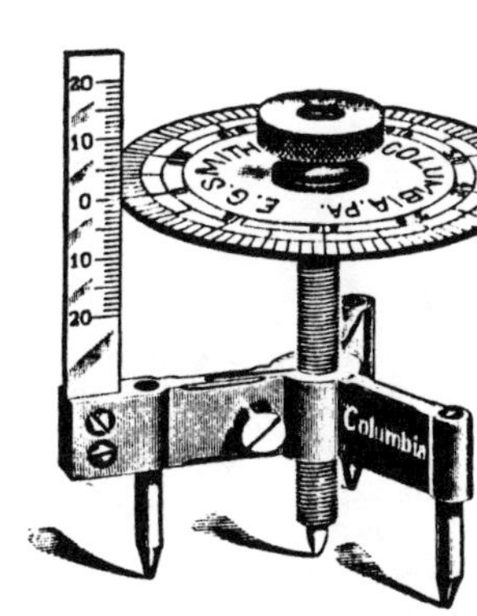

1895

1894

All from American Machinist

Machinery 1909

Machinery 1900

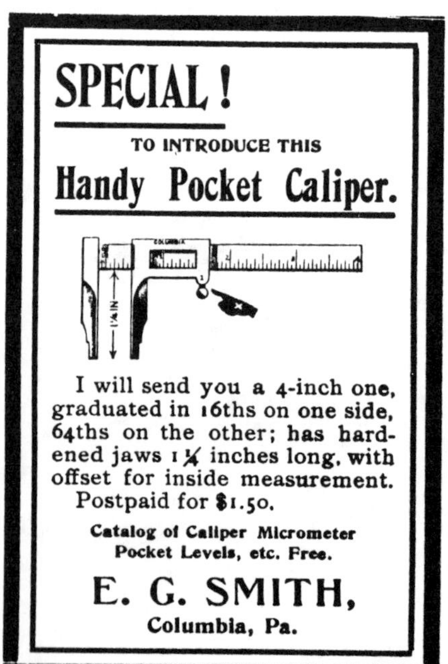

Machinery 1900

Machinery 1900

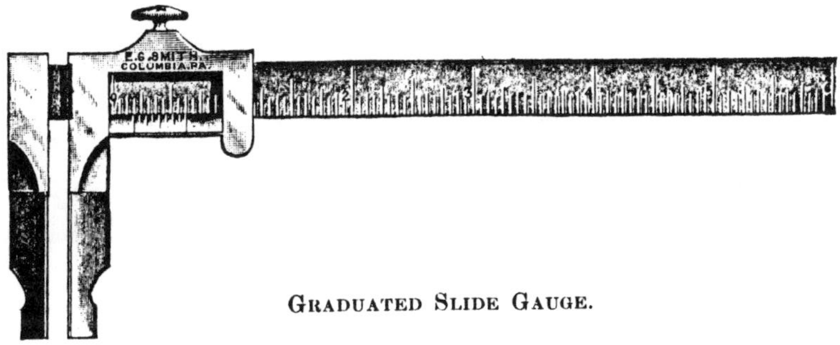

GRADUATED SLIDE GAUGE.

1892

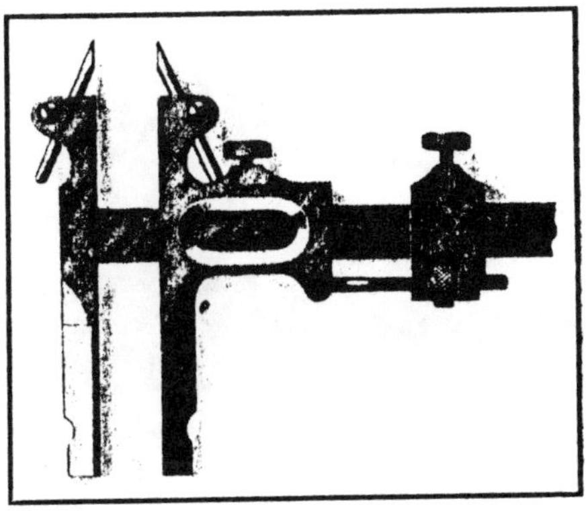

1905

IMPROVED CALIPER.

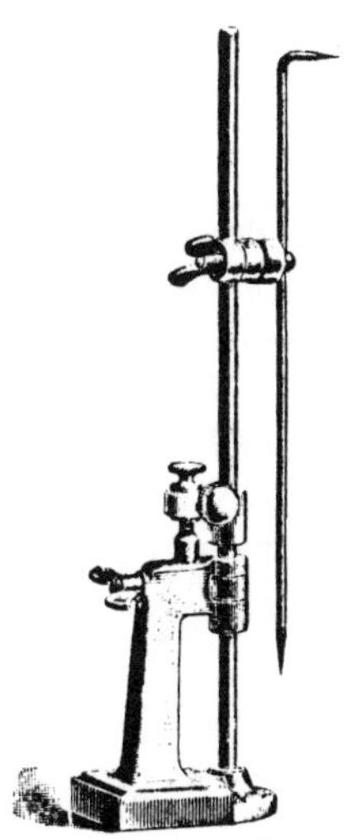

NEW SURFACE GAUGE.

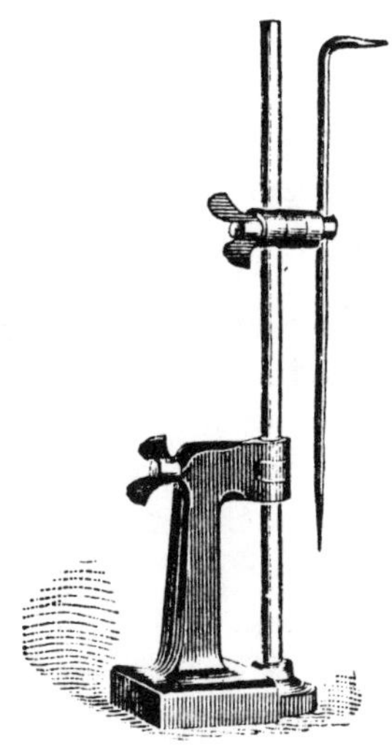

An Extension Surface Gauge.

1891

All from American Machinist

E.G. Smith

1910

Views showing Various Applications of Universal Angle Gage

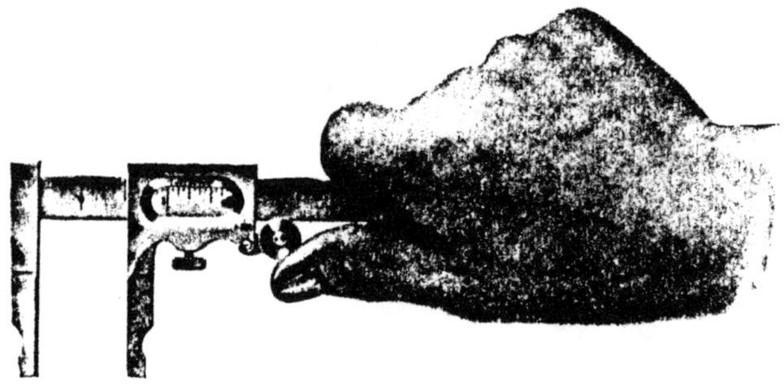

1909

A Friction Roller Attachment for Making Fine Adjustments on the Beam Caliper

SMITH, WILLIAM W. Chicopee, MA

Inventor and probably maker of spring calipers patented October 1, 1850 (No. 7689). His patent was witnessed by Joshua Stevens and E.V.B. Holcomb, both of whom were working at the Mass. Arms Co., Chicopee Falls, MA.

SOMERS, J.L.& C.A. Brooklyn, NY

Makers of an odd combination caliper patented December 12, 1865 (No. 51517) by Benjamin G. Martin of Philadelphia, PA. The caliper combines "arms" for inside measurements and "legs" for outside measurements with a registering scale common to both inside and outside measurements. The odd part is a pointed bar freely swinging on the caliper pivot. The bar has a pointer and a protractor scale and when the caliper is placed on its extended flat feet, the swinging bar becomes an angle indicator or pendulum plumb. When not in use the heavy German silver head of the swinging bar is held fast by a small spring clip.

The firm is reported to have also made spring calipers.

SPALDING, FRANK (also spelled Spaulding in some records) See VICTOR SEWING MACHINE CO. and BROWN & SHARPE MFG. CO.

SPENCER, WILLIAM D. Middletown, CT

Maker of machinist's depth gauges for which he received a patent, March 13, 1888 (No. 379423). The tools are very well made.

SPRINGFIELD LEVEL & TOOL CO. Springfield, MA

Listed in the Springfield city directory from 1895 to 1907 as "manufacturers of spirit levels and machinists tools". Nothing else is known at this time.

SPRINGFIELD MACHINE SCREW CO. Springfield, MA

Makers, in 1900, of a "cheap" spring caliper which is similar, in many respects to the J. STEVENS CO. coil spring caliper patented in 1880.

Diamond Pocket Inch Caliper.

The Springfield Machine Screw Company, Springfield, Mass., are putting on the market the caliper shown herewith. It is made from a high grade spring steel, finely tempered, making, it is remarked, an attractive and

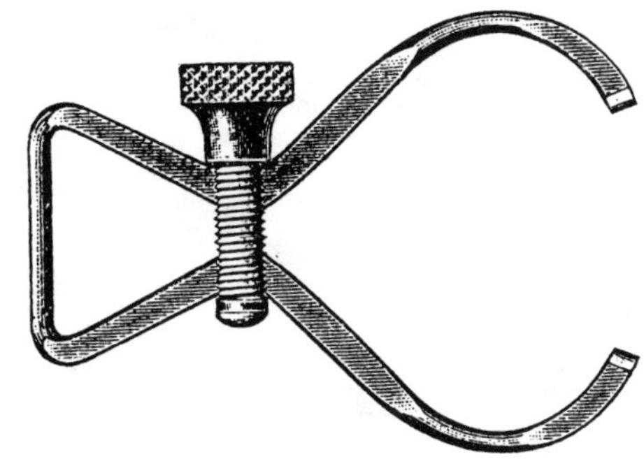

Diamond Pocket Inch Caliper.

cheap spring caliper as well as being very handy.

SQUIER, J.B. Boston, MA

Maker, about 1890, of a machinist's taper gauge calibrated on one side in 64ths and on the other side in 100ths.

STAAB, S. Baltimore, MD

Maker of inside micrometers marked "PAT PENDING". Period of operation is unknown.

STAFFORD & WHIPPLE Hopedale, MA

Founded by Benjamin E.D. Stafford and Fenner E. Whipple who had jointly patented a combination surface and planer gauge on December 16, 1886 (No. 352584). In 1887, the tool was manufactured and advertised by ARTHUR B. PHILLIPS. By 1904 Stafford & Whipple had formed a partnership and were making and selling the gauge under their own name. See PHIILLIPS, ARTHUR B., for an illustration of the gauge.

STANLEY RULE & LEVEL CO. New Britain, CT

This well known firm made a wide variety of tools, mostly for woodworkers. However, some machinist's tools were offered. It was the maker, from 1863 to 1879, of the #43 machinist's plumb and level and, from 1867 to 1892, of Nicholson's patent levels, patented May 1, 1860 (No. 28104). A gear cutting rule patented by Charles B. Long on June 20, 1865

(No. 47436), was offered in its 1870 catalog. Stanley also made the "Eclipse" #34 and 34V levels offered from 1897 to 1984 the #38 and #39 machinist's levels from 1877 to 1897 and the #38½ and 39½ levels from 1898 to 1958. Its "Improved Trammel Points" for "millwrights, machinists, carpenters, and all mechanics" were offered as early as 1872. The #5 and #6 machinist's trammel points were made from 1909 to 1935. The 1924 catalog offered machinist's trammel points, machinist's "Eclipse" levels and "Victor" iron levels.

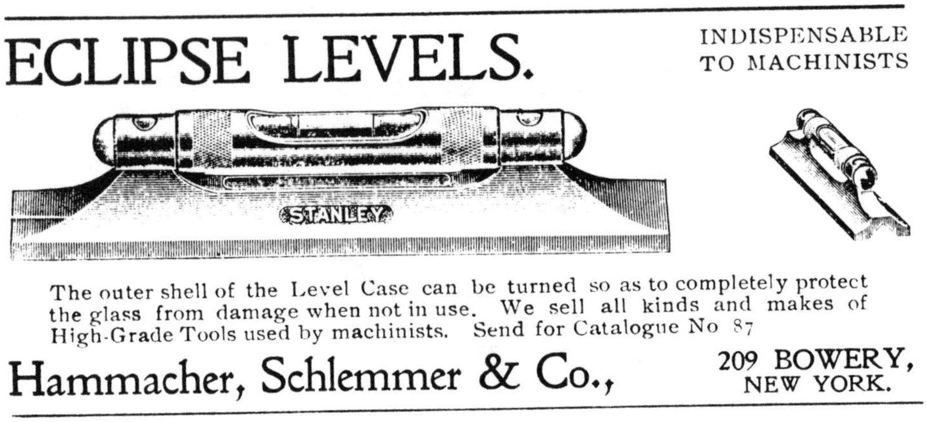

Machinery 1900

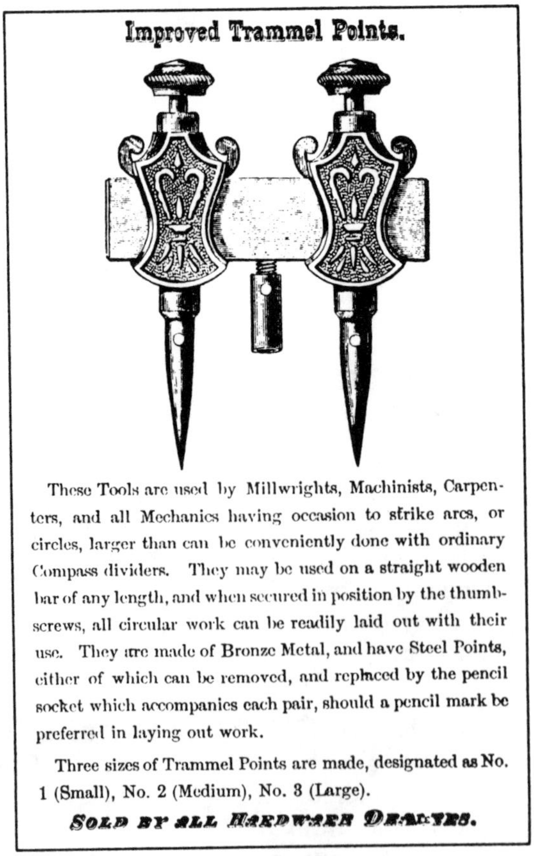

Stanley Pocket Catalogue, dated 1872

STAR TOOL CO. Middletown, CT

Formed about 1867, the firm was the maker of calipers and dividers patented February 22, 1876 (No. 173802) by T. McDonough and of machinist's squares marked with the words "TOOL CO" inside a five pointed star. It also offered a wide selection of steel machinist's rules in an 1880 catalog.

STAR INSIDE CALIPER.
No. C 409

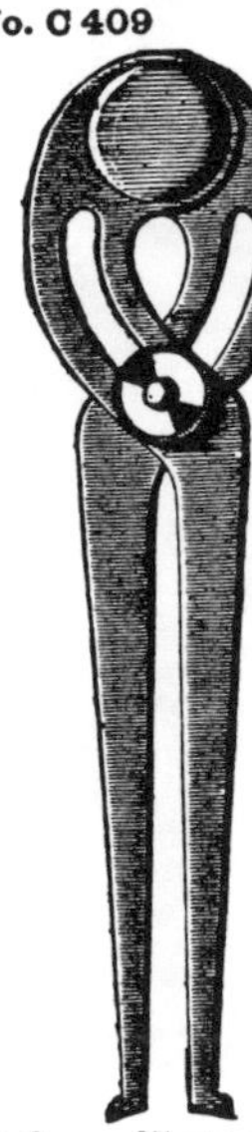

3 inch....65 cts. each.
5 " ....85 " "

STAR OUTSIDE CALIPERS.
No. C 410

3 inch .... 65 cts. each.
5 " .... 85 " "

STAR DIVIDERS.
No. C 411

3 inch..65 cts. each.
5 " ..85 " "

1884

STANDARD STEEL RULES.

Length,	1	2	3	4	6	9	12	18	24	36 in
Price,	$0.25	35	40	60	85	1 25	1 60	2 50	3 50	7 50 each.

The rules in this list are divided five ways in parts of inches as follows:

	No. 1 Graduations.	No. 2 Graduations.	No. 3 Graduations.	No. 4 Graduations.	No. 6 Graduations.
1st cor.	10, 20, 50, 100	10, 20, 50, 100	16, 32, 64	64	32 whole length.
2d cor.	12, 24, 48	12, 24, 48	16	32	48 "
3d cor.	16, 32, 64	16, 32, 64	16	16	50 "
4th cor.	14, 28	8	8	8	64 "

12 in. Steel Rule of No. 5 graduation...$2 50
24 in " " " 5 00

No. 5. Graduations—1st cor. 16, 32, 64
2d cor. 11, 14, 15, 17, 18, 19, 20, 21, 22, 23, 24, 25
3d cor. 26, 27, 28, 29, 30, 31, 33, 34, 35, 36, 37, 38
4th cor. 39, 40, 41, 42, 43, 44, 45, 46, 47, 48, 49, 50, 100

1880

STARK TOOL CO. Waltham, MA

Founded by John Stark in 1862, with small machine tools as the primary product. In 1912, they advertised themselves as makers of fine tools.

STECKENREITER MFG. CO. Chicago, IL

Makers, in 1903, of a "speed measuring machine" which was a small hand-held instrument for measuring the surface speed of a rotating part.

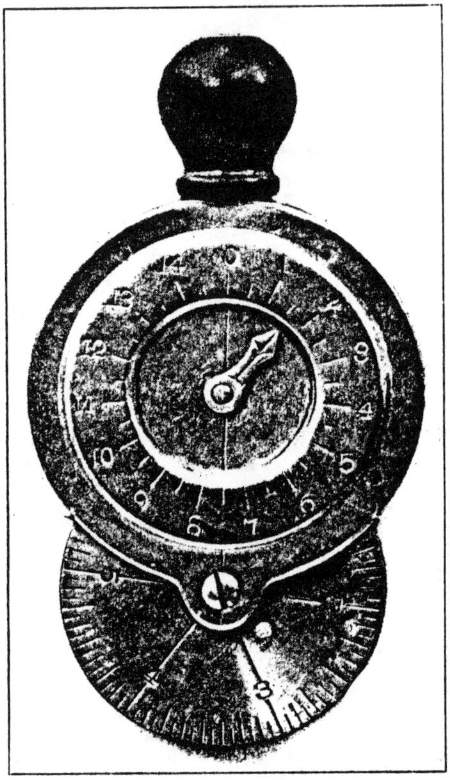

1903

A SPEED MEASURING MACHINE.

STERLING WATCH TOOL CO. Rochester, NY

Advertised, in 1901, a "watch charm" micrometer which measured up to 3⁄16". It was gold plated to wear as a watch charm but was claimed to be accurate and usable as a micrometer.

STILES & POLLARD Middletown, CT

A partnership of Norman C. Stiles and Charles E. Pollard, the firm was an offshoot of THE STILES & PARKER PRESS CO., Middletown, CT, where Pollard was superintendent. It was a maker of adjustable machinist's blocks introduced in 1889. The blocks were patented March 9, 1889 (No. 399926) by Charles E.Pollard. *(continued next page)*

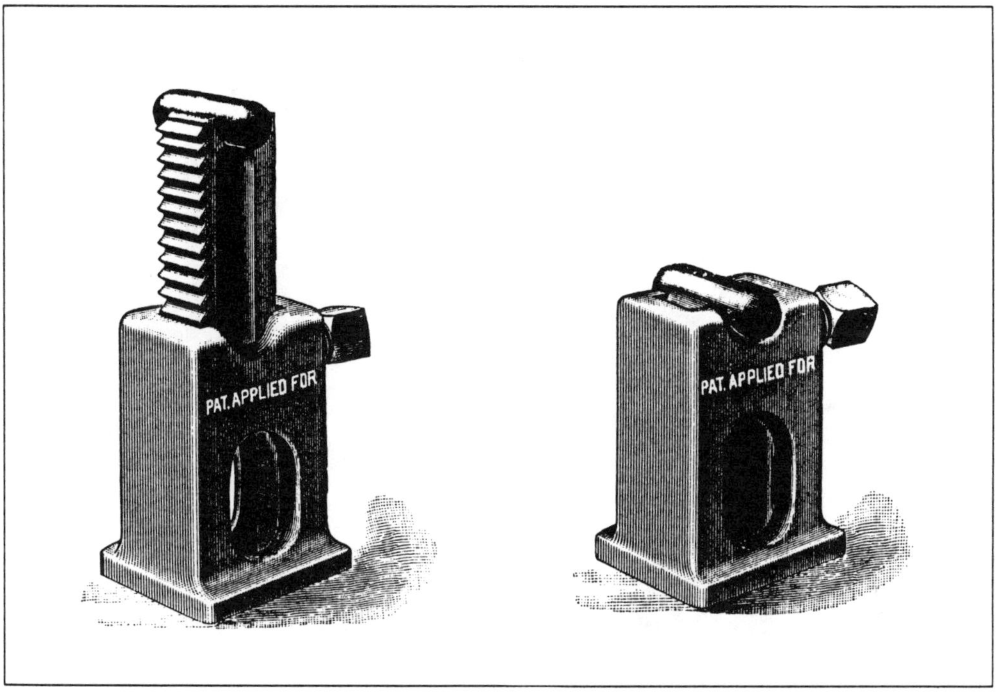

American Machinist 1889

SUPERIOR INDICATOR CO. Rochester, NY

Maker of a machinist's test indicator similar to the "Ideal" type but with the needles enclosed in clear plastic bubbles. The tool was probably made in the 1930's or 1940's.

SWEET, JOHN E. Syracuse, NY

John Edson Sweet (1832-1916) was a professor at Sibley College, Syracuse, NY and has been credited with developing the standard caliper gauge in the mid 1870's. Various types of home-made caliper gauges had been made for a long period. Sweet, however, appears to have been the first to develop standard instruments intended as an article of manufacture. In 1877, Sweet developed a 12" measuring machine to accurately set the gauge sizes. He patented the machine on March 10, 1885 (No. 313552) and it was built and sold by the SYRACUSE TWIST DRILL CO. from 1883 until as late as 1910. Sweet was also president of the STRAIGHT LINE ENGINE CO., Syracuse, NY, from its formation in 1880 until his death on May 8, 1916.

SYRACUSE TWIST DRILL CO. Syracuse, NY

Maker of bench micrometers (Sweet's Measuring Machine), patented March 10, 1885 (No. 313552) by John E. Sweet. Production began in 1883 with an improved version introduced in 1885. The same units were advertised as late as 1910. In 1885, the company introduced a new comparator for checking dimensions against standards, patented by Jakob Hurst on November 30, 1886 (No. 353561). *(continued next page)*

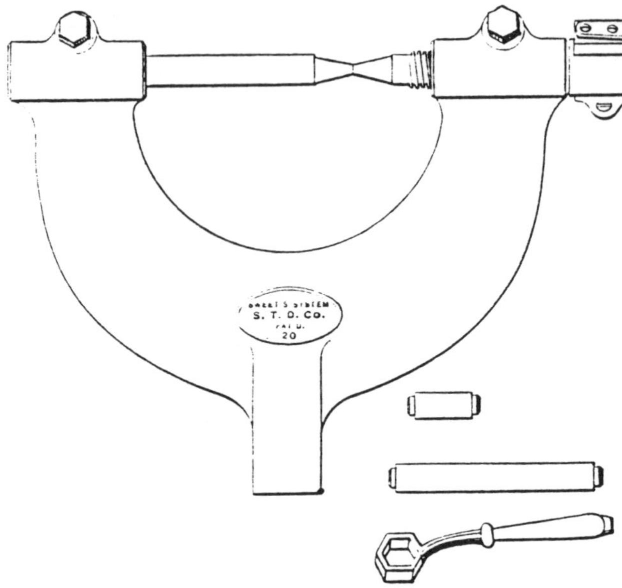

Fig. 1.

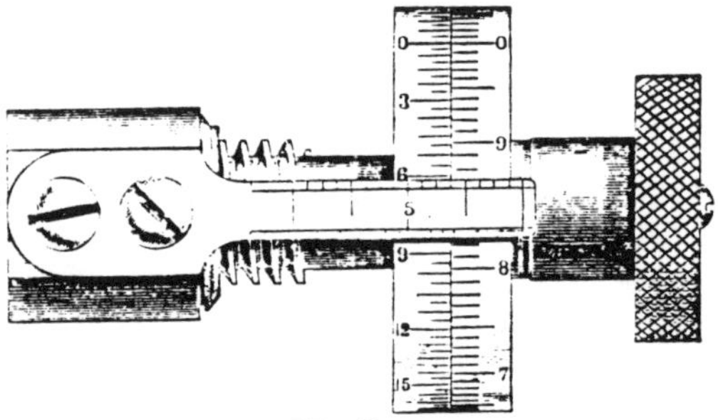

Fig. 2.

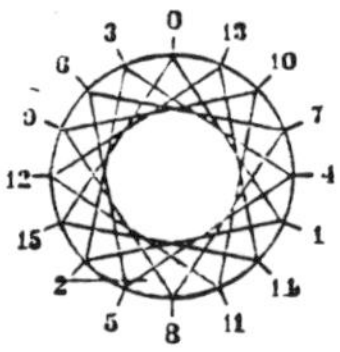

Fig. 3.

American Machinist 1885

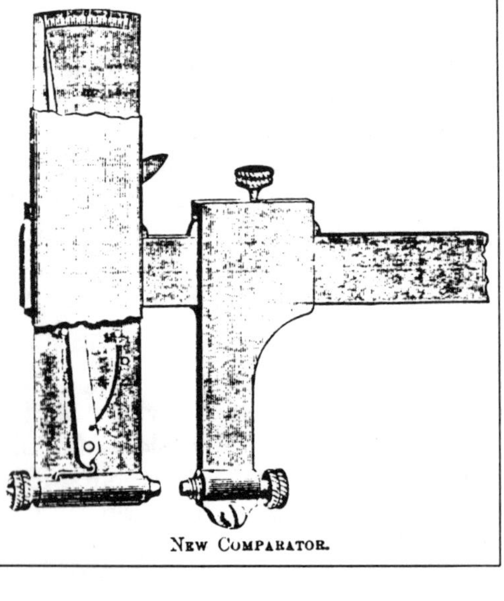

New Comparator.

American Machinist 1885

1897

TABOR MFG. CO. Philadelphia, PA

Founded by Harris Tabor (1843-1907) in 1885 when he resigned as superintendent of the WESTINGHOUSE MACHINE CO., Pittsburgh, PA, and began making moulding machines of his own design. His company was also the maker of the Tabor revolution counter introduced in 1886. The counter was patented on September 8, 1885 (No. 325991) by Harris Tabor and was made by the TABOR MFG. CO. until the product line was sold to the GRANT MFG. &MACHINE CO., Bridgeport, CT, in 1907.

1886

TAFT-PIERCE MFG. CO. Woonsocket, RI

Maker of a variety of precision measuring tools. Its catalog #10, issued in 1921, offered sine bars, taper test gauges, V-blocks, parallels, universal right-angle irons, measuring irons, bench plates, and universal squares. The following tool patents were assigned to the company:

William Simmons, universal right-angle iron, July 27, 1915 (No. 1148196)
William Commons, surface plate, October 26, 1915 (No. 1157762)
Elmer Bryant, limit-gauge, September 24, 1918 (No. 1279645)
Elmer Bryant, sine-bar fixture, May 13, 1919 (No. 1303633)
Elmer Bryant, plug-gauge, October 28, 1919 (No. 14743, re-issue)
Elmer Bryant, precision square, October 19, 1920 (No. 1355808)
Elmer Bryant, toolmaker's adjustable knee, October 19, 1920 (No. 1355809)
Elmer Bryant, parallel for machine tools, July 5, 1921 (No. 1383979)
Albert Martell, plug-gauge, October 11, 1921 (No. 1393233)
Elmer Bryant, ring thread gauge, October 31, 1922 (No. 1433528)
Elmer Bryant, adjustable limit gauge, March 13, 1923 (No. 1448356)
Elmer Bryant, clamp block, April 28, 1925 (No. 1535570)
Albert Martell, adjustable plug gauge, September 22, 1925 (No. 1554226)
(see illustration next page)

JAMES L. TAYLOR MFG. CO. Bloomfield, NJ

This company issued a pamphlet in 1905 describing the LaFrancis patent combination drill block and gauge (patented byAnthony La Francis, January 30, 1906) and the Taylor quick-adjusting screw clamps.

UNIVERSAL SQUARES

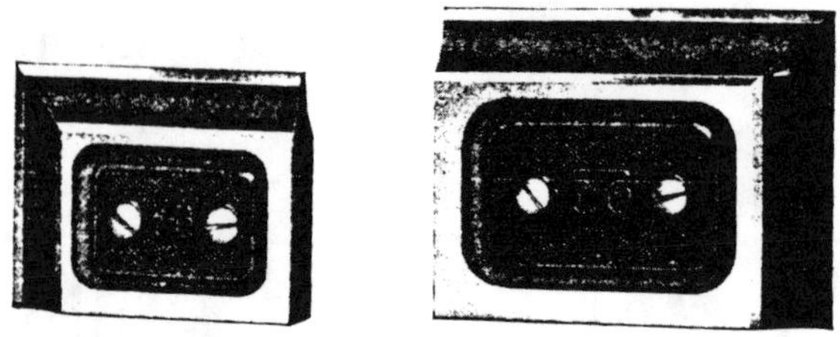

Taft-Peirce Universal Squares are designed to fill the need for a high precision tool of this character.

Made from alloy tool steel drop forgings, each tool is hardened, ground and lapped.

The rectangular construction with two knife edges and two flat sides finished to a high degree of accuracy, makes a variety of applications possible.

The center of the square is covered with a hard rubber pad providing an excellent grip, and preventing heat transference in handling.

1929

TAYLOR-RICE ENGINEERING CO. Wilmington, DE

Successors to AMERICAN STANDARD GAUGE & TOOL WORKS in 1895. It continued to make the same standard gauges for a number of years.

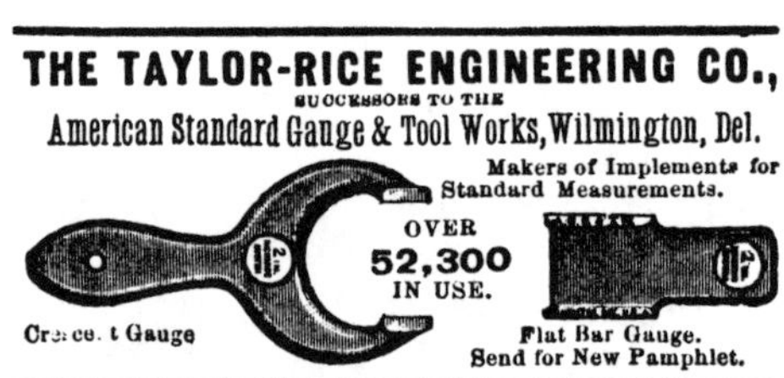

1895

TAYLOR-SHANTZ CO. Rochester, NY

Maker of depth gauges, dial gauges, surface gauges, and snap gauges in 1929. Little is known of the company or the products.

THEWLIS & CO. Boston, MA

Maker of small machinist's squares. The example which has been examined appears to have been made c1880-1890.

TICKELL & DYESON Cleveland, OH

Maker, in 1888, of an inside micrometer patented March 5,1887 (No. 359295), by John Tickell. The company advertised it as available in "sizes from 2" to 40". Only the small sizes are presently known. *(continued next page)*

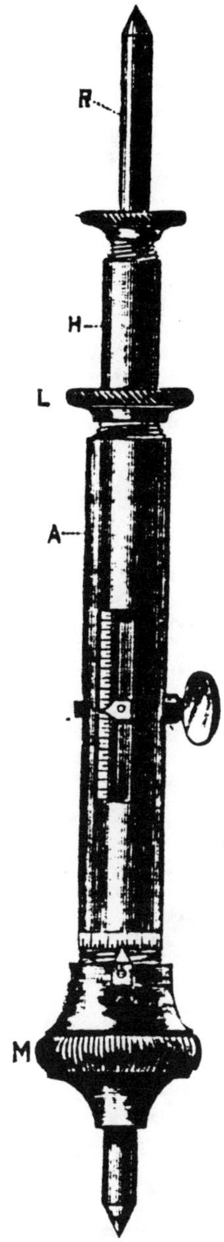

1888

Tickell's Inside Micrometer Gauge.

TIP TOP TOOL CO. Worcester, MA

Maker, in 1907, of "a new adjustable size block or packing gauge". The tool is very similar to a planer gauge. *(continued next page)*

New Adjustable Size Block or Packing Gauge.

The adjustable size block or packing gauge shown in the illustration is a new type of mechanics' tool designed for use with planer, milling machine, boring mill and lathe, and, with other machine tools, its special purpose being to procure accurate adjustment of the cutting tool, where a packing is to be planed into a lathe rest, planer head, saddle or any of the many other places where inside packing is used. The taper base has on its upper surface a 60-degree dovetailed slot, in which are movable for its whole length the step piece and the adjusting nut bracket. The step piece is broached out to contain the clamping nut, under which is a spring washer to give the necessary slight friction. The adjusting nut bracket carries a clamping nut with spiral spring inside to give the friction. The adjusting screw is fastened to the step piece, the adjusting nut working upon it. By tightening the clamping nut of the bracket and loosening that of the step piece the adjusting nut can be used for procuring fine adjustments, one turn of the nut giving a vertical movement of 5-1000 in. By loosening both clamping nuts the step piece may be moved instantly to any position on the base. If a packing is to be planed into a part the required hight of the cutting tool may be obtained on the first chip. A second trial is unnecessary, nor is it possible to take off too much metal and spoil the work. In setting two roughing heads in a double head planer much time wasted in running the saddle back and forth for the adjustment of the second tool may be saved by the use of the size block. If the work is higher than the range of the block it is placed on a parallel. If a number of pieces of one thickness are to be planed separately, and, as often happens, the size is odd, or the pieces too high for standard size blocks, the adjustable size block, placed on a parallel of sufficient hight, will accomplish the desired adjustment of the tool. If a piece is to be bored out to some odd measurement and must leave the mill accurately sized, the gauge is set to the desired distance from table to boring bar, the machine is set so that the gauge just slides under the bar and the result must be correct. In setting straddle milling cutters for any required distance apart on small work the tool is convenient. Lathe hands find it useful where they have much squaring to do for accurate inside measurements in narrow places. The range of adjustment by movement of the dovetailed slot is from 3-16 in., the smallest, to 1½ in., though, as already stated, the scope of the tool may be increased indefinitely by the use of parallels. It will be seen that all the adjusting parts are below the top surface. All important surfaces are ground, and the step piece is hardened and ground parallel with the bottom surface. The tool is manufactured by the Tip Top Tool Company, 78 Vernon street, Worcester, Mass.

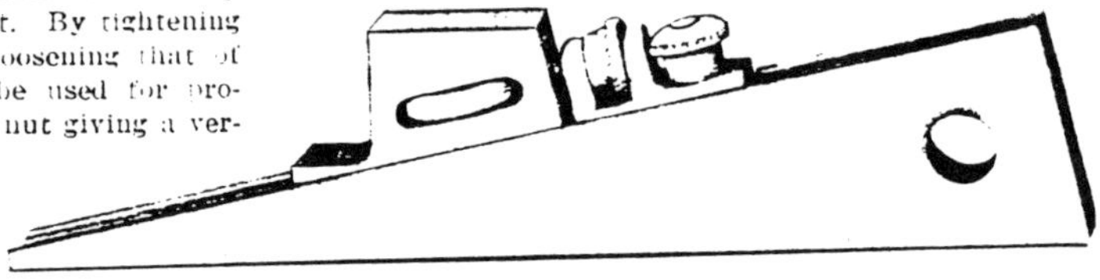

A New Adjustable Size Block or Packing Gauge.

Iron Age, Nov. 14, 1907

THOMPSON, W.L., Springfield, IL

Maker, in 1884, of a machinist's depth gauge "composed of a graduated tube and a flat base— a disk works in the tube, carrying a measuring wire, which projects through the base and also carries a pointer, which indicates the depth measured. The spring on the side of the tool acts as an automatic friction stop, holding the measuring wire in any position that it may be forced to while measuring." The range was $1\frac{1}{2}$" and the gauge was calibrated in 100ths of an inch on one side of the slot and 64ths of an inch on the other. *(continued next page)*

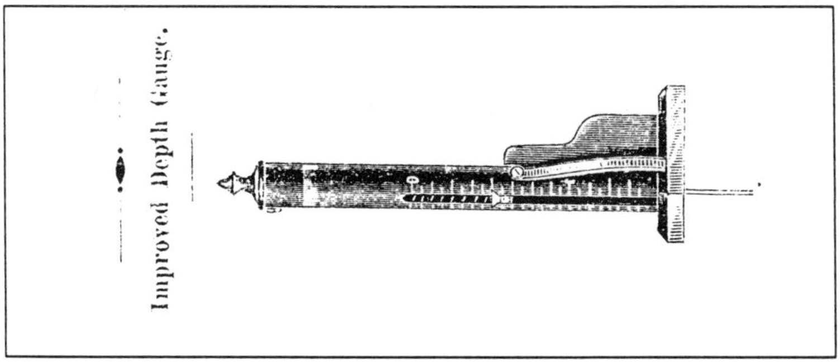

American Machinist 1884

TUBULAR MICROMETER CO. St. James, MN

Maker of "TUMICO" micrometers, vernier height gauges, depth gauges and a variety of small tools for machinists beginning about 1946. The company purchased the former REED SMALL TOOL WORKS micrometer line in 1946. Boxes with the Tubular Micrometer Co. label pasted over a REED SMALL TOOL WORKS label are fairly common. About 1960 it merged with GEORGE SCHERR CO., an importer of machinist's tools, and became SCHERR-TUMICO CO.

ULSTER MACHINE CO. Providence, RI

Founded about 1885 by R.D. Williams who served as president. The firm was the maker, in 1882, of a revolution counter patented September 26, 1882 (No. 264900) and, in 1885, of adjustable calipers patented October 20, 1885 (No. 328861). Both patents were issued to William G. Richards. In 1892, the same revolution counter was advertised by the A.B. PITKIN MAC'Y CO. Even the ad copy was identical except for the name of the company. This may have been a company name change only.

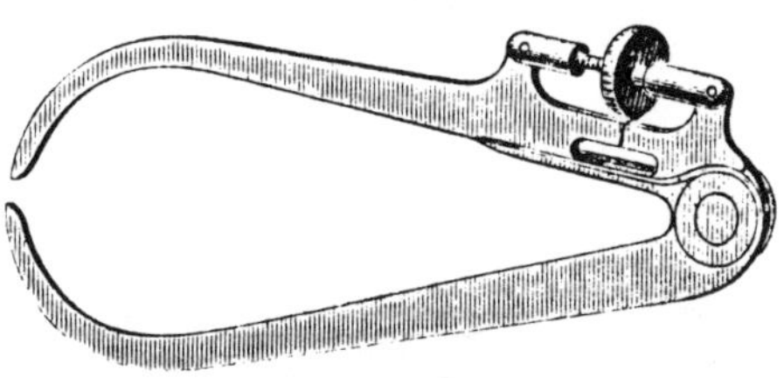

ADJUSTABLE CALIPER.

American Machinist 1885

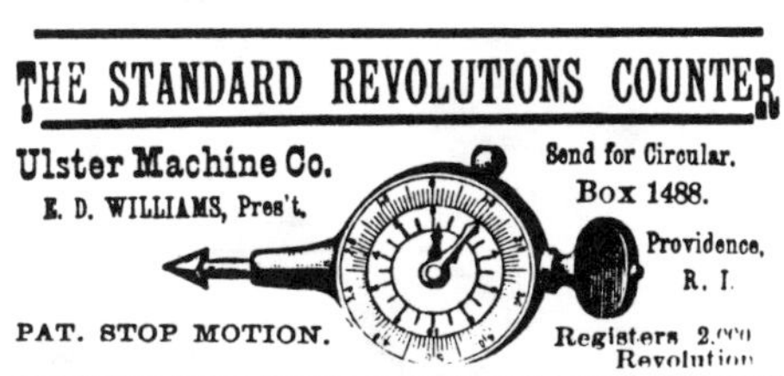

1885

UNION CALIPER CO. Fitchburg, MA, later, Orange, MA

Founded by Emory E. Ellis (?-1924) on May 28, 1908, to make a line of machinist's calipers. Ellis had been superintendent of the SAWYER TOOL MFG. CO., Fitchburg, MA, in 1901 and became a director and part owner when Sawyer sold out in 1902. Ellis apparently left the SAWYER TOOL MFG. CO., when it was sold to Carl A. Hubbel in 1908. Ellis moved the company to Orange, MA, in March, 1911. In 1913 the UNION CALIPER CO. purchased the BATES MFG. CO. of Fitchburg, MA, adding the Bates line of machinist's rules to its products. It continued to use the Bates name for some time as rules, combination squares, and thickness gauges marked BATES MFG. CO., ORANGE, MASS. are quite common.

Its 1915 catalog listed the following machinist's tools: a variety of calipers and dividers basically the same as the STARRETT "Yankee" calipers; firm joint calipers, a unique quick adjusting line of calipers and dividers based on the Emory Ellis patent of June 30, 1914 (No. 1102146, assigned to Union Caliper Co.), nail sets and punches, thread gauges, thickness gauges, a line of steel rules, key seat rule blocks, combination squares, and hack saw frames. The catalog also listed Hill pattern drill holders and machine vises, stating that they were the same as formerly manufactured by the Hill-Standard Mfg. Co. of Anderson, IN, which it had purchased in May, 1911.

Company officers in 1915 were: Emory Ellis, president; James C. Bates, general manager, and after his death Herbert H. Eames; and Kathleen Jackson, secretary. James C. Bates was probably the same man as J. Clayton Bates who started the BATES MFG. CO. in 1913. In 1916 the name of the company was changed to:

UNION TOOL CO. Orange, MA

The 1916 catalog offered calipers, steel rules, combination squares, tap wrenches, reamers, gauges, clamps, slide caliper rules, punches, scribers, trammels, center finders, and protractors.In 1930 the company had capital of $400,000 and 100 employees. H.A. Ellis was president, replacing E.E. Ellis who had died in 1924, and P.E. Pederson was secretary. The company eventually became a subsidiary of the MILLERS FALLS CO., Greenfield, MA, and was sold off in 1975. *(continued on next two pages)*

UNION BEVEL PROTRACTOR
No. 610

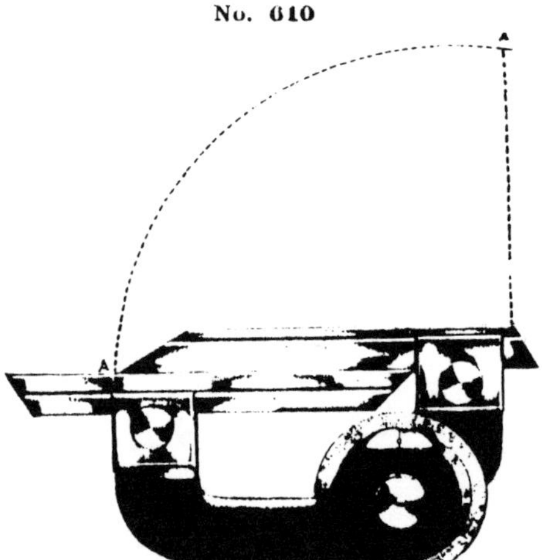

Shaver Patent

FIGURE No. 1 shows the Protractor closed, ready to be opened from A to A, describing all acute angles from 0 to 90 degrees. Please note this is done without the aid of any auxiliary attachments, also that this is the only Protractor offered which is so constructed that the blade will close down flat on the base or work surface.

This Protractor is indeed a universal instrument, and stands for all that the word "universal" means.

Years of study by competent designers have contributed to the perfection of this very fine tool, and the mechanical profes-

No. 610
Shaver Patent

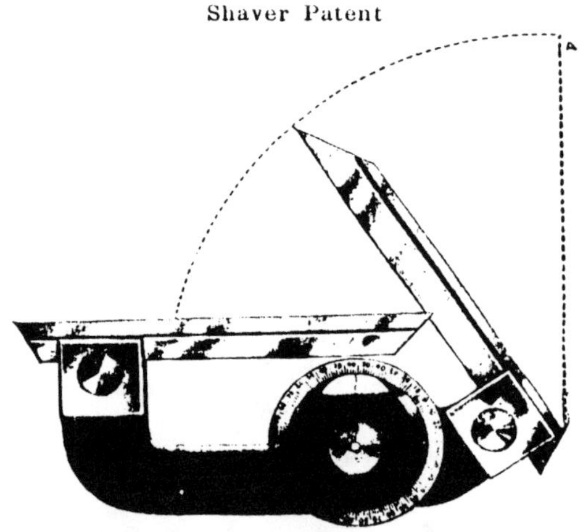

FIGURE No. 2 shows the Protractor opening from position shown in Figure No. 1. Note how the adjustable base follows up the the blade, completing any angle from the starting-point at the base line to the first 90 degrees.

sion will appreciate the ingenuity shown in its construction.

This Bevel Protractor is complete in itself. It needs no auxiliary attachments.

As furnished, the body is made of tough cast metal. Blade and base clamps are steel hardened. Six-inch adjustable base and 6-inch blade are both hardened.

UNION COMBINATION SET

No. 611

The most complete Combination Set ever offered. Made with our 12-inch Combination Square No. 600, with hardened blade, Square Head and Center Head, Scratch-awl, Level, all complete, together with Bevel Protractor No. 610, shown and described on Pages 54, 55, 56, 57, 58, and 59 of this catalogue. The Protractor is reversible and universal in all senses of the word. In considering the purchase of a Combination Set, the mechanic who wants the better kind cannot afford to consider any other tool.

Weight each, 39 oz. $8.00

UNION DIE-MAKERS' PROTRACTOR AND DEPTH GAUGE

No. 333

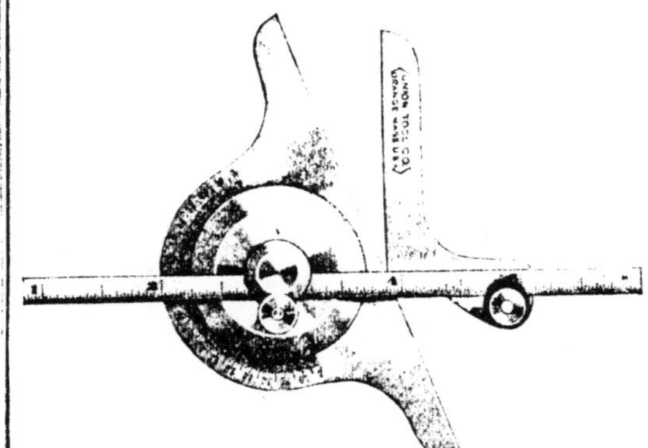

A new type Bevel Protractor, having many new features that particularly adapt it to tool and die work. The design is neat, accurate and light, making it sensitive on delicate particular work.

The base or work surface is 4 inches long, accurately ground, and in it is mounted a 2 inch graduated turret, rigidly locked at will in any desired position by a draw bolt and nut.

Through a 3-16-inch milled slot in the turret, slides a narrow, tempered, graduated rule, which may be locked in any position with the positive taper-locking device.

The turret and rule locking devices operate independently of each other, which permits the instrument being used as a depth gauge at a right or any other angle, making this tool valuable for use on die and tool work in testing the draft of dies and patterns, or work of a like nature.

An auxiliary arm that may be removed at will is adjustably mounted on the rule, making an excellent taper gauge for accurately measuring tapers and all acute angles.

Weight each, 5 oz. $5.00

ILLUSTRATING POSITIVE TRANSFER FEATURES OF UNION QUICK-ADJUSTING SPRING CALIPERS AND DIVIDERS

This positive transfer feature, mechanics will find very valuable for the measurement over flanges, thickness of panels, internal chambers having flanges which must be measured over, or in fact any place where accurate transfer of sizes must be taken over flanges or shoulders.

To operate, just take the size as you would with any ordinary calipers, getting it exact, note the notch in which the adjusting bar is set, raise the adjusting bar, remove the calipers from the work and set it back to the notch noted. A very simple, positive process. No danger of mistakes.

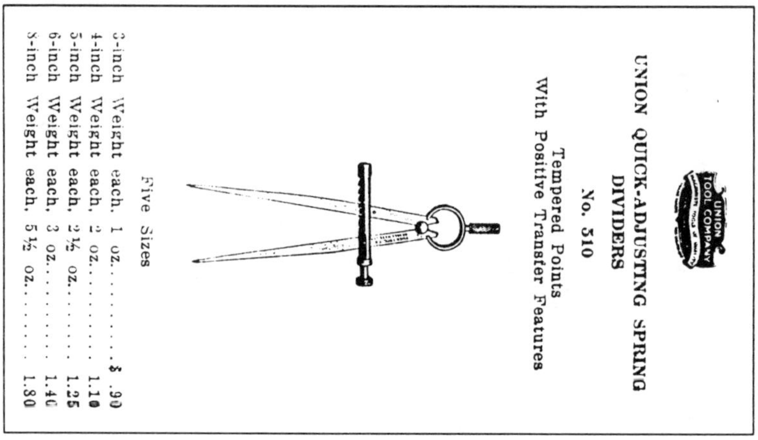
UNION QUICK-ADJUSTING SPRING DIVIDERS
No. 510
Tempered Points
With Positive Transfer Features

Five Sizes

3-inch Weight each, 1 oz.	$ .90
4-inch Weight each, 2 oz.	1.10
5-inch Weight each, 2½ oz.	1.25
6-inch Weight each, 3 oz.	1.40
8-inch Weight each, 5½ oz.	1.80

UNION TOOL CHEST CO., INC. Rochester, NY

Maker of machinist's tool chests from c1919 to c1946.

UNION TOOL MFG. CO. Chicago, IL

Maker, in 1904, of a universal rule slide patented January 16, 1904 (No. 750204) by R.A. Lachmann. The rule slide could be used to convert a steel rule into a trammell, caliper setting gauge, height gauge, or beam caliper. *(continued next page)*

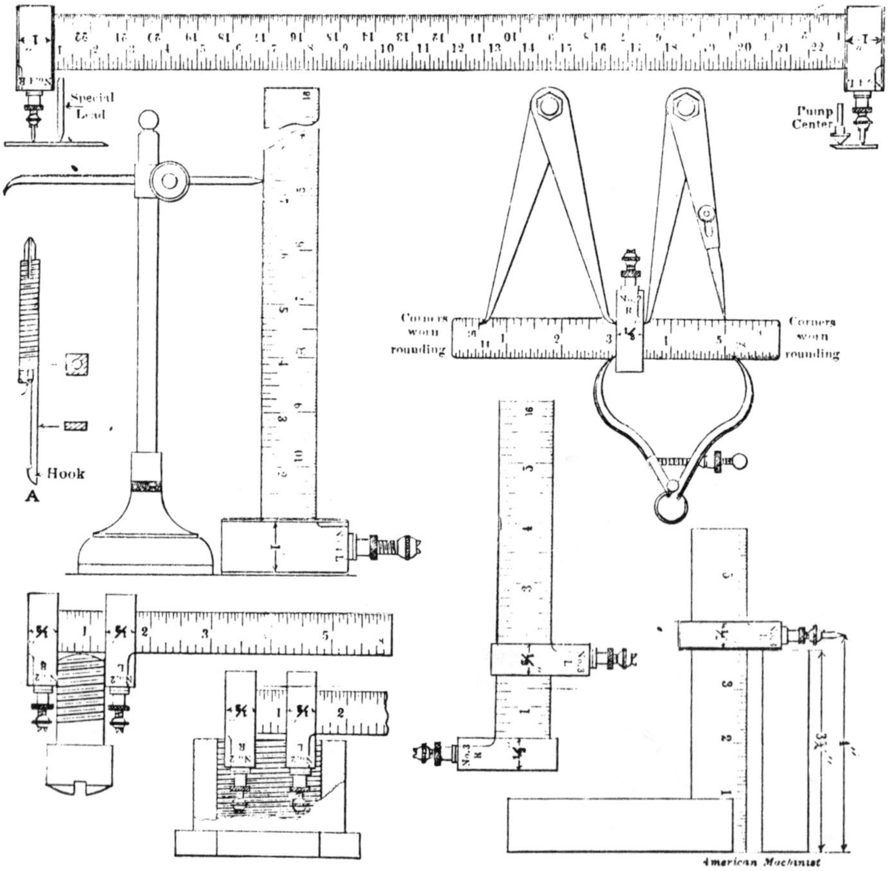

TYPICAL APPLICATIONS OF UNIVERSAL RULE SLIDE.

1904

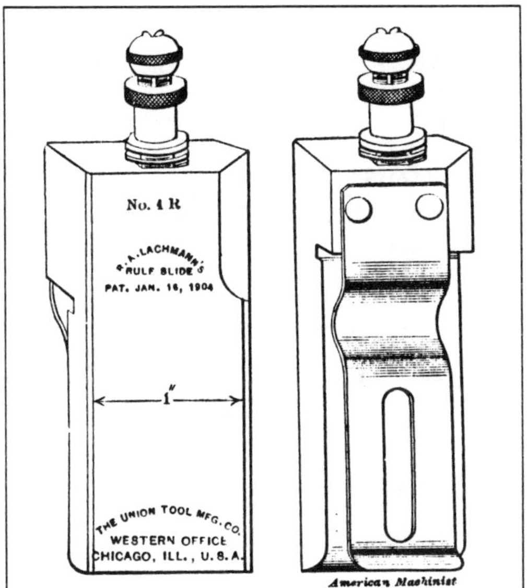

FIG. 1. UNIVERSAL RULE SLIDE.

1904

U.S. LOCK CO. Boston, MA

Maker of a 2" caliper square, probably in the 1890's. Only a single example is now known.

UNIVERSAL BORING MACHINE CO. Hudson, MA

Founded by Charles A. Clarke in 1907 to make large horizontal boring machines. In addition to the machine tool line the company offered a line of very well made precision machine aligning levels from c1921 to 1929.

USHER , JOHN T. Elmira, NY

Inventor and maker of the first inside micrometer. He received his patent August 24, 1886 (No. 347844).

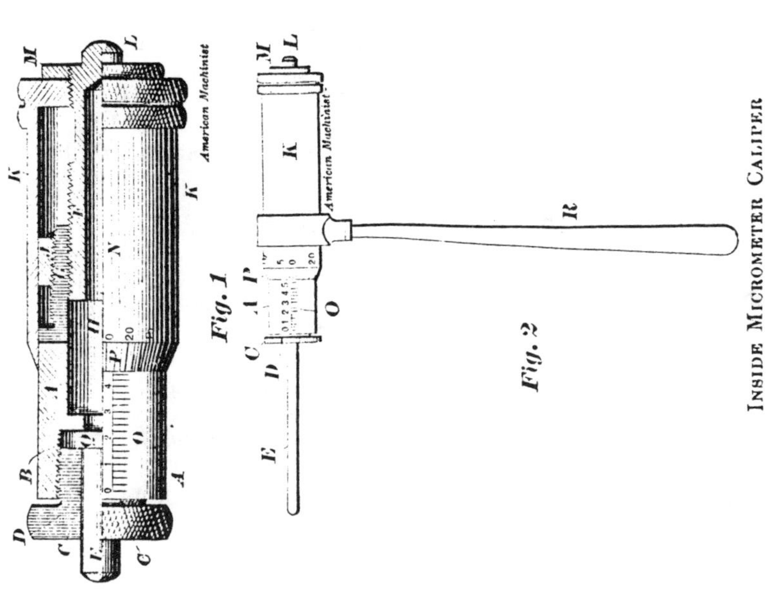

1886

VAN KEUREN CO. Boston, MA

Its 1922 catalog offered gauge blocks, plug gauges, measuring wires, and lapped surface plates. By 1936, it was also offering micrometers.

VEEDER MFG. CO. Hartford, CT

Introduced a new speed indicator in 1908. This same unit was made unchanged until the 1960's. The company name, however, had been changed to the VEEDER-ROOT CO.

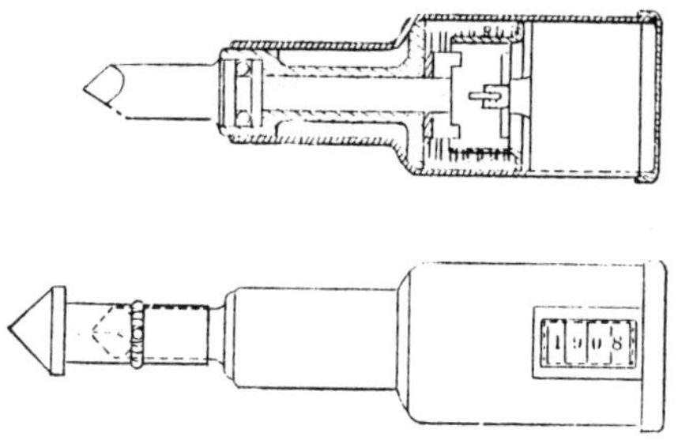

A NEW SPEED INDICATOR

1908

VICTOR MFG. CO. Athol, MA

Maker, in 1885, of a bevel protractor patented June 2, 1885 (No. 318984), by Jaspar J. Hedges. The firm, or at least the product, was taken over by the STANDARD TOOL CO., Athol, MA, about 1890. (illustration dated 1889)

VICTOR BEVEL PROTRACTOR.

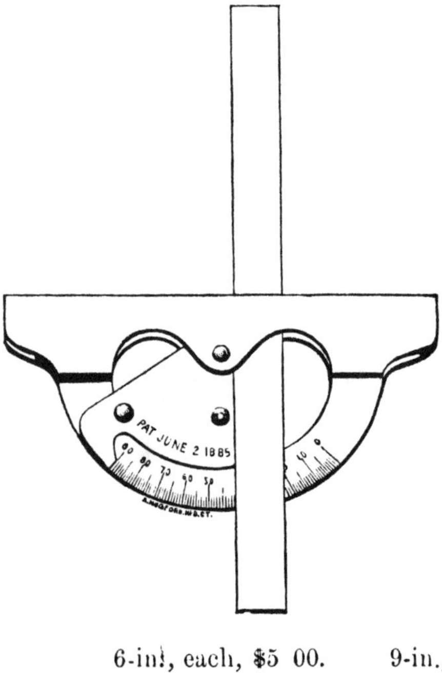

The blade swings between the two working faces of the head, or stock, and when clamped at any degree the angle may be set off to the right or left, without any further adjustment, by simply turning the tool over.

By means of the index point and divided quadrant the blade may be set at any angle from 0 to 90, and firmly clamped.

The blade slides its full length on the working edge of the sector, thus giving a long bearing surface and not liable to wear. The blade may also be drawn back, leaving the face of the head clear for use on surface plates and on platens of planers for setting the head at any angle—a very valuable feature. An extra blade three inches long, and brought to a point by bevelling off one side at an angle of 30 degrees, is included with the tool, and is very convenient for inside dove tailing. Instead of setting the index point at the centre of the division mark this tool is adjusted so as to show the full size of the mark forward of the index. The quadrant is plainly and correctly divided to degrees. This is a practical tool, made entirely of steel, well finished all over, convenient in size, weighing only four ounces, durable in every part, and sold as low as the quality of the tool will admit.

6-in., each, $5 00. 9-in., each, $5 50. Nickelling, 75c. each extra.

VICTOR SEWING MACHINE CO. Middletown, CT

Formed to make sewing machines, the company began making micrometers about 1878. The micrometers were covered by the George M. Pratt patent of April 23, 1878 (No. 202866) for a "U" shaped frame and inserted hardened points. This was one of the key micrometer patents as it had proven impossible to harden points which were integral with the micrometer frame.In 1883 the business was closed. The BROWN & SHARPE MFG.CO. of Providence, RI, bought its micrometer business including all stock and patents. Frank Spalding, who had been in charge of the micrometer business for Victor, went to BROWN & SHARPE to head up their micrometer department.

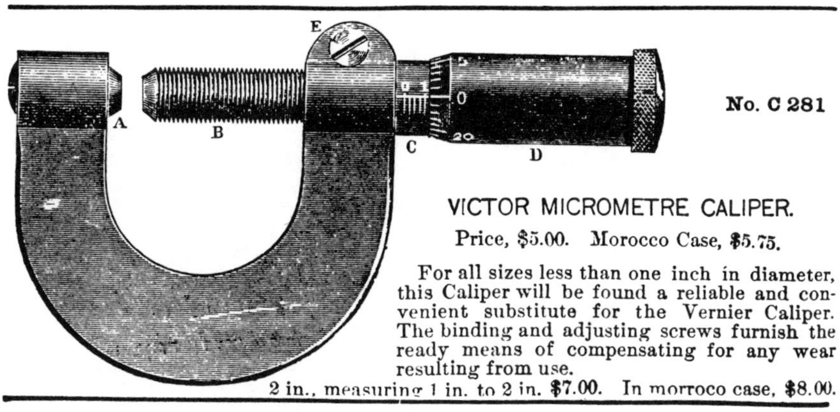

1883

WALKER MACHINE CO., S.A. Boston, MA

Operated by S.A. Walker, the company was the maker of surface gauges patented July 16, 1912 (No. 1032999) by John B. Flint.

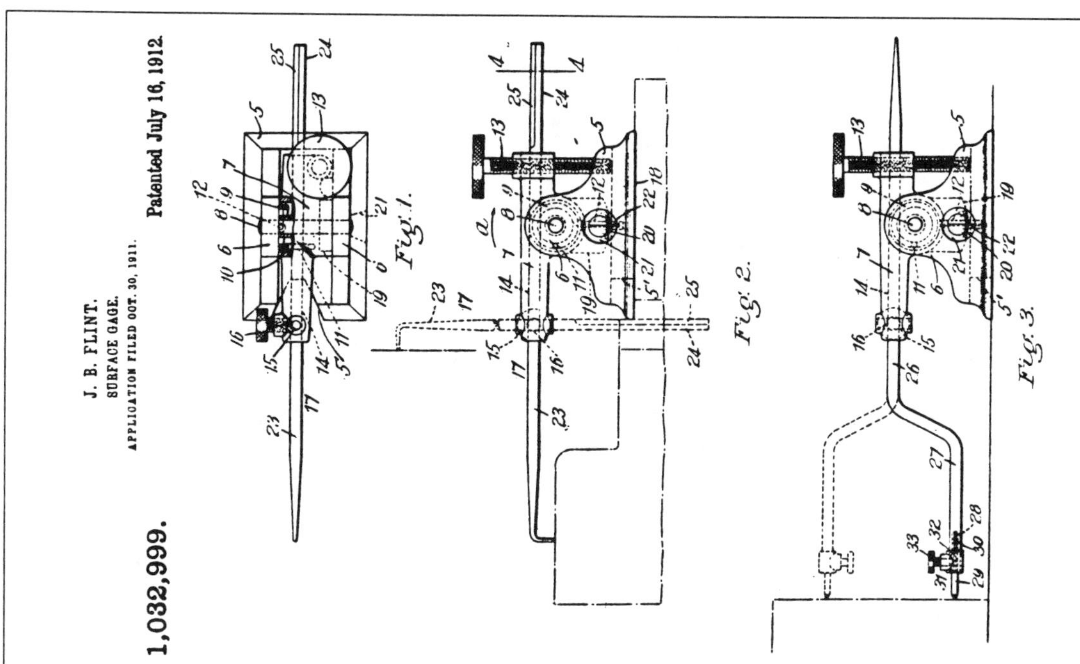

WALKER TOOL CO. Milwaukee, WI

Introduced, in 1902, attachable points "adapted to use on rods, calipers, dividers, trammels, etc."

CHANGEABLE POINTS FOR CALIPERS.

The cut shows a handy little attachment for the familiar inside calipers of the machinist whereby the points of either or both legs are changed at will. The cut

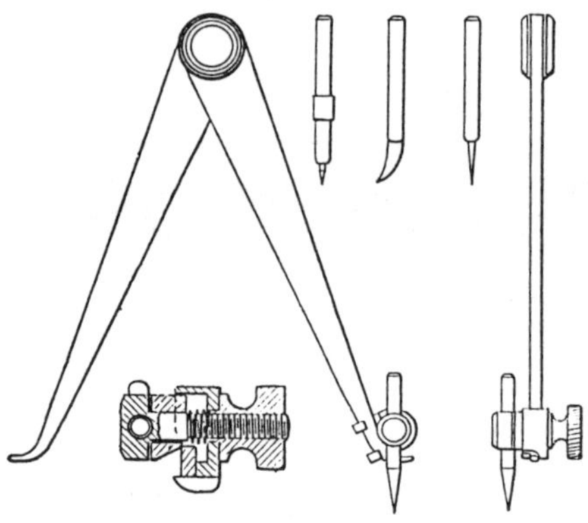

CHANGEABLE POINTS FOR CALIPERS.

shows a needle point, a pencil lead holder and a reverse hook point, either of which may be slipped into the clamping device and at once be attached securely and in the desired position to the caliper leg. Tightening the knurled thumb nut tightens both the hooks which clasp the leg, and the holder for the point. The latter may of course be turned in any direction. The hooks will attach or can be made to attach to any of the familiar forms of caliper legs. This device is made by the Walker Tool Company, Milwaukee, Wis.

Machinery, May 1902

WALKER TOOL CO., E. Erie, PA

Formed in 1883 by Edwin Walker (1847-1917) to make, among other things micrometer surface gauges and marking gauges which he had patented March 15, 1887 (No. 359336). In 1887, it was offering adjustable face planes, chisels, and micrometer surface and marking gauges. Walker sold out in 1888 and the company was dissolved in 1892. *(see illustrations next two pages)*

WALKER & CO., O.S. Worcester, MA

Formed in 1902 as a maker of grinding machines and magnetic chucks, the company also made machinist's parallels and size blocks. These were offered as late as 1936.

WALTHAM DIAL GAUGE CO. Waltham, MA

Maker, in 1925, of machinist's dial indicators.

Walker's Micrometer Surface Gauge.

Nickel Plated.

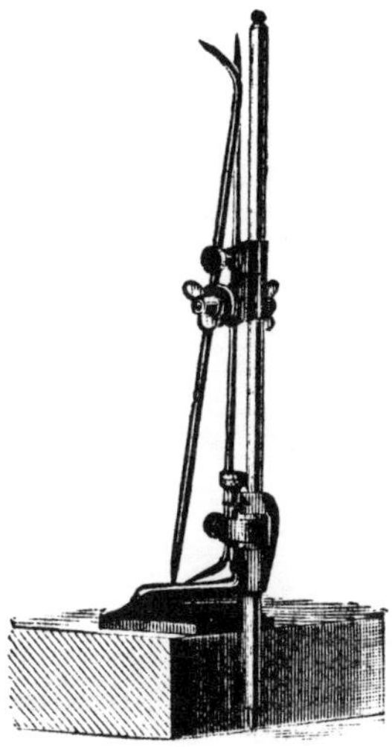

Fig. 1.

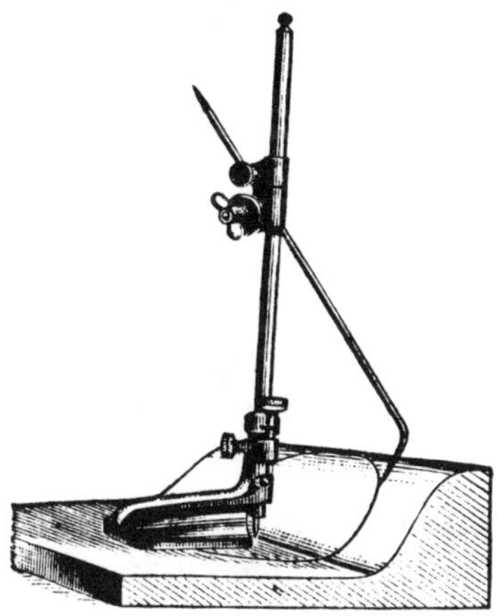

Fig. 2.

Fig. 3.

Fig. 1 shows the gauge in use as a depth gauge. As will be seen, the base is of a convenient shape for handling, and has an overhanging projection, through which the stem slides freely. In a rectangular opening in this projection is placed a block, through which the stem also passes. This block is split and can be clamped firmly to the stem by a knurled-head clamping screw. The block is also tapped out to receive the vertical adjusting screw, which is journaled above, and by means of which the block is adjusted vertically, of course carrying the stem with it when clamped. This provides the adjustment, when it is used either as a depth gauge or as a surface gauge.

The lower end of the stem is provided with a piece of pointed steel wire, which constitutes the point of contact when gauging depths, and is also used in the manner indicated in Fig. 2.

This figure shows the manner of using the gauge for scribing a circle, where ordinary dividers utterly fail to work. In this application of the tool, the centre of the hole being located and a centre mark made, the point of the stem is placed in the mark and held there. Then the scriber being set to the desired radius and clamped, the head is left loose on the stem, when, upon being swung around, it rises and falls in a vertical line over the centre, thus striking a true circle, regardless of the shape of the surface.

Then, if it is desired to lay off the projection of a circle at an angle with the surface upon which it is struck, it may be done by blocking up under the base so as to make the stem stand at the required angle.

The stem may be inserted into the base from below if desired, and the gauge then used, as shown by Fig. 3, where a line is being drawn in the recess parallel to the base of the piece, without the use of a marking table or surface plate, the adjustment being, of course, available when in this shape as well as when in the other forms.

The needles are clamped independently of each other, and the screw by which they are clamped does not turn when they are moved to different angles. A part of each of these screws is squared, and the washers which are under the wing nuts fit over these squares, and are thus prevented from turning, thereby avoiding the tightening or loosening of the nuts when the needles are moved. The upper head with needles can be used separately as a tram, dividers, inside and outside calipers, etc.

Walker's Micrometer Marking Gauge.

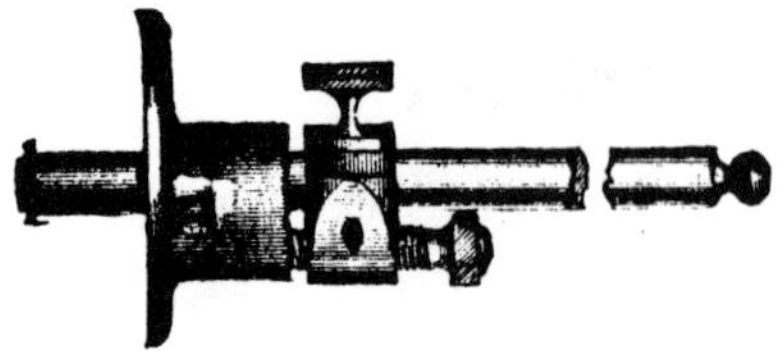

This is a superior tool for all classes of wood or iron workers. The double head is set at any distance by the small split head, which clamps it to the rod. The gauge head is then adjusted with the micrometer screw, which is much quicker and more accurate than the old way of rapping the rod up or down.

This gauge is provided with a circular steel cutter, tempered, and will gauge either right or left handed without following the grain of the wood; will also gauge into a corner, or up to a shoulder and in a hole one-half-inch diameter or over, and in various places that the common gauge cannot be used.

The gauge bar is 8 inches by $\frac{5}{16}$ steel.

Full nickel plated.......................... each, $1 25.

WARD'S SONS, Edgar T. Boston, MA

Maker of a machinist's center finder c1890. Little is known of the company.

WARWICK TOOL CO. Middletown, CT

Makers, in the 1860's, of a simple surface gauge. It appears to have been one of the first commercial makers of surface gauges.

WATERSTON, J.M. Detroit, MI

Maker of drill grinding gauges in 1927. See GROVE MFG. CO.

WATERMAN & KELLEY Pawtucket, RI

Maker, from 1883 to c1887, of a surface gauge which had been jointly patented by Stukey E. Waterman and Frederick H. Kelley on March 13, 1883 (No. 273784).

WAY CO., A.F. Hartford, CT

Maker of "Loway" size block "within the reach of every mechanic's pocketbook" and "Loway" sine bars and sine squares. The company operated in the 1920's.

WEISS, KARL Waterbury, CT

Maker, in 1909, of a speed indicator which was "a combination of a watch movement and speed indicator, which automatically throws the worm gear out of mesh with the worm at the expiration of any desired time, up to three minutes".

WEISS BROTHERS Brooklyn, NY

Maker of a speed indicator introduced in 1892 and patented November 22, 1892 (No. 486651), by Louis T. Weiss. The indicator was designed to ring a small bell when the pointer was at zero. This allowed the user to zero the unit by holding it to the rotating shaft until the bell rang. He could then start timing. The speed indicator was made until at least 1912. *(continued next page)*

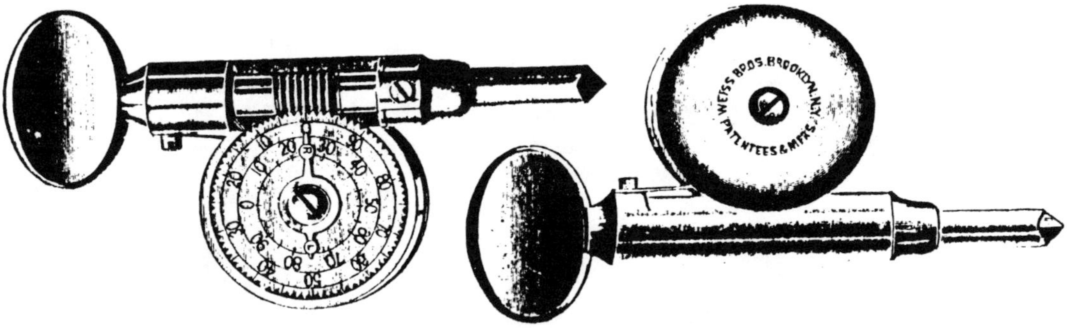

NEW SPEED INDICATOR.

1892

T.F. WELCH & CO. Boston, MA

Its 1902 catalog lists bench lathes, trammel points, gauges, jeweler's tools, etc. It claimed to be a manufacturer but may have been a dealer only. None of its products have been observed.

WELLES, F.A. Milwaukee, WI, later, Waukesha, WI

Frederick A. Welles (?-1916) was listed as a toolmaker in the Milwaukee city directories from 1882 to 1890+. In 1888, he began making a line of machinist's calipers for which he was granted a patent, May 7, 1889 (No. 402876). He later offered a combination bar caliper and trammel patented May 20, 1890 (No. 428191), and surface gauges patented January 20, 1891 (No. 444865).

In December, 1894, Welles wrote:

> "I take pleasure in saying that there are better indications of good business in the near future than a year ago. This year's business was not as large as usual but to offset this, I have received a good many letters from users of my tools who are much pleased with them. I have added several new machines of my own design, which not only facilitate the manufacture, but improve the quality of my tools."

Welles moved to Waukesha, WI, in June, 1908. In August, 1911, he moved back to Milwaukee and incorporated:

WELLES CALIPER CO. Milwaukee, WI

By 1913 the WELLES CALIPER CO., now listed as "associated with A.J. MACHEK & CO." was owned by Arthur J. Machek who was listed as a machinist in the City Directory. Machek had started making tubular micrometer frames in 1905. His 1913 catalog listed firm joint calipers, hermaphrodite calipers, calipers and dividers with inserted points, steel rules, and tubular micrometer frames. See also: MEIGS-POWELL Co. and PERDIEU TOOL MFG. CO. *(see illustrations next two pages)*

WESSMAN, C.W. Chicago, IL

Maker of "Wessman's Combination Beam Caliper", patented October 7, 1884 (No. 306123). At least one speciman of this caliper has survived complete with the cardboard box containing instructions. The unique feature of the tool is an outer jaw which slides between two stops. In one position the jaw is set for inside measurements and in the other position is set for outside measurements.

Welles Firm Joint Calipers

Made of Tool Steel.

Are the most satisfactory on the market.

OUTSIDE.

Fig. 288.

INSIDE.

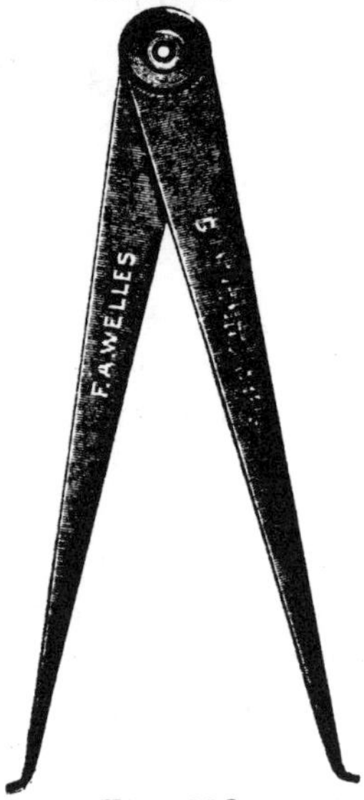

Fig. 289.

Inches	3½	5	6½	8	10	12	14	18	24	36
Prices	$0.50	.60	.70	.80	.90	1.00	1.60	1.80	2.40	6.00

Welles Adjustable Calipers

OUTSIDE.

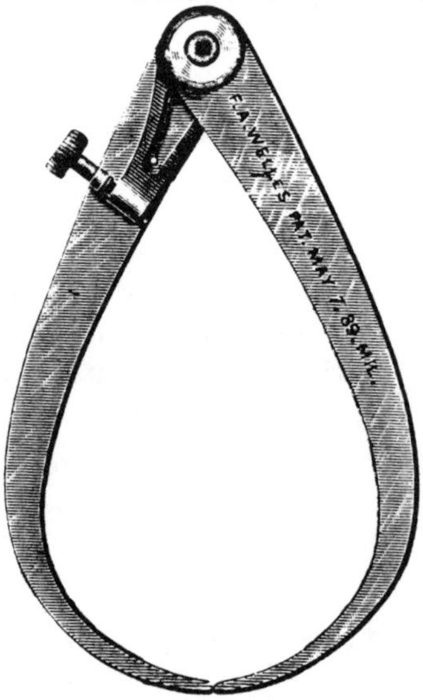

Fig. 290.

INSIDE.

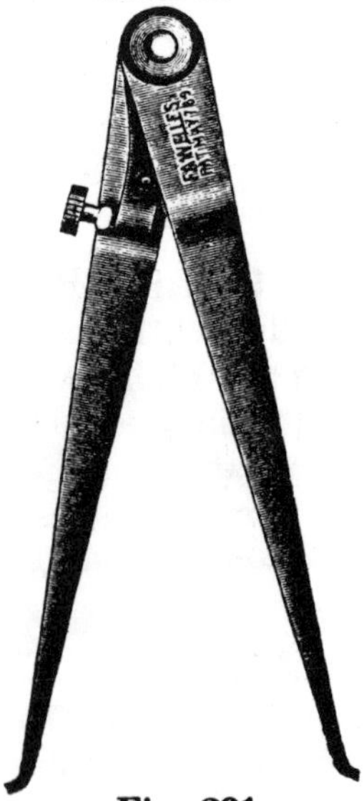

Fig. 291.

Inches	3½	5	6½	8	10	12	14	18	24	36
Prices	$1.15	1.35	1.60	1.80	2.00	2.20	2.70	3.20	4.20	8.00

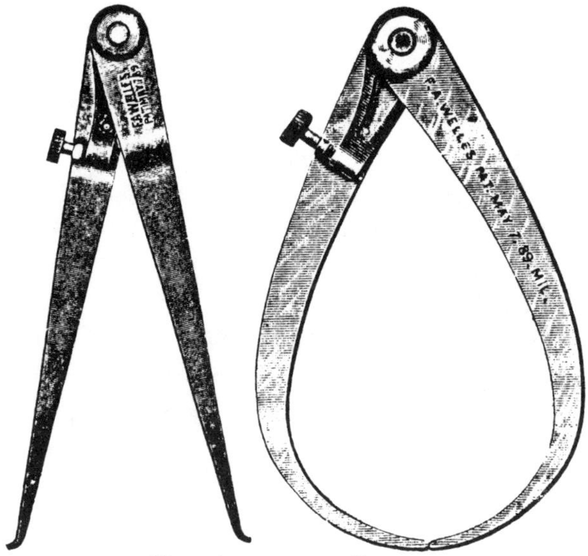

NEW ADJUSTABLE CALIPER.

1890

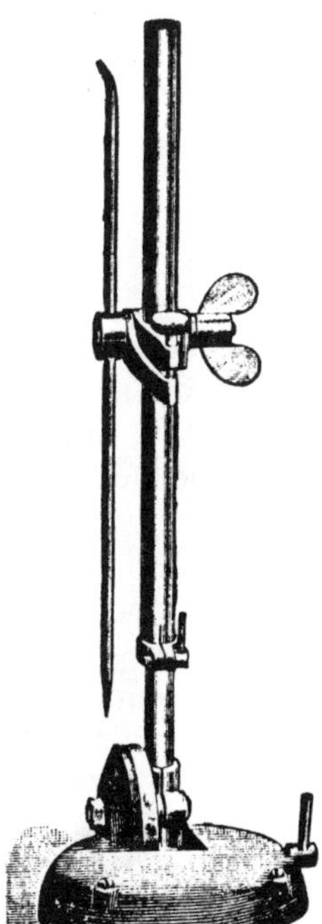

Pat. May. 20, 1890.

BAR CALIPER.

1892

WHITEHEAD, George Providence, RI

Patentee of a screw thread micrometer on December 27, 1887 (No. 375688). He was superintendent at the BROWN & SHARPE MFG. CO., Providence, RI, at the time of the patent. The micrometer, the first of the style still in use, was made by BROWN & SHARPE for many years.

WIET-GOETHE Sacramento, CA

Maker of a combination gauge-tool patented March 21, 1905 (No. 785245) by Eugene Wiet. The tool is a combination center-gauge, protractor, bevel, and rule.

WIGHT, HIRAM P. Worcester, MA

Inventor and maker of machinist's calipers, patented June 25,1878 (No. 205225).

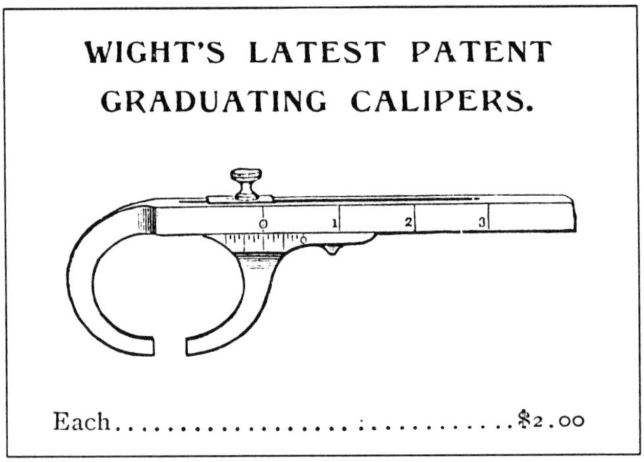

1889

WILKINSON & CO., A.J. Boston, MA

The company's primary products were reported to be dental instruments but the company was making machinist's tools as early as 1874. It introduced a new "indicating caliper" (we would now call it a beam micrometer) in 1883. The device was covered by the Merrick Barnes patent of January 30, 1883 (No. 271401). In 1885, the company introduced a shifting screw micrometer with a 0" to 3" range, and a bench type indicating caliper, both made under Merrick Barnes patents of January 30, 1883 (No. 271401), July 3, 1883 (No. 280339) and August 25, 1885 (No. 324804) . The first two patents were for a means of covering the screw thread and the third was for a "shifting" micrometer which allowed a longer range of measurment. These are noteworthy as the first micrometers made with a covered screw thread. It also offered a sheet metal gauge made under the Barnes' patents and the Benjamin H. Bristol patent of July 18, 1882 (No. 261201). By 1890, it was also offering a micrometer which included a ratchet stop patented May 22, 1883 (No. 278094) by George W. Church. The BROWN & SHARPE MFG. CO. Providence, RI, acquired the Barnes patents about 1890-1895. (*continued next two pages*)

IMPROVED INDICATING CALIPER WITH COVERED SCREW.

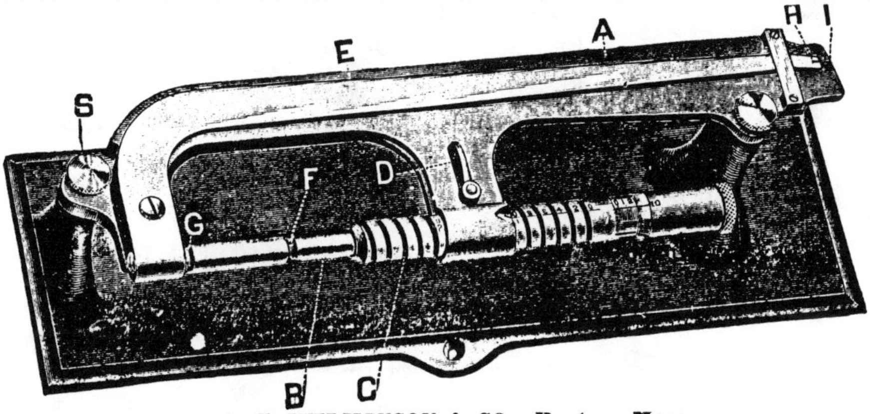

A. J. WILKINSON & CO., Boston, Mass.

Makers of the Improved Indicating Caliper, designed and used as a STANDARD TEST GAUGE in GUN, PISTOL, WATCH FACTORIES, &C. Send for Illustrated Catalogue.

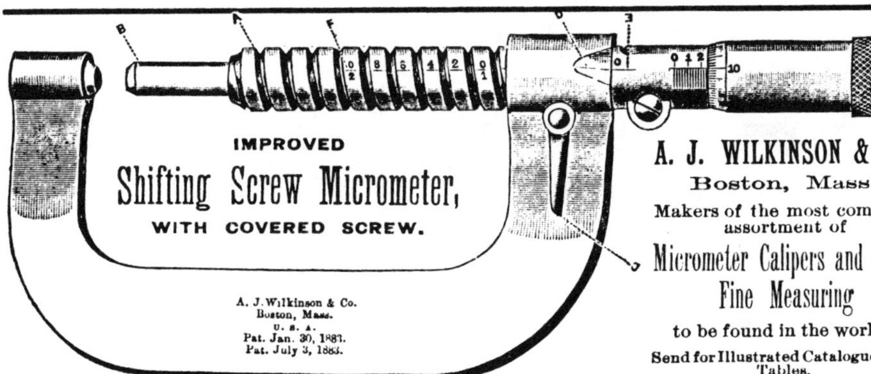

A. J. WILKINSON & CO.
Boston, Mass.,

Makers of the most complete assortment of

Micrometer Calipers and Fine Measuring Tools

to be found in the world.

Send for Illustrated Catalogue with Tables.

IMPROVED RATCHET POCKET MICROMETER.

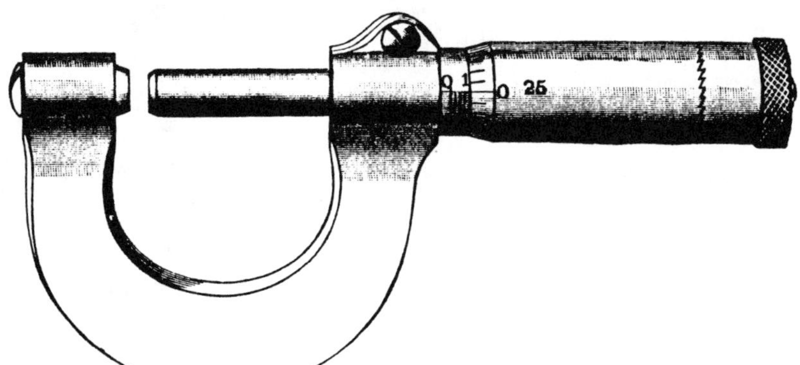

1885

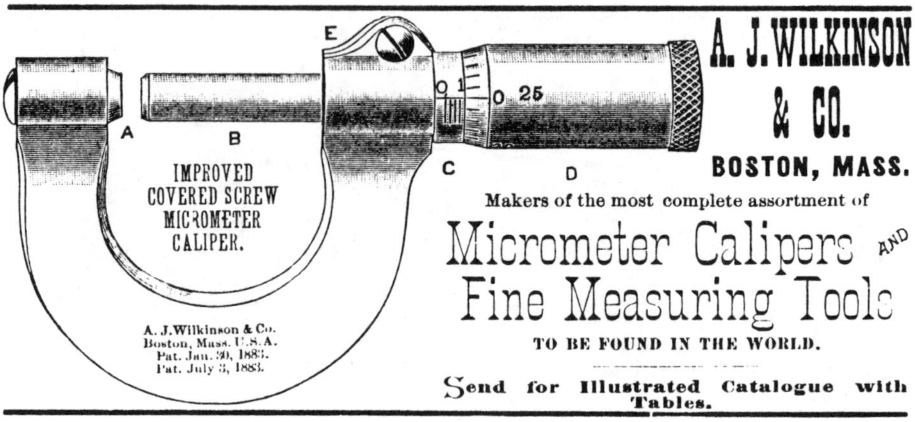

1885

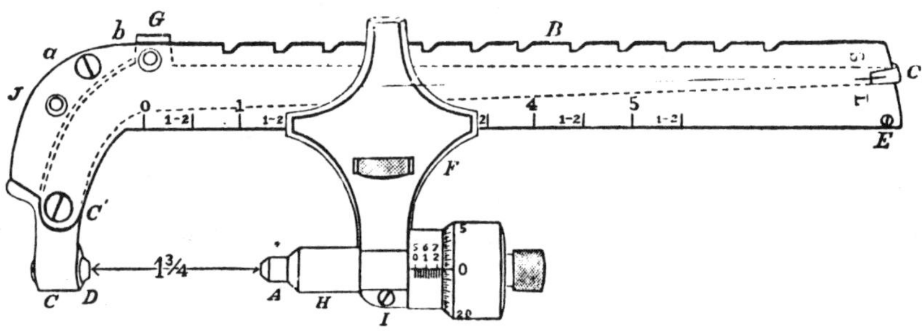

NEW INDICATING CALIPER.—(One-third full size.)

1882

ESTABLISHED
HARDWARE AND TOOLS.
1842. A. J. WILKINSON & CO., 1842.
No. 2 Washington street, Boston, Mass.,

Offer a superior Stock of Tools, in which may be found Bangor Squares; Center and Universal Squares; U. S. Scales; Protractors; Vennier Callipers; Thread Gauges; Surface Gauges; Stubbs' Fine Tools, Files, Drill Wire, and Cast Steel; Morse's Twist Drills, Sockets, and Chucks; Lathes, with or without foot motion; Hand Tools for turning; Addis Carving Tools; Monk's Molders' Tools; Pond's Carriage Tools; Bailey's Iron Planes; Wilkinson's Polished Planes (unequaled); John Martin's Files. Agents for Horton's, Chushman's, and Beache's Chucks; Pattern Letters; Rolled Brass, German Silver, and Wires; Wellington Mills Emery, Emery Cloth, Emery Wheels; Chests of Tools, all sizes, etc., etc.

Their reputation for a full assortment, superior quality, and all the new inventions, will be fully sustained. Orders promptly filled. Illustrated Catalogues sent on the receipt of 50 cents.

26 to 2

American Artisan, January 1868

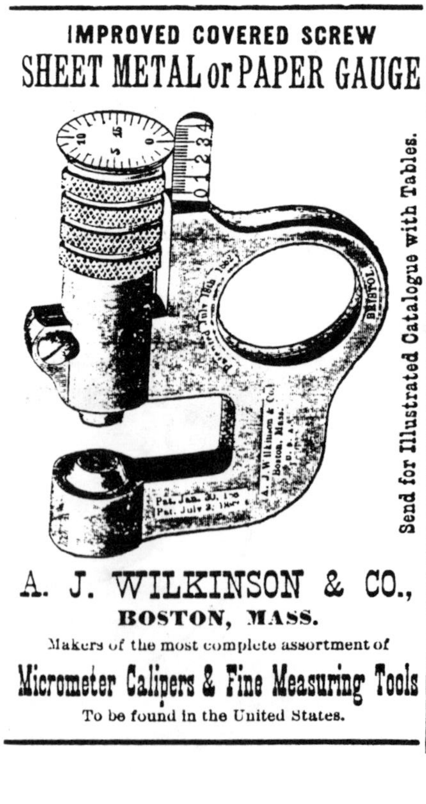

1885

WILKINSON & CO. Plattsburg, NY

Maker of a combination calipers and wire gauge, jointly patented by John D. Wilkinson and E.O. Boyle on October 27,1868 (No. 83577).

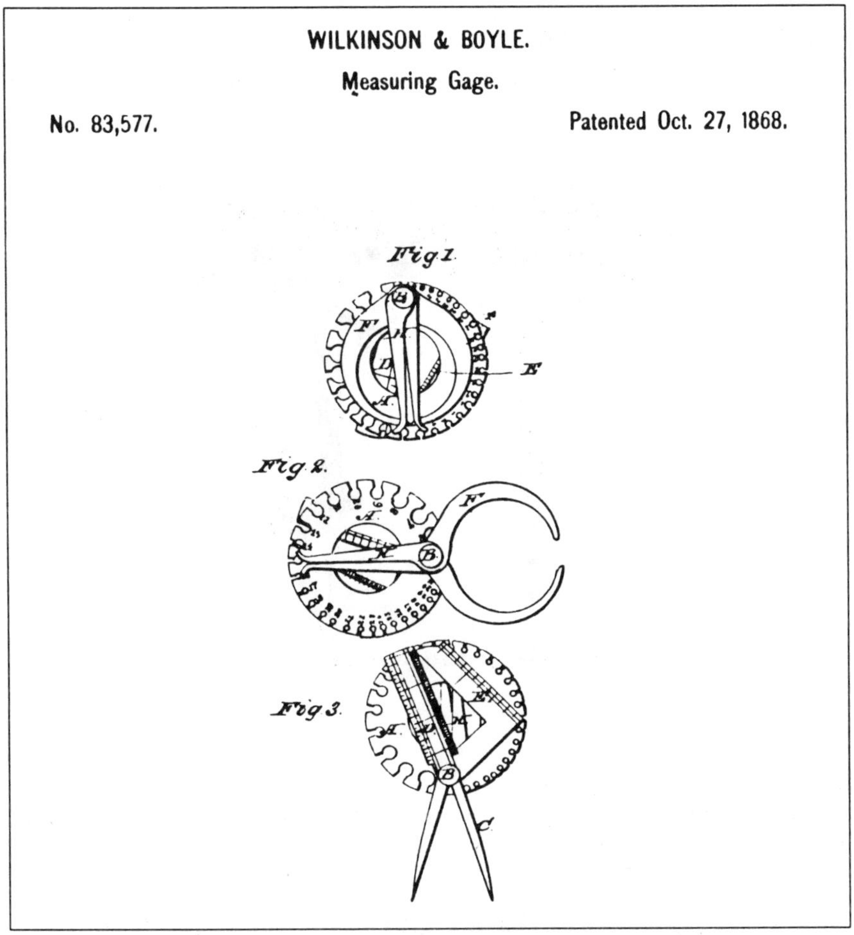

WILSON TOOL CO. Athol, MA

Maker of steel machinist's rules. Period of operation is unknown at this time.

WING, J.T. Detroit, MI

Maker of speed indicators shaped to look like a pair of wings, presumably as a play on his name.

WINSOR & JERAULD MFG. CO. Providence, RI

The F.O. Jaques micrometer patents of October 27, 1908 (Nos. 902492 and 902493) were assigned to this company. No products of the company have been noted. See CENTRAL TOOL CO.

WISMACH & CO . Milwaukee, WI

Maker, in 1916, of rectangular gauge blocks. These blocks were furnished in sets and in $\frac{1}{64}$" increments.

L.J. WOHNLICH & CO. Detroit, MI

Operated by Louis J. Wohnlich, the firm made "Diamond" trammel points which it introduced in 1886.

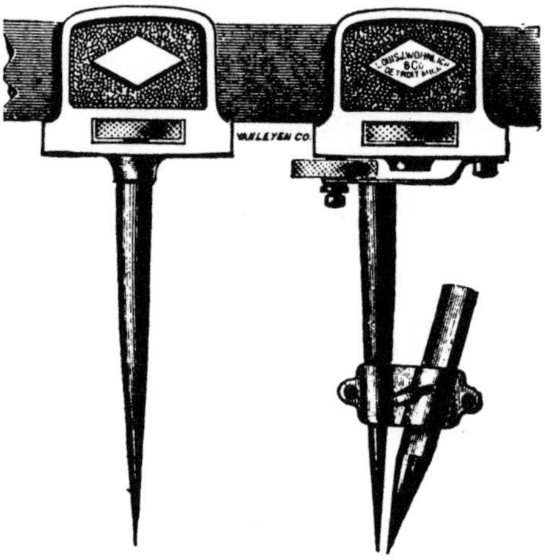

1886

The "Diamond" Trammel Points.

WOODMAN, R. Boston, MA

Maker of speed indicators, patented September 12, 1876 (No. 182177), by Daniel Davis, Jr. and Edward Wright. By 1901, Woodman had reorganized as:

WOODMAN MFG. AND SUPPLY CO., R. Boston, MA

Woodman continued to make speed indicators as late as 1910. By then they were known as the Woodman and Hudson speed indicators. *(see illustrations next page)*

WOODS MACHINE CO., S.A. Boston, MA

The Carl G. Osteman patent of July 25, 1893 (No. 502199) for a machinist's bevel was assigned to this company. The Osteman patent was later acquired by the SAWYER TOOL CO., Athol, MA, which made the tool for many years. The same tool was also made by the L.S. STARRETT CO., Athol, MA, after the patent had expired.

WRIGHT, EDWARD Worcester, MA

Edward Wright (1835-1916) was listed as a machinist in the 1867 city directory and as a maker of woolen machinery in 1877. On September 24, 1867, Wright received a patent for spring calipers (No. 69292) which he, and others, made for many years. The calipers were first made by the L.W. POND MACHINE CO. of Worcester, MA, where Wright was probably employed. In 1880, Wright formed a partnership and began operating as:

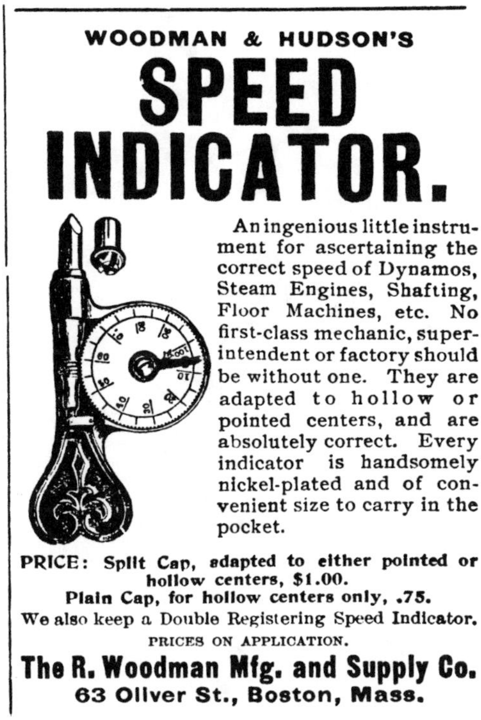

1900

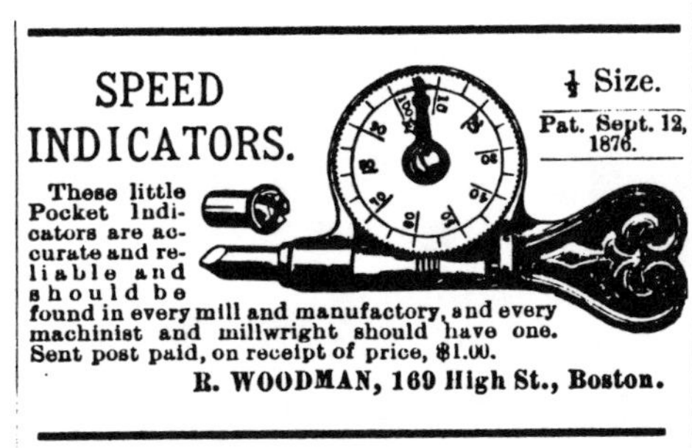

1885

See entry for Woodman Mfg. & Supply Co., previous page

WRIGHT & CO., EDWARD Worcester, MA

This new company lasted only a year until the formation of:

WRIGHT MACHINE CO. Worcester, MA

Edward Wright was president and the new company's principal products were cotton and woolen machinery. By 1883, he was also making the patented calipers under his own name, although he still referred to them as the WRIGHT-POND calipers as late as 1891.

In 1887 he added a speed nut feature patented January 4, 1887 (No. 355430), by Jacob Wright and assigned to the WRIGHT MACHINE CO. It is interesting to note that this speed nut design was still in use on calipers made by the LUFKIN RULE CO., in the 1960's.

The WRIGHT MACHINE CO. was succeeded by Anthony-Bates Machine Co. in 1896. Wright, in 1894-1896 was operating the Wright Hardware Mfg. Co. and the Wright Wrench Co., both of Worcester, MA. *(continued next page)*

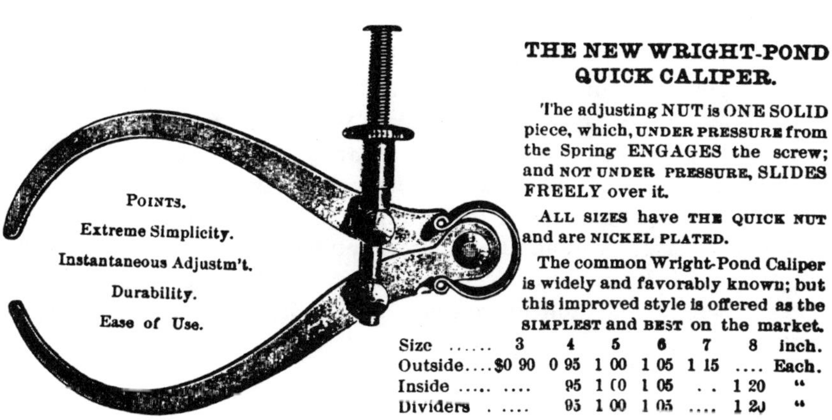

1891

Wright Machine Co.

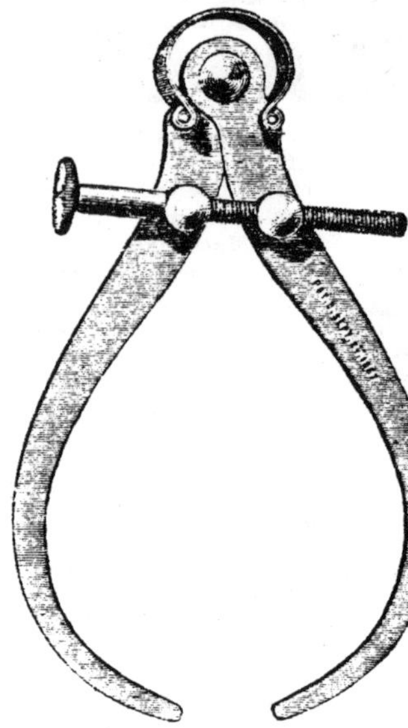

1881

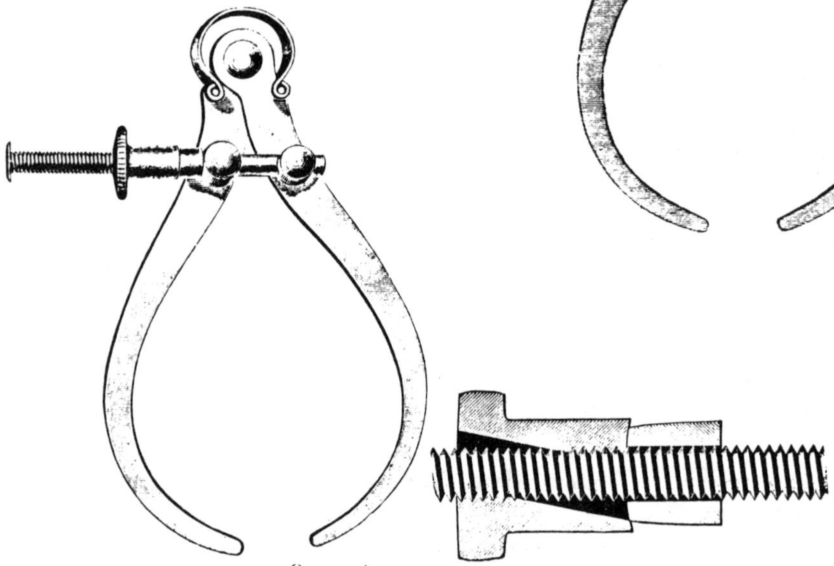

QUICK-ADJUSTING CALIPER.

1887

WYKE & CO., J. East Boston, MA

Wyke began making, in 1885, universal screw-cutting center, depth, angle and twist drill gauges which he had patented September 11, 1883 (No. 284702). These tools were advertised for sale as late as 1898. Hardened machinist's scales were offered from 1885 to 1897. Wyke introduced a combination surface gauge in 1886. A key-seating/splining gauge covered by his patent of October 27, 1891 (No. 461995) was introduced in 1891.

In an 1893 business report, Wyke noted that he had sold 16,000 of the universal screw cutting, center, depth, angle and twist drill gauges in the last five years. His 1897 catalog offered: rules, end measures, combination rules, combination surface gauges, metal T squares, pocket drawing squares, key seat or splining gauges, and depth or angle gauges. A new, adjustable surface gauge was introduced in 1899.Wyke went out of business in 1911. *(continued next five pages)*

Wyke's Improved End Measure Combination Rule.

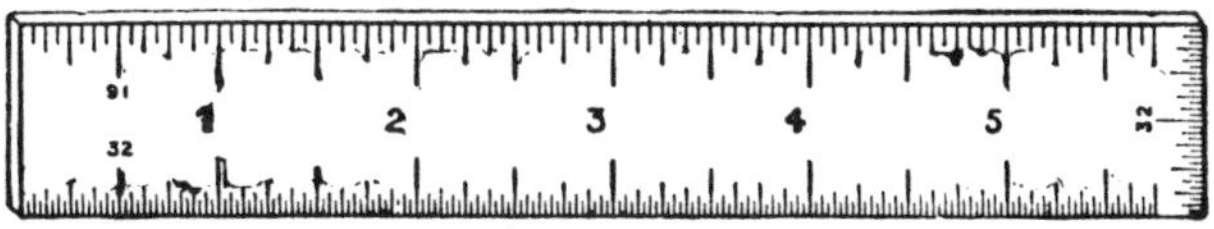

The above cut represents our improved end measure combination rule. It is of exact standard length and width, and warranted correct by micrometer. It is intended for use on work that is limited in space, and where the ordinary scale cannot be used; such as the bottoms of grooves, and short work between lathe centres and shoulders, collars, etc. It is also useful in many situations constantly occurring to the machinist. The Wyke scale is superior to other end measure scales, being graduated entirely across the end (not partially across as in other scales), thus the mechanic can measure grooves or other spaces from left to right without reversing the scale or taking his eye from the lines of measure; giving him a more correct judgment. It is graduated on four corners to his No. 3 graduation, and across one end it is graduated on one side to 32ds of an inch.

These rules can also be used as standard gauges of length and width by which calipers and other tools can be set accurately. They are especially useful to have around planing, shaping, milling, and other machines, as by their use the tools can be set to exact distances when planing surfaces, projections, etc., of uneven heights and widths, also when planing out keyseats, splines, etc., to exact size. They are, as their name implies, a combination tool; being a short rule, a long rule, and a gauge combined. Machinists owning a set of these tools have the following standard sizes, but by placing the rules side by side, or side to end, an almost unlimited number of sizes can be obtained.

6 inches long, 1 inch wide	each,	$0 85
5 " " 7/8 " "	"	75
4 " " 3/4 " "	"	65
3 " " 5/8 " "	"	55
2 " " 1/2 " "	"	40
The full set complete in neat wood box		3 70
Box alone		50

1889

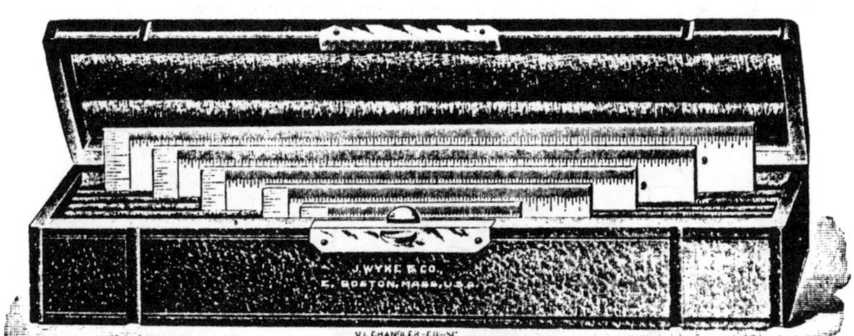

CASE OF STEEL RULES.

1891

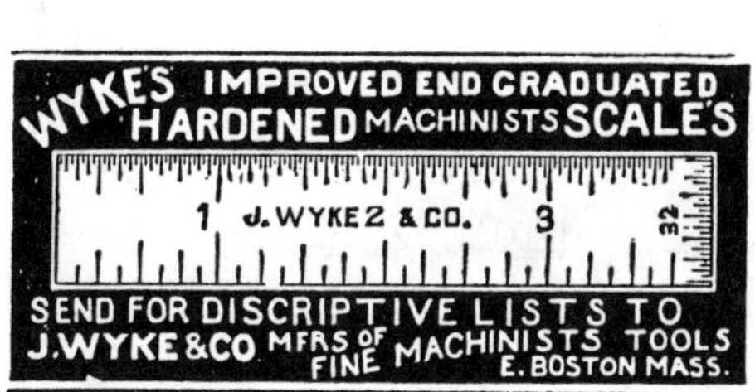

1897

1886

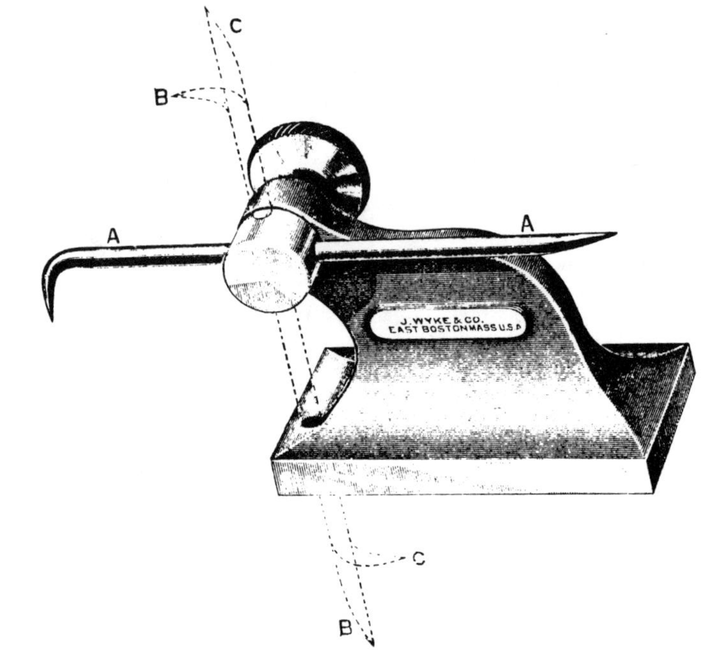

COMBINATION SURFACE GAUGE.

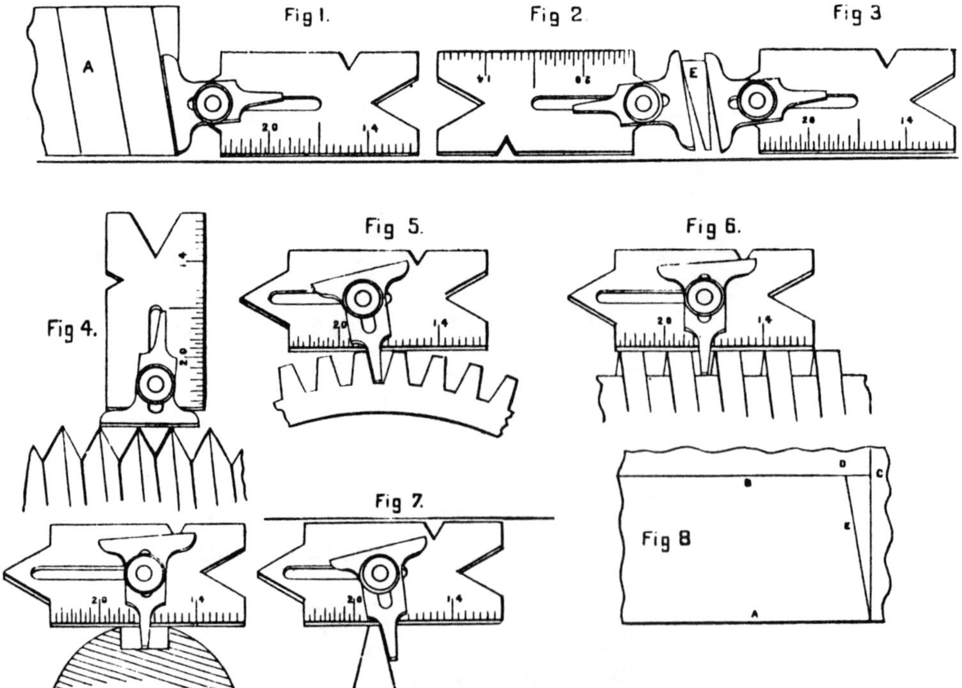

1883

WYKE'S SCREW-CUTTING GAUGE.

Wyke's Keyseat or Splining Gauge.

J. Wyke & Co., East Boston, Mass., are introducing a tool for the use of engineers and machinists in general, as illustrated herewith. It is designed for setting shafting on the planing, shaping, milling or drilling machine, when about to have slots, splines, or keyseats cut into them; also for setting the tools, whether for planing or milling out the slots. The cut gives an excellent idea of the operation of the tool. The manufacturers claim that the tool

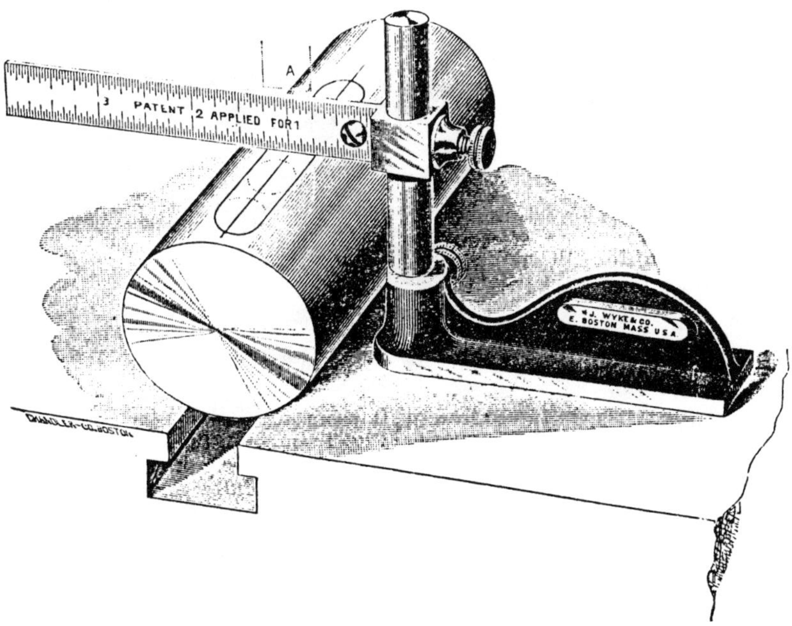

Wyke's Keyseat or Splining Gauge.

enables the artisan to work rapidly, correctly and mechanically, and that it is so simple any one can understand it.

May 7, 1891

Adjustable Surface Gage.

We illustrate an adjustable scribing block, surface and depth gage. The needle passes transversely through the usual clamping bolt, which in this case is fitted to a hole passing through the end of a flat piece which fits over two fixed studs that project from the upper surface of the block, and below it is a flat spring which keeps the piece always pressed

ADJUSTABLE SURFACE GAGE.

snugly upwards against a knurled nut, so that there can be no lost motion.

A hole is drilled through the base in such position that the needle may pass through it, and the tool thus becomes an adjustable depth gage. The base of the gage is 3¾ inches long, 1⅛ inches wide, with finished working surfaces and edges, japanned center, and is made by J. Wyke & Co., East Boston, Mass.

1899

XANDER, J.G. Reading, PA

Xander introduced, in 1916, a "universal electric test indicator." The unit was made to mount on a surface gauge and was designed to "save the operator from eye strain while watching for the contact of the surface gauge needle with the work. With this tool, the moment the point touches the highest point of the work, either internal or external, an electric light flashes, indicating the direction in which the work must be moved."

YORK CO., S.M. Cleveland, OH

Maker of combination inside/outside calipers very similar to those made by the L.S. STARRETT CO.

YOUNG-FISCHER INCLINOMETER CO. Milwaukee, WI

Introduced a two dial inclinometer in 1921. The device was supposed to take the place of a plumb, level, sine-bar, and protractor in machine shops.

APPENDIX I

EARLY AMERICAN STEEL RULES

by Ambrose Webster

The following is excerpted from the April 12, 1894 edition of American Machinist *magazine.* NOTE: Material in square brackets has been added by K. Cope.

The machinist of today who is using the fine Brown & Sharpe steel scales and micrometer calipers has little conception of the difficulties of making accurate measurements which confronted the machinist of 1840.

When the writer commenced his term of four years apprenticeship, there were no steel scales in the market, and we were obliged to set our calipers to the proper dimensions by laying them on an ordinary boxwood rule; trying to have the ends of the caliper coincide with the division lines. It was utterly impossible, of course, to secure uniform measurements in this way.

In 1848 there were, in the Springfield Armory, some home or hand-made steel scales, graduated on a device consisting of a wooden slide to which the scales were clamped, this slide being moved by a screw on which was a wooden index, to the periphery of which was tacked a ribbon of paper on which the proper graduations were spaced with dividers. [Webster described making some rules in 1851].

Previous to this date Mr. S.W. Putnam of Fitchburg, Mass. had done something in this line. Mr. Putnam evidently desired to get a great range of subdivisions ranging from 8 to 34 to the inch, and as quarters of an inch were indicated by long lines, it seemed to him to be advisable to use these long lines or "quarters" in all the graduations of an inch. Consequently the number of divisions in the "quarters" varies somewhat; as, in the case of the division 14, there are two "quarters" containing four divisions and two containing three divisions each. The term "quarters" is not correct, but we naturally drift into the use of that term. [Salmon W. Putnam, with his brother John, began making machine tools and steam engines about 1838.]

Mr. Charles L. Hildreth, superintendent of the Lowell Machine Shop writes:

> "In 1835 it was impracticable for the workmen employed at these shops to readily procure scales and rules which were in exact accord with one another. One of our superior mechanics, Samuel Brown, made in 1835 a standard screw for use in graduating measuring scales. This screw was made to agree with the standard brass scale of Cary, of London, and it took nearly a year to perfect it. Mr. Brown then made two short duplicate standard screws for use in graduating small scales. Mr. W.D. Blanchard has now in his possession a steel scale 4 inches long which he graduated about 1846 by the aid of one of the standard screws."

Mr. Hildreth also writes:

> "The Lowell Machine Shop made standard scales long before they were undertaken to be manufactured in Bangor." [He is probably referring to Samuel Darling of Bangor, ME who began making rules in 1852.]

Mr. Joseph Brown, of the firm of Brown & Sharpe, Providence, R.I., commenced the manufacture of steel rules and scales and vernier calipers in the year 1850, and the first printed circular with price list was issued March 1, 1851. The following is a copy of the first circular issued by him:

U.S. STANDARD RULES

Manufactured by
JOSEPH R. BROWN
115 South Main St., Providence, R.I.

The subscriber has recently invented and put into operation a machine for dividing rules and scales of all descriptions, in the most accurate manner. Having been convinced by long experience that it is impracticable to make a screw sufficiently accurate to be used in a dividing engine, he has constructed a machine upon an entirely new principle which, he is happy to say, produces the most satisfactory results. It is operated entirely by power, which render it automatic, thereby obviating the imperfections arising from the employment of hand labor. The attention of engineers, machinists, draftsmen, and all others wanting a very accurate instrument, is respectfully invited to these rules.

In 1852, Samuel Darling and Edward H. Bailey began the manufacture of steel rules in Bangor, ME. This partnership continued a year, then Mr. Darling purchased Mr. Bailey's interest and took another partner named Michael Schwartz, and continued the business under the name of Darling & Schwartz until 1866, when the firm was dissolved, and Mr. Darling formed a partnership with Joseph R. Brown and Lucien Sharpe, under the name of Darling, Brown & Sharpe, and soon moved the works to Providence, R.I.

In the early manufacture of scales it was natural to resort to a screw for feeding the carriage, but Mr. Brown's mechanical genius led him to see the difficulties of making a perfect screw, and he resorted to another process for moving the carriage, which was for a long time a secret, and may still be, as the graduating machine of Brown & Sharpe is rarely or never shown to visitors. [Another source states that the "secret machine" was actually designed and built by Samuel Darling and came with him to Providence when the merger took place in 1866. This was confirmed by Mr. Henry Sharpe, grandson of Lucien Sharpe and still employed at Brown & Sharpe. Mr. Sharpe mentioned that the Darling machine was built mostly of old saw blades from the sawmills around Bangor. Darling had not bothered to remove the teeth so there was no doubt about the source of the material.] Among the many things which have led to the present high skill and precision in machine shops, I consider that steel scales and micrometer calipers stand in the first rank.

APPENDIX II

THE STORY OF THE COMBINATION SQUARE

by L.S. STARRETT

The following is excerpted from the January, 1915 edition of MACHINERY *magazine.*
NOTE: Material in square brackets has been added by K. Cope.

In 1876, while making patterns, I was impressed with the need of improvements in try-squares to fill the place of a kit with different lengths of fixed blades for pattern-makers' and carpenters' use on various grades of work. The idea of a sliding blade struck me as the right principle to work on. After I conceived the design I made patterns, got castings, and a blade and had the first sample made up by a good machinist, who, not believing in an adjustable square blade, said he "wouldn't give a damn for it." But that did not discourage me. I showed it to Samuel Colt of Hartford, the noted revolver manufacturer. He grabbed it out of my hand and exclaimed: "That is a good thing. That will sell. Did you get it up? I want one." [Starrett received a patent for the combination square on May 5, 1879.]

He being a man after my own heart this was encouraging. I made a contract with a Connecticut company who made old style try-squares, etc. [Stanley Rule and Level?] to make 5,000 as a starter. After awhile I got the first lot and showed them to the hardware men, all of whom admitted it to be a good thing, but said that as it was not introduced there was no demand. They said "if you will get them introduced, we will buy them." I saw the point. So I employed agents, giving them a generous commission to canvass the manufacturers and sell direct to employees, in many cases taking orders to be delivered after the next payday. The squares sold like hot cakes, and so my agents did well for themselves and well for me. But I could not get them made fast enough to fill orders.

My first one-sheet circular headed by a cut of the combination square gave the following description: "The old style try-square superseded by an instrument that is more than a substitute for a whole kit, from the shortest to the full length blade, and more convenient than a whole kit can be. It is also a miter square, a plumb and level, a depth and mortise gage, a steel rule and straightedge, a scriber or scratch gage, and a gage for transferring measurements."

When this circular was shown to the manager of the concern I had contracted with he exclaimed: "No, you don't! No, you don't!" meaning superseding the old fashioned try-square, being somewhat jealous, you see. How much this had to do with my not getting the squares as fast as I wanted I don't know, but being a good friend to the foreman I interested him in pushing things to get out a lot on time. The management could not understand why their foreman was so intensely interested in getting the work along as fast as he did.

However, as the work was not as good as I wanted, I fitted up a small shop in Athol and started to make them myself, hiring the best workmen I could find. But to get the blades graduated was what bothered me most. I tried to place a contract with a firm making scales but was turned down. I could not buy a graduating machine for love or money. So after deciding on the best method of graduating I got hold of an idea and perfected my own machines and system, and not only made my square blades, but put on the market [in 1882] an improved line of spring-tempered steel rules that are now acknowledged and accepted as standard the world over. Among the first scales I made was a "very thin scale". I christened this "Flexible" to distinguish it from the heavier spring-tempered scale. This has since been copied by all of our competitors who make scales.

Soon after I put out my squares I added a center head [patented June 29, 1880, No. 229283] to slip off and on to the same blade to use in finding the center of circular work; also a bevel protractor [patented August 7, 1883, No. 282583] to use with the same blade, all of which have added much to the value of the tool, and since my patents have expired these tools have been copied at home and abroad by many of our competitors who no doubt think imitation is no crime.

APPENDIX III

THE STARRETT/ ATHOL MACHINE CO. LITIGATION

NOTE: Material in square brackets has been added by K. Cope.

WORCESTER WEST CHRONICLE, February 1, 1883

"Sometime since we noticed that the suit pending between Newton and Bellows of the Athol Machine Co. and L.S. Starrett had been decided in favor of the latter: now we learn that another suit, brought by Mr. Starrett against the said Newton and Bellows and the Athol Machine Co., for an infringement of his patent squares has also been decided in his favor by Judge Lowell, and an injunction ordered against the defendents. By this last decision we presume Mr. Starrett will be intitled to heavy damages, in addition to having his patent sustained, which must be very gratifying to himself and his friends."

THE ATHOL TRANSCRIPT, February 6, 1883

"In the United States Court at Boston, Monday, Judge Lowell presiding, was heard the case of Laroy S. Starrett vs Athol Machine Co. et. al. This was a suit in equity for the infringement of two patents for improvements in try-squares. The defense was a general denial, and that the Athol Company was not liable under a contract by which the company's power, machinery, etc. were let out to another firm for the manufacture of tools, that firm agreeing to pay as rent 10% of their net sales. The plaintiff insists that the whole contract is a device to shield the Athol Company as infringers, but this is not proved. There is, however, evidence tending to show that the company has made tools for the express purpose of manufacturing these try-squares and that a few were made before the contract went into effect. An injunction should go against all the defendents, but in accounting the plaintiff must prove that the company is liable to his in profits of damages. Interlocutory decree for complainant." [An "Interlocutory" decree is, in law, a decree not final or definitive, and an interlocutory order is an order made during the progress of a suit, upon some incidental matter which arises out of the proceedings.]

WORCESTER WEST CHRONICLE, June 7, 1883

"Some time since we published the decision of Judge Lowell, given Jan. 31st, in the case of L.S. Starrett against D.A. Newton and Stephen H. Bellows, wherein the Judge sustained Mr. Starrett's patent for Try Squares and ordered an injunction against the defendents. Since then, Newton and Bellows have attempted to evade the injunction of the court, by continuing to make the squares, whereupon, Mr. Starrett took legal action against them. This week Judge Lowell has decreed that Newton and Bellows are clearly guilty of violating the injunction in the manufacture of Try Squares and fined them fifty dollars each and costs. In making the centre head, for which an injunction was also granted, the defendents have since cut away one half of it, making only one side available and thereby evading Mr. Starrett's patent for the centre head. But this does not affect his valuable patent for try squares, for which they were fined."

WORCESTER WEST CHRONICLE, July 2, 1885

"Another contest decided. Some over a year ago, Mr. L.S. Starrett informed the Standard Tool Co. of this town, that a Bevel Protractor they were making infringed his patent, and requested them to stop making the same, whereupon Mr. S. H. Bellows, one of said Co., claimed that the invention was his own, and caused an interference suit to be instituted against Mr. Starrett. This case, after being hotly contested at great expense, has just been decided by the Commissioner of Patents at Washington, D.C., in favor of Mr. Starrett."

APPENDIX IV

EVOLUTION OF THE SPRING CALIPER

by E.R. Markham

The following is excerpted from the October 2, 1902, edition of American Machinist *magazine.*
NOTE: Material in square brackets has been added by K. Cope.

At page 909 of the *American Machinist*, in the article "From a Designer's Notebook," is a cut showing a form of "rapid transit" nut, which calls to my mind a number of incidents and changes that have taken place in the evolution of the spring caliper, which I thought might be of interest to your readers.

The J. Stevens Company—afterwards the J. Stevens Arms & Tool Company—of Chicopee Falls, Mass., manufactured for years a spring caliper patterned after English tools which were sold in this country as the "Stubbs" calipers.

I can remember as a boy of six years of age [this would have been in 1866] going into their blacksmith ship and watching Grandpa Spalding—as he was known to the village boys—forging calipers and dividers. The legs and bows were made, as will be seen by a glance at Fig. 1, in one piece. [Illustration on page 310.] There was at this time no drop or power hammer in the shop; all drawing of stock was done with hammer and sledge, and the shapes were produced with special hand tools and forms rigged up as jumpers. The calipers after forging were sent to the machine shop to have the necessary machine work done on them, after which they were returned to the blacksmith shop to be hardened and tempered. Little did I think as a lad of six years that twenty-one years after I should have charge of making this very caliper in this shop—but I am getting ahead of my story.

After manufacturing this caliper for many years, the company put on the market what was known as their patent helical coil caliper, a cut of which is given in Fig. 2 [T.C. Page and G.W. Hadley patent of February 8, 1870, No. 99698]. This was made from steel wire; the spring coiled and the legs flattened and bent to shape. It was necessary to cut a slot nearly the entire length of the screw, leaving it solid at each end; the slot was necessary, as the screw had to straddle the legs. This tool never met the expectations of the company as a revolutionizer of the caliper business.

About the year 1878, Charles P. Fay, working as an apprentice for the above-mentioned concern, invented and had patented [December 20, 1881, No. 251200] a caliper of the form shown in Fig. 3. He afterward sold the patent right to a Mr. Hyde, of Springfield, Mass.; it was known as the Hyde patent screw caliper. The operation of the caliper will be seen by a glance at the cut. Two coil springs which were inserted in the cross-bar kept the legs against the screws shown at each end of the bar, but as there was no provision for taking up wear in the joints, or back lash in the screw, the popularity of this tool was short lived.

Shortly after this, Mr. Fay invented and had patented [June 2, 1885, No. 319215] a detachable bow caliper, which differed from the Pond caliper made in Worcester [made under the Edward Wright patent of September 24, 1867, No. 69292], in that the bow was placed above the joint of the caliper and opened it by compressing the upper end of the legs, as shown in Fig. 4. As the pressure of the bow kept the legs together, all that was necessary for a fulcrum was a pin fitting in a semi-circular shaped slot in each leg. A cut of the fulcrum pin—or joint stud—is shown in Fig. 5, which also shows a view of the legs and slot to receive the pin.

Mr. Fay started the manufacture of this caliper in Springfield, Mass., in 1885. I went to work for him in the fall of the same year; shortly afterward I was appointed foreman of his shop. During the summer of 1886 he placed on the market a cheaper form of his caliper and divider known as the "Yankee"; his object in making this caliper was to meet the cheaper forms of goods sent to this country from Germany.

The quick-adjusting nut put on the Fay caliper and divider and on the Yankee, when ordered, is shown in Fig. 6 [covered by Samuel Dover patent of November 7, 1882, No. 267166, and Charles P. Fay patent of January 26, 1886, No. 334764]. This nut, after being made to shape and drilled and tapped in the screw machine, was placed in a jig and hollow milled on two opposite sides to receive a ring. After elongating the hole in the small end of the nut, it was sawed in two and filed, to allow the small end to close in order to open the opposite end that contained the thread, and to release it from the screw. A ring was soldered in opposite sides of each part of the nut, which was then put together and polished.

At some time previous to the putting of the Fay quick-adjusting nut on the market the nut mentioned under the heading "From a Designer's Note Book" was brought to our notice; this form of nut is shown in Fig. 7. At about this time—I cannot recall the precise date—another form of quick-adjusting nut, known as the Richardson rapid nut [Frederick Thomas patent of March 22, 1881, No. 239072, assigned to C.F. Richardson], was added to the list, and was used by the Stevens Company in connection with their calipers, when ordered.

In 1887 I accepted a position with the J. Stevens Arms & Tool Company in charge of their fine tool department. They had previously brought out a detachable bow caliper, whose legs were joined together by means of a half-and-half joint, the bow being connected with the legs below the joint and operated similar to the Pond caliper bow [J. Stevens patent of March 9, 1886, No. 337531]. Fig. 8 shows the joint and bow. Shortly after I became acquainted with them the joint was changed and a compression bow substituted for the style formerly used, on account of the possibility of getting a much evener tension [J. Stevens and O.D. Warfield patent of October 26, 1886, No. 351474]. The form of quick-adjusting nut used by them is shown in Fig. 9.

They afterwards put on the market a cheaper form of caliper, known as the Leader. They continued to manufacture the solid forged caliper up to the time I left their employ in 1895, but on a very small scale, as the detachable bow goods gave much better satisfaction, and the tension of the bow was much evener, especially on the large sizes.

In 1887 L.S. Starrett, of Athol, Mass., bought out the Fay and Yankee caliper and divider, and with some slight improvements is manufacturing them now.

The Brown & Sharpe Manufacturing Company of Providence, R.I., has recently gone into the manufacture of detachable bow calipers and dividers. While there have been some changes in forms of joints and styles of quick-adjusting nuts, the detachable bow caliper made by the three leading manufacturers of these goods is substantially the same as Mr. Fay brought out nearly twenty years ago.

When Mr. Fay sold out his business to Mr. Starrett he went to Athol to oversee the manufacture of the goods. Nine years later he left Mr. Starrett's employ, and returned to the J. Stevens Arms & Tool Company. He is now the works manager and part owner of the concern.

It is very interesting to note the growth of that company, that a few years ago employed about thirty hands, today employs, I am informed, about 1,400 hands. Fifteen years ago their product was carried to the express and freight offices every night on a wheelbarrow. I think it safe to say they find it necessary to use some different method of conveying their goods to the railway at the present time. *(illustration next page)*

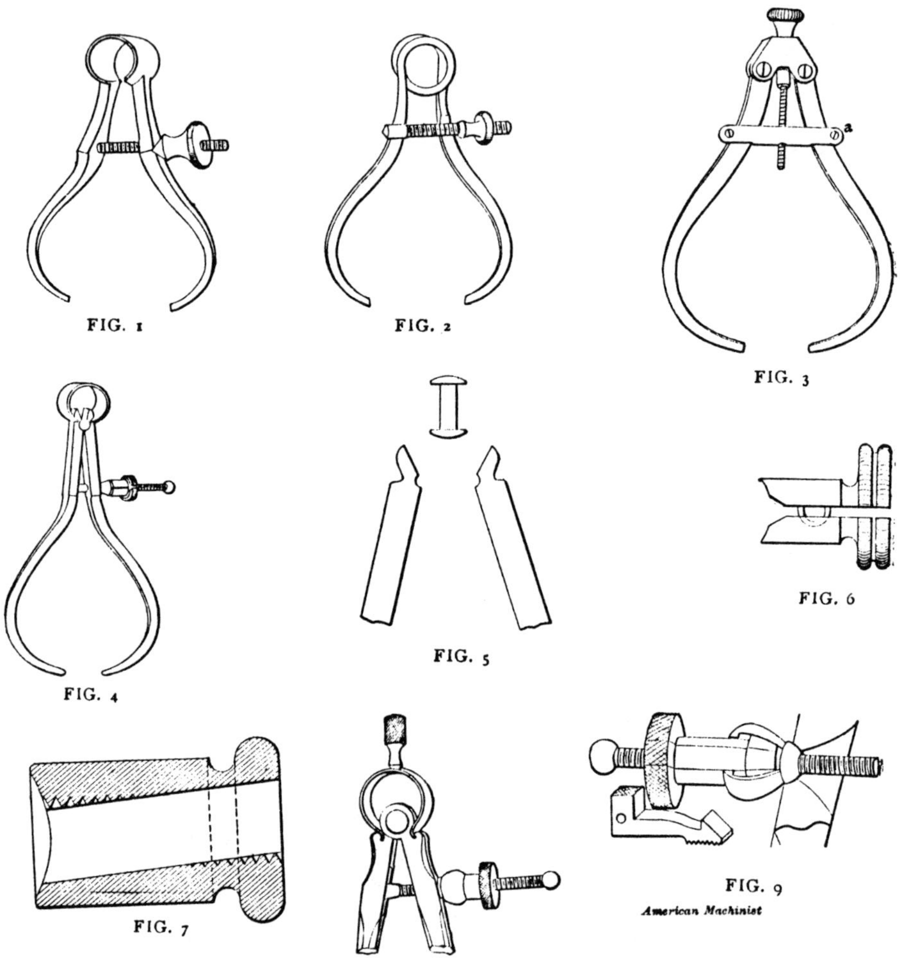

EVOLUTION OF THE SPRING CALIPER.

APPENDIX V

THE KOOPMAN CALIPERS

by Ferd. Wertheim

The following article is excerpted from the December 4, 1902 edition of American Machinist *magazine.*

Micrometers and other refined measuring appliances are in everyday use, but the spring caliper and the firm-joint kind, recently much improved, are being called for and produced in greater quantity than ever. They have their own field, and the changes in shop methods have not brought about their retirement.

I have thought that a description of several calipers, which used to be held in high esteem by toolmakers and the best machinists, and are superior to anything on the market today would interest *American Machinist* readers.

The tools illustrated were brought out about ten years ago by A.T. Koopman, of Chicago, and patents were applied for. [No patents, however, have been found.]

Fig. 1 shows the strap-joint caliper. This was made of saw steel, and the points were really finished—one did not have to doctor them before they were fit for use, as with most of those now for sale. The joint was perfect—the steel straps and solid bronze ring giving a friction absolutely free from back lash—and so made as to be readily adjusted. One could hold these calipers by inserting the small finger in the bronze ring, or they could be hung up thereby. These strap- joint calipers were notable for their extreme stiffness and lightness.

Fig. 2 shows the combination caliper. The construction of the joint is shown in Fig. 3. The caliper points are finished to a thin, flat point, and hardened, making the tool valuable for the most accurate measurements. The usual binding clamps or thumb-screws are done away with, and all the points are tapered at one end, fitting sockets in the head, friction being depended on to hold them.

I am not sure how this construction stood the rough-and-tumble usage of the shop, but offer the suggestion that it is a very good way to make the better grade of drafting instruments.

The folding wing W is especially neat. The quadrant on ordinary dividers or combination calipers sticks out "like a sore thumb". It is often in the way of the fingers when using the dividers for stepping off; it takes up too much room in the tool box, and requires springs and check-nuts to keep it steady. The folding wing W opens and closes automatically with the tool, and a fine positive adjusting movement of the leg is secured by the eccentric nut C. This nut has a taper bearing to take up supposable wear.

These tools are as simple as any on the market now. They were brought out by a small and poor concern, but if manufactured today by a firm equipped for the work, their excellence should assure a ready and abundant sale.

The importance of the workman's small tools, both to himself and to the shop, through the results so largely dependent upon them, justifies almost unlimited discussion of them, and many should be interested in the tools to which I here call attention. *(illustration next page)*

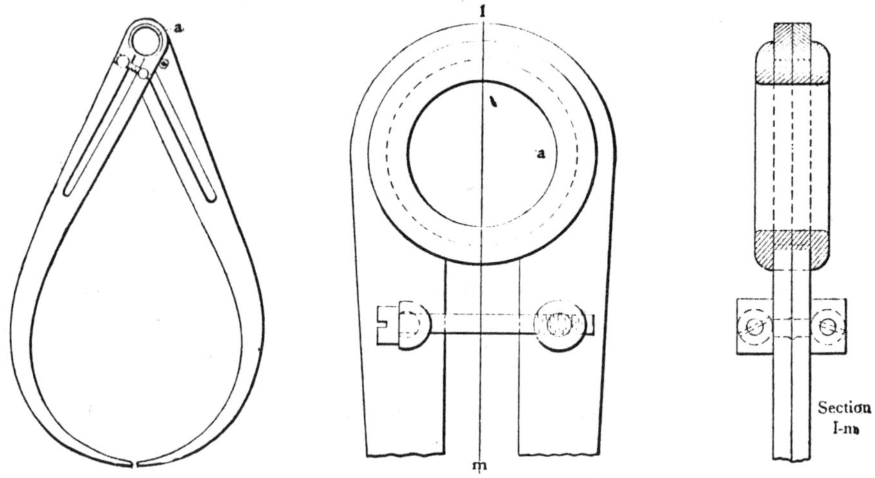

FIG. 1

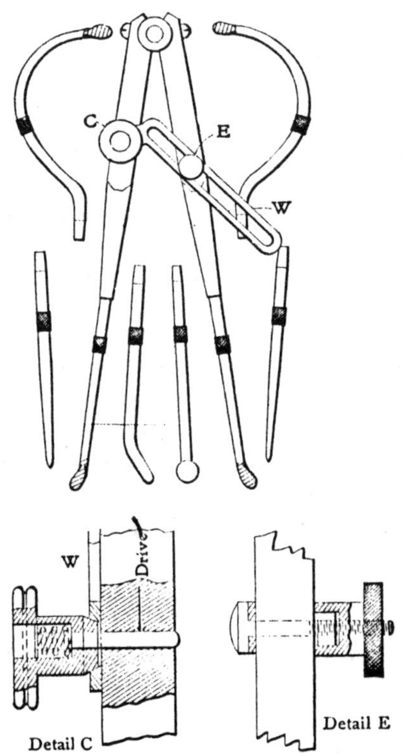

FIG. 2

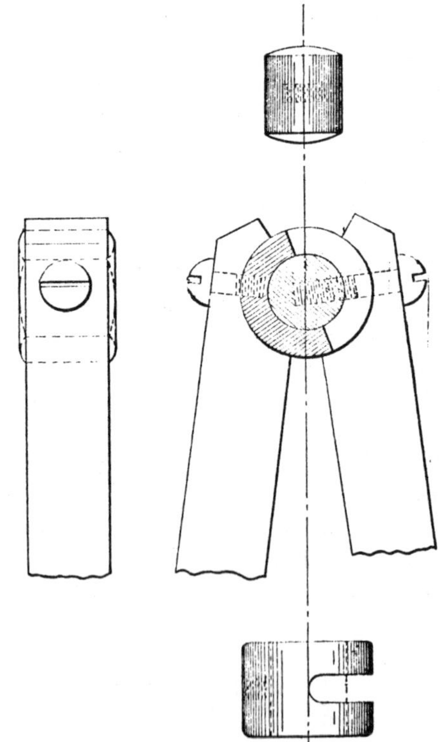

FIG. 3

THE KOOPMAN CALIPERS.

American Machinist

APPENDIX VI

THE DEVELOPMENT OF THE MICROMETER CALIPER

The following exchange of letters between F.A. Strong and Lucien Sharpe was published in American Machinist *magazine.*

Bridgeport, Conn. Dec. 8, 1887

Mess. Brown & Sharpe, Providence, R.I.

Gentlemen: Will you kindly gratify the curiosity of the undersigned relative to the origin of the micrometer caliper as orginally made by you. I, of course, know that the principle involved is old, but can you give me the name of the inventor and the approximate date of invention—if it is proper to speak of it as an invention—of the form first made by you? We have with us a man who claims to have first made and used this tool as a sheet metal gauge, that he showed it to Mr. Sharpe, and from that the B.& S. micrometer was developed. How much truth or fiction there is in this, I would deem it a great favor to learn from you.

Respectfully,
F.A. Strong, Asst. Supt.
The Eaton, Cole & Burnham Co.

Providence, R.I., Dec. 10, 1887

Mr. F.A. Strong, Asst. Supt.
Eaton, Cole & Burnham Co.'s Works
Bridgeport, Conn.

Dear Sir: Your favor of the 8th inst. is received, and we take great pleasure in telling you the history of the micrometer caliper as far as our knowledge extends.

Some time in the first seven months of 1867 a man connected with the Bridgeport Brass Co., brought a gauge here the size and form of the smallest tool shown on the blue-print herewith inclosed (Fig. 1), and desired to have some made.

It was invented by some one employed in that establishment, we were told. The reading of thousandths of an inch was done by means of a spiral line of same pitch as the screw, 40 to the inch, running round the cylindrical head of the screw, crossed by 25 lateral lines, dividing this head into 25 equal parts. The pointer which is attached to the body of the instrument shows where the lines intersect, and if you count the number of screw thread lines the pointer has passed over, added to the number of lateral lines, the number of thousandths of an inch the caliper is open can be obtained; but as the cylinder is so completely covered by the crossed lines, rendering it impossible to put any figures upon it, we did not at the time consider the instrument of any practical value, as it could not be easily read.

In August, 1867, our Mr. Brown and the writer visited the Paris Exposition and there saw a micrometer caliper like the other tool shown on blue-print (Fig. 2) and brought one home with us. It was at that time well known in France as "Systeme Palmer", or Palmer's plan of gauge, from the name of the inventor.

As a gauge was wanted for measuring sheet metal, we adopted Palmer's plan of division and the Bridgeport man's size of gauge, adding the clamp for tightening the screw and the adjusting screw for compensating the wear of ends of points where the metal is measured and produced our "Pocket Sheet Metal Gauge", as shown on first page of inclosed circular. As to whom the credit is due for producing the "Pocket Sheet Metal Gauge", you can determine from the above statement of facts.

We never should have made such a gauge as was shown us by the Bridgeport man in 1867 to sell on our own account, as it would be too troublesome to read to be saleable. If we had not happened to find the Palmer gauge, and thereby found a practical way to read thousands of an inch, no gauges would have been made. If we had never seen the Bridgeport device we should have found the Palmer at Paris, and without doubt have made such gauges, but possibly would have made a larger one first. The immediate reason of making the "Pocket Sheet Metal Gauge" was the suggestion coming from the Bridgeport Brass Co., of the want of a gauge of the size of the sample shown us for use of the brass trade.

Beside the micrometer shown in the circular, we shall soon have one to measure ½ inch and another 2 inch.

Yours truly,
Brown & Sharpe Mfg. Co.
L. Sharpe, Treas.

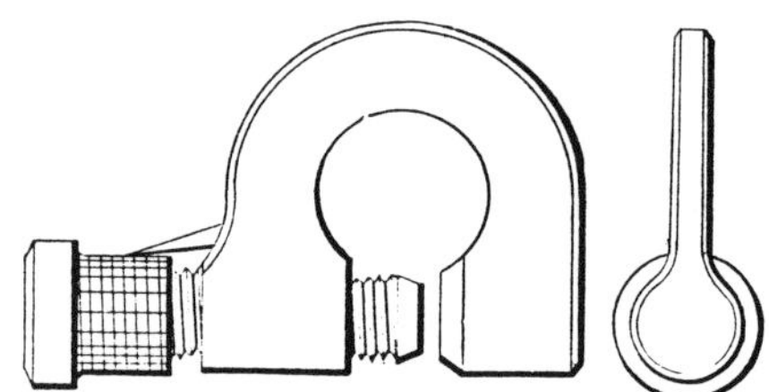

Fig.1

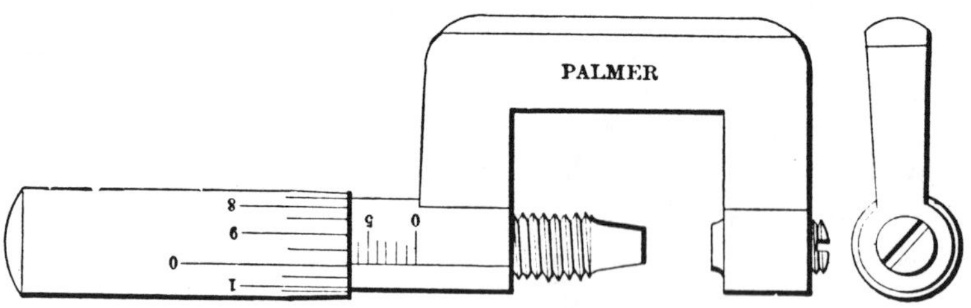

Fig.2